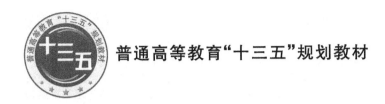

普通高等教育"十三五"规划教材

Engineering Mathematics
—Functions of a Complex Variable and Mathematical Methods for Physics

工程数学
——复变函数与数学物理方法

石　霞　默会霞　钱　江　杨建奎　**编著**

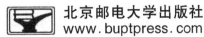

北京邮电大学出版社
www.buptpress.com

内 容 简 介

本教材是一本用于同名课程双语教学的英文教材，编者按照国家教育部对本课程的基本要求，参考了多本有关的英文教材，结合多年的教学实践编著而成。本教材包含复变函数和数学物理方法两部分。复变函数部分的内容包括：复数与复变函数的基本概念、复变函数的导数和积分、解析函数的定义和性质、解析函数的幂级数表示、留数定理及其应用等。数学物理方法部分的内容包括：三类典型方程的导出和定解问题的定义、二阶线性偏微分方程的分类和化简以及达朗贝尔公式求解波动方程、分离变量法求解定解问题和特征值问题、贝塞尔函数和勒让德多项式、傅里叶变换和拉普拉斯变换。

本教材可作为理工科院校非数学专业本科生的双语教材，也可供有关工程技术人员参考。

This book contributes to the bilingual teaching of engineering mathematics. The selection of the contents is in accordance with the fundamental requirements for this course and based on years of practice in the course of the authors. The contents of this book include two main parts, that is, complex variables and mathematical methods for physics. Complex variables part includes the basic concepts of the complex number and complex function, the derivative and integral of the complex function, the definition and properties of the analytic function, the power series expansion of the analytic function, the residue theorem and its application. Mathematical methods for physics part includes the three typical equations and the definition of the well-posed problem, the classification and simplification of the second order linear partial differential equations and the D'Alembert formula for string oscillation, the method of separation of variables for the well-posed problems and the eigenvalue problem, Bessel function and Legendre polynomial, Fourier integral transformation and Laplace integral transformation.

This book can be used as the bilingual textbook for undergraduate students in the science and technological schools whose majors are not mathematics, and may also be suitable to the related engineers and technicians.

图书在版编目(CIP)数据

工程数学：复变函数与数学物理方法 ＝ Engineering Mathematics：Functions of a Complex Variable and Mathematical Methods for Physics / 石霞等编著. -- 北京：北京邮电大学出版社，2017.9(2024.7重印)

ISBN 978-7-5635-5264-1

Ⅰ. ①工… Ⅱ. ①石… Ⅲ. ①工程数学—高等学校—教材 Ⅳ. ①TB11

中国版本图书馆 CIP 数据核字(2017)第 214378 号

书　　　名	工程数学——复变函数与数学物理方法
著作责任者	石　霞　默会霞　钱　江　杨建奎　编著
责 任 编 辑	马晓仟
出 版 发 行	北京邮电大学出版社
社　　　址	北京市海淀区西土城路 10 号(邮编：100876)
发　行　部	电话：010-62282185　传真：010-62283578
E-mail	publish@bupt.edu.cn
经　　　销	各地新华书店
印　　　刷	河北虎彩印刷有限公司
开　　　本	787 mm×1 092 mm　1/16
印　　　张	16.25
字　　　数	419 千字
版　　　次	2017 年 9 月第 1 版　2024 年 7 月第 5 次印刷

ISBN 978-7-5635-5264-1　　　　　　　　　　　　　　　　　定价：35.00 元

・如有印装质量问题，请与北京邮电大学出版社发行部联系・

前　　言

为了适应日益增强的国际交流,培养更优秀的人才,双语教学已经成为中国高等教育课程改革的一个重要发展趋势。近十几年来,编者在北京邮电大学国际学院进行了本课程(工程数学)的双语教学实践。按照本课程的教学大纲,参考多本国内外优秀教材,经过几年的教学实践和不断完善,编者在讲义的基础上编写了本教材。

本教材包括复变函数和数学物理方法两部分。虽然目前部分高等工科院校将复变函数和数学物理方法放在一起称为工程数学,但是将这两部分内容放在一起写入工程数学的教材较少,而适用于双语工程数学教学的英文教材目前还没有,这给本课程的双语教学带来了很大的困难。基于此,我们下决心编写一本适合高等理工科院校特色专业学生学习的英文"工程数学"教材,以满足开设工程数学的高等理工科院校对本课程双语教学的需要。

本书的所有编者都在我校主讲了多年的双语工程数学课程,有丰富的教学经验,了解学生在学习双语工程数学过程中需要注意的问题。本书的第一、四、九、十章由石霞副教授编写,第二、七、八章由钱江副教授编写,第三、十一章由默会霞副教授编写,第五、六章由杨建奎副教授编写,全书由石霞副教授审阅校对。

本教材在编写的过程中得到了北京邮电大学、北京邮电大学国际学院的教改项目的资金支持,也得到了北京邮电大学出版社的大力支持,同时也得到了同事和研究生的帮助,在此一并表示衷心感谢!

由于编者水平有限,书中的缺点和疏漏在所难免,恳请广大读者批评指正。

编　者

Preface

This book is a textbook for the undergraduates in the science or engineering who are taking a course on *Functions of a Complex Variable and Mathematical Methods for Physics* in English in China.

This book essentially includes two parts.

The first part, Chapter 1 ~ Chapter 5, is Functions of a Complex Variable. The functions of a complex variable is the branch of mathematical analysis that investigates functions of complex numbers. The function of a complex variable was developed in the nineteenth century, mainly by Augustin Cauchy (1789—1857), and later his theory was made more rigorous and extended by such mathematicians as Peter Dirichlet (1805—1859), Karl Weierstrass (1815—1897), and Georg Friedrich Riemann (1826—1866). Complex analysis has become an indispensable and standard tool of the mathematician, physicist and engineer.

The content of this part is organized as follows:

In Chapter 1, complex number and its operations are reviewed, and the definition of a complex function and its continuity are also presented.

In Chapter 2, the concept of the derivative of a complex function and the definition of analytic function are introduced, and the necessary and sufficient conditions on differentiability and analyticity are presented. Elementary functions are showed in this chapter.

In Chapter 3, the concept of the contour integral of a complex function and Cauchy integral formula are given.

In Chapter 4, the complex series representations, including Taylor series and Laurent series, of an analytic function are presented.

In Chapter 5, the definition of the residue and Cauchy residue theorem are introduced. And the application of residues on three types of definite integrals is presented.

The second part, Chapter 6 ~ Chapter 11, is Mathematical Methods for Physics. This part contributes to partial differential equations (PDEs) of mathematical physics. It is an important and specialized course of math, which is widely used in math, physics and engineering. Basically, our objective is to describe even predict the physical phenomena and evolution by mathematical methods. We don't take too much time on explaining how we establish the model of equations, but pay more attention to how we can solve the equations technically. In this part, we mainly introduce the method of separation of variables and two special functions derived in the process of separation of variables for some special PDEs.

The content of this part is organized as follows:

In Chapter 6, the basic concepts and definitions on partial differential equations are introduced, three typical mathematical models (wave equation, heat equation and Laplace equation) and well-posed problem are showed.

In Chapter 7, the classification, simplification and standard forms for linear second order PDEs are demonstrated, and the integral method on characteristics for solving hyperbolic and parabolic equations are showed in Chapter 8.

In Chapter 9, the method of separation of variables on finite regions and the Sturm-Liouville eigenvalue problems are presented.

In Chapter 10, two special functions, Bessel function and Legendre polynomial are introduced through practical problems and the properties of these functions are also described.

In Chapter 11, the definition, properties on Fourier and Laplace integral transformations and some applications on solving the definite problems are showed.

Contents

Part I Functions of a Complex Variable

Chapter 1 Complex Numbers and Complex Functions 3

1.1 Complex number and its operations 3
   1.1.1 Complex number and its expression 3
   1.1.2 The operations of complex numbers 6
   1.1.3 Regions in the complex plane 13
Exercises 1.1 14
1.2 Functions of a complex variable 15
   1.2.1 Definition of function of a complex variable 15
   1.2.2 Complex mappings 17
Exercises 1.2 20
1.3 Limit and continuity of a complex function 21
   1.3.1 Limit of a complex function 21
   1.3.2 Continuity of a complex function 26
Exercises 1.3 27

Chapter 2 Analytic Functions 29

2.1 Derivatives of complex functions 29
   2.1.1 Derivatives 29
   2.1.2 Some properties of derivatives 31
   2.1.3 A necessary condition on differentiability 31
   2.1.4 Sufficient conditions on differentiability 34
Exercises 2.1 36
2.2 Analytic functions 37
   2.2.1 Analytic functions 37
   2.2.2 Harmonic functions 39
Exercises 2.2 41

2.3 Elementary functions ... 41
2.3.1 Exponential functions ... 41
2.3.2 Logarithmic functions ... 42
2.3.3 Complex exponents ... 45
2.3.4 Trigonometric functions ... 46
2.3.5 Hyperbolic functions ... 48
2.3.6 Inverse trigonometric and hyperbolic functions ... 49
Exercises 2.3 ... 50

Chapter 3 Integral of Complex Function ... 52

3.1 Derivatives and definite integrals of functions $w(t)$ ... 52
3.1.1 Derivatives of functions $w(t)$ ... 52
3.1.2 Definite integrals of functions $w(t)$ ... 53
Exercises 3.1 ... 56
3.2 Contour integral ... 56
3.2.1 Contour ... 56
3.2.2 Definition of contour integral ... 58
3.2.3 Antiderivatives ... 66
Exercises 3.2 ... 73
3.3 Cauchy integral theorem ... 75
3.3.1 Cauchy-Goursat theorem ... 75
3.3.2 Simply and multiply connected domains ... 76
Exercises 3.3 ... 80
3.4 Cauchy integral formula and derivatives of analytic functions ... 81
3.4.1 Cauchy integral formula ... 81
3.4.2 Higher-order derivatives formula of analytic functions ... 84
Exercises 3.4 ... 87

Chapter 4 Complex Series ... 89

4.1 Complex series and its convergence ... 89
4.1.1 Complex sequences and its convergence ... 89
4.1.2 Complex series and its convergence ... 90
Exercises 4.1 ... 93
4.2 Power series ... 93
4.2.1 The definition of power series ... 93
4.2.2 The convergence of power series ... 95
4.2.3 The operations of power series ... 97
Exercises 4.2 ... 97
4.3 Taylor series ... 98
4.3.1 Taylor's theorem ... 98

  4.3.2 Taylor expansions of analytic functions ······ 100
Exercises 4.3 ······ 104
4.4 Laurent series ······ 105
  4.4.1 Laurent's theorem ······ 105
  4.4.2 Laurent series expansion of analytic functions ······ 109
Exercises 4.4 ······ 111

Chapter 5 Residues and Its Application ······ 113

5.1 Three types of isolated singular points ······ 113
Exercises 5.1 ······ 118
5.2 Residues and Cauchy's residue theorem ······ 118
Exercises 5.2 ······ 123
5.3 Application of residues on definite integrals ······ 123
  5.3.1 Improper integrals ······ 124
  5.3.2 Improper integrals involving sines and cosines ······ 125
  5.3.3 Integrals on $[0, 2\pi]$ involving sines and cosines ······ 128
Exercises 5.3 ······ 130

Part II Mathematical Methods for Physics

Chapter 6 Equations of Mathematical Physics and Problems for Defining Solutions ······ 135

6.1 Basic concept and definition ······ 135
  6.1.1 Basic concept ······ 136
  6.1.2 Linear operator and linear composition ······ 138
  6.1.3 Calculation rule of operator ······ 140
6.2 Three typical partial differential equations and problems for defining solutions ······ 141
  6.2.1 Wave equations and physical derivations ······ 141
  6.2.2 Heat (conduction) equations and physical derivations ······ 143
  6.2.3 Laplace equations and physical derivations ······ 144
6.3 Well-posed problem ······ 145
  6.3.1 Initial conditions ······ 146
  6.3.2 Boundary conditions ······ 146

Chapter 7 Classification and Simplification for Linear Second Order PDEs ······ 148

7.1 Classification of linear second order partial differential equations with two variables ······ 148
Exercises 7.1 ······ 149
7.2 Simplification to standard forms ······ 149
Exercises 7.2 ······ 156

Chapter 8 Integral Method on Characteristics ... 158

8.1 D'Alembert formula for one dimensional infinite string oscillation ... 158
Exercises 8.1 ... 160
8.2 Small oscillations of semi-infinite string with rigidly fixed or free ends, method of prolongation ... 160
Exercises 8.2 ... 162
8.3 Integral method on characteristics for other second order PDEs, some examples ... 162
Exercises 8.3 ... 165

Chapter 9 The Method of Separation of Variables on Finite Region ... 166

9.1 Separation of variables for (1+1)-dimensional homogeneous equations ... 167
  9.1.1 Separation of variables for wave equation on finite region ... 167
  9.1.2 Separation of variables for heat equation on finite region ... 170
Exercises 9.1 ... 172
9.2 Separation of variables for 2-dimensional Laplace equations ... 174
  9.2.1 Laplace equation with rectangular boundary ... 174
  9.2.2 Laplace equation with circular boundary ... 177
Exercises 9.2 ... 180
9.3 Nonhomogeneous equations and nonhomogeneous boundary conditions ... 181
Exercises 9.3 ... 192
9.4 Sturm-Liouville eigenvalue problem ... 192
Exercises 9.4 ... 198

Chapter 10 Special Functions ... 199

10.1 Bessel function ... 199
  10.1.1 Introduction to the Bessel equation ... 199
  10.1.2 The solution of the Bessel equation ... 201
  10.1.3 The recurrence formula of the Bessel function ... 204
  10.1.4 The properties of the Bessel function ... 207
  10.1.5 Application of Bessel function ... 210
Exercises 10.1 ... 213
10.2 Legendre polynomial ... 214
  10.2.1 Introduction of the Legendre equation ... 214
  10.2.2 The solution of the Legendre equation ... 216
  10.2.3 The properties of the Legendre polynomial and recurrence formula ... 218
  10.2.4 Application of Legendre polynomial ... 221
Exercises 10.2 ... 223

Chapter 11　Integral Transformations 224

　11.1　Fourier integral transformation 224
　　11.1.1　Definition of Fourier integral transformation 225
　　11.1.2　The properties of Fourier integral transformation 228
　　11.1.3　Convolution and its Fourier transformation 230
　　11.1.4　Application of Fourier integral transformation 231
　Exercises 11.1 235
　11.2　Laplace integral transformation 236
　　11.2.1　Definition of Laplace transformation 236
　　11.2.2　Properties of Laplace transformation 238
　　11.2.3　Convolution and its Laplace transformation 241
　　11.2.4　Application of Laplace integral transformation 242
　Exercises 11.2 244

References 245

Part I
Functions of a Complex Variable

Chapter 1 Complex Numbers and Complex Functions

In this chapter, we survey the algebraic and geometric structure of the complex number system. We assume various corresponding properties of real numbers to be known. Then we consider functions of a complex variable and develop the theory of limit and continuity.

1.1 Complex number and its operations

1.1.1 Complex number and its expression

It is well known that the equation $x^2 = -1$ is unsolvable in the real number system. So an imaginary unit i is introduced as a root of the above equation, that is, $i^2 = -1$.

Definition 1.1.1 Suppose that x and y are two real numbers, then
$$z = x + iy \tag{1.1.1}$$
is called a ***complex number***, where x and y are known as the ***real and imaginary parts*** of z, and we can write
$$\text{Re } z = x, \quad \text{Im } z = y. \tag{1.1.2}$$
When $x = 0$, $y \neq 0$, $z = iy$ is called a ***pure imaginary number***. The y axis is then referred to as the ***imaginary axis***.

When $y = 0$, $x \neq 0$, $z = x + i0$ is a real number, which means that complex number is the extension of the real number.

Two complex numbers $z_1 = x_1 + iy_1$ and $z_2 = x_2 + iy_2$ are ***equal*** whenever they have the same real parts and imaginary parts. A complex number is ***equal to zero*** if and only if its real and imaginary parts are zero simultaneously.

Complex numbers can also be defined as ordered pairs (x, y) of real numbers that are to be interpreted as points in the complex plane. It is natural to associate any nonzero complex number $z = x + iy$ with the directed line segment, or

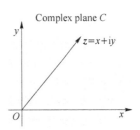

Fig. 1.1.1

vector, from the origin to the point (x,y) that represents z in the complex plane. In fact, we often refer to z as the point z or the vector z. In Fig. 1.1.1 the number $z=x+iy$ is displayed graphically as both point and radius vector.

The vector interpretation of complex number is especially helpful in extending the concept of absolute values of real numbers to the complex numbers.

Definition 1.1.2 The **modulus**, or absolute value, of a complex number $z=x+iy$ is defined as the nonnegative real number $\sqrt{x^2+y^2}$ and is denoted by $|z|$; that is,

$$|z|=\sqrt{x^2+y^2}. \tag{1.1.3}$$

Geometrically, the number $|z|$ is the distance between the point (x,y) and the origin, or the length of the vector representing z. It reduces to the usual absolute value in the real number system when $y=0$.

It also follows from definition 1.1.2 that the real numbers $|z|$, $\operatorname{Re} z=x$, and $\operatorname{Im} z=y$ are related by the equation

$$|z|^2=(\operatorname{Re} z)^2+(\operatorname{Im} z)^2. \tag{1.1.4}$$

Thus

$$\operatorname{Re} z \leqslant |\operatorname{Re} z| \leqslant |z| \quad \text{and} \quad \operatorname{Im} z \leqslant |\operatorname{Im} z| \leqslant |z|. \tag{1.1.5}$$

Let r and θ be polar coordinates of the point (x,y) that corresponds to a nonzero complex number $z=x+iy$, since $x=r\cos\theta$ and $y=r\sin\theta$, the number z can be written in polar form as

$$z=r(\cos\theta+i\sin\theta). \tag{1.1.6}$$

This is also called the **trigonometric expression** of a complex number.

Definition 1.1.3 The real number θ in (1.1.6) represents the angle, measured in radians, that z makes with the positive real axis when z is interpreted as a radius vector (Fig. 1.1.2), and it is called the **argument** of the complex number z and is denoted as $\theta=\arg z$.

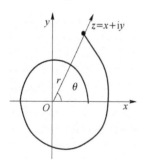

Fig. 1.1.2

If $z=0$, the coordinate θ is undefined; so it is always understood that $z\neq 0$ whenever $\arg z$ is discussed. As in calculus, θ has an infinite number of possible values, including negative ones, which differ by integral multiples of 2π. Those values can be determined from the equation $\tan\theta=\dfrac{y}{x}$, where the quadrant containing the point corresponding to z must be specified.

The **principal value** of $\arg z$, denoted by $\operatorname{Arg} z$, is that unique value Θ such that $-\pi<\Theta\leqslant\pi$, then

$$\arg z=\operatorname{Arg} z+2n\pi \quad (n=0, \pm 1, \pm 2, \cdots). \tag{1.1.7}$$

Also, when z is a negative real number, $\operatorname{Arg} z$ has value π, not $-\pi$.

Note When $z=x+iy\neq 0$, the principal value of the argument of z can be determined by the following formula

$$\operatorname{Arg} z = \begin{cases} \arctan \dfrac{y}{x}, & x \geqslant 0, \\ \arctan \dfrac{y}{x} + \pi, & x < 0, y \geqslant 0, \\ \arctan \dfrac{y}{x} - \pi, & x \leqslant 0, y < 0. \end{cases} \quad (1.1.8)$$

Example 1.1.1 Find the modulus and argument of $z = -1 - i$.

Solution It is obvious that $|z| = \sqrt{(-1)^2 + (-1)^2} = \sqrt{2}$.

This complex number lies in the third quadrant, has principal argument $-\dfrac{3\pi}{4}$. That is,

$$\operatorname{Arg}(-1-i) = -\dfrac{3\pi}{4}.$$

It must be emphasized that, because of the restriction $-\pi < \Theta \leqslant \pi$ of the principal argument Θ, it is not true that $\operatorname{Arg}(-1-i) = \dfrac{5\pi}{4}$.

And then

$$\arg(-1-i) = -\dfrac{3\pi}{4} + 2n\pi, \quad (n = 0, \pm 1, \pm 2, \cdots).$$

Example 1.1.2 Find the modulus and argument of $z_1 = 2 - 2i$ and $z_2 = -3 + 4i$.

Solution $|z_1| = \sqrt{2^2 + (-2)^2} = 2\sqrt{2}$, $|z_2| = \sqrt{3^2 + 4^2} = 5$.

Since $\operatorname{Re} z_1 = 2 > 0$, then $\operatorname{Arg} z_1 = \arctan(-1) = -\dfrac{\pi}{4}$, and

$$\arg z_1 = -\dfrac{\pi}{4} + 2n\pi, \quad (n = 0, \pm 1, \pm 2, \cdots).$$

Notice that $\operatorname{Re} z_2 = -3 < 0$, and $\operatorname{Im} z_2 = 4 > 0$, then

$$\operatorname{Arg} z_2 = \arctan\left(-\dfrac{4}{3}\right) + \pi = \pi - \arctan\dfrac{4}{3},$$

and

$$\arg z_2 = \pi - \arctan\dfrac{4}{3} + 2n\pi, \quad (n = 0, \pm 1, \pm 2, \cdots).$$

According to Euler's formula

$$e^{i\theta} = \cos\theta + i\sin\theta, \quad (1.1.9)$$

we can rewrite the trigonometric expression (1.1.6) more compactly in **exponential form** as

$$z = re^{i\theta}. \quad (1.1.10)$$

The choice of the symbol $e^{i\theta}$ will be fully motivated later in next subsection and its use will, however, suggest that it is a natural choice.

According to (1.1.10), the number $-1-i$ in Example 1.1.1 has exponential form

$$-1-i = \sqrt{2}\exp\left[i\left(-\dfrac{3\pi}{4}\right)\right]. \quad (1.1.11)$$

With the agreement that $e^{-i\theta} = e^{i(-\theta)}$, this can also be written $-1-i = \sqrt{2}\,e^{-i\frac{3\pi}{4}}$. Expression (1.1.11) is, of course, only one of an infinite number of possibilities for the exponential form of $-1-i$,

$$-1-\mathrm{i}=\sqrt{2}\exp\left[\mathrm{i}\left(-\frac{3\pi}{4}+2n\pi\right)\right], \quad (n=0, \pm 1, \pm 2, \cdots). \tag{1.1.12}$$

Example 1.1.3 Express the following complex numbers in their exponential form:

(1) $z=1+\mathrm{i}$;   (2) $z=\sin\frac{\pi}{5}+\mathrm{i}\cos\frac{\pi}{5}$.

Solution (1) Since $r=|z|=\sqrt{1^2+1^2}=\sqrt{2}$, $H=\mathrm{Arg}\,z=\arctan 1=\frac{\pi}{4}$, the exponential form of z is $z=\sqrt{2}\exp\left(\mathrm{i}\frac{\pi}{4}\right)$.

(2) Since $r=|z|=1$, and

$$\sin\frac{\pi}{5}=\cos\left(\frac{\pi}{2}-\frac{\pi}{5}\right)=\cos\left(\frac{3\pi}{10}\right), \quad \cos\frac{\pi}{5}=\sin\left(\frac{\pi}{2}-\frac{\pi}{5}\right)=\sin\left(\frac{3\pi}{10}\right),$$

then the exponential form of z is $z=\exp\left(\mathrm{i}\frac{3\pi}{10}\right)$.

Definition 1.1.4 The *complex conjugate*, or simply the conjugate, of a complex number $z=x+\mathrm{i}y$ is defined as the complex number $x-\mathrm{i}y$ and is denoted by $\bar{z}$; that is,

$$\bar{z}=x-\mathrm{i}y. \tag{1.1.13}$$

The number $\bar{z}$ is represented by the point $(x,-y)$, which is the reflection in the real axis of the point (x,y) representing z (Fig. 1.1.3).

Note that

$$\bar{\bar{z}}=z \text{ and } |\bar{z}|=|z| \tag{1.1.14}$$

for all z.

It is also seen from Fig. 1.1.3 that

$$\mathrm{Arg}\,z=-\mathrm{Arg}\,\bar{z}. \tag{1.1.15}$$

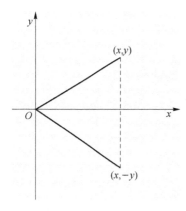

Fig. 1.1.3

1.1.2 The operations of complex numbers

Definition 1.1.5 The *sum* z_1+z_2 and the *product* $z_1 z_2$ of two complex numbers $z_1=x_1+\mathrm{i}y_1$ and $z_2=x_2+\mathrm{i}y_2$ are defined as follows:

$$z_1+z_2=(x_1+x_2)+\mathrm{i}(y_1+y_2) \tag{1.1.16}$$

$$z_1 z_2=(x_1 x_2-y_1 y_2)+\mathrm{i}(y_1 x_2+x_1 y_2) \tag{1.1.17}$$

Observe that the right-hand sides of these equations can be obtained by formally manipulating terms on the left as if they involved only real numbers and by replacing i^2 by -1 when it occurs.

Various properties of addition and multiplication of complex numbers are the same as for real numbers. We list here the more basic of these algebraic properties.

(1) *commutative laws*

$$z_1+z_2=z_2+z_1, \quad z_1 z_2=z_2 z_1. \tag{1.1.18}$$

(2) *associative laws*
$$(z_1+z_2)+z_3=z_1+(z)_2+z_3), \quad (z)_1z_2)z_3=z_1(z_2z_3). \tag{1.1.19}$$

(3) *distributive law*
$$z(z_1+z_2)=zz_1+zz_2. \tag{1.1.20}$$

There is associated with each complex number $z=(x,y)$ an *additive inverse* $-z=(-x,-y)$, satisfying the equation $z+(-z)=0$.

For any nonzero complex number $z=(x,y)$, there is a number z^{-1} such that $zz^{-1}=1$. According to (1.1.17), the *multiplicative inverse* of $z=(x,y)$ is

$$z^{-1}=\left(\frac{x}{x^2+y^2},\frac{-y}{x^2+y^2}\right), \quad (z\neq 0). \tag{1.1.21}$$

Definition 1.1.6 *Subtraction* and *division* are defined in terms of additive and multiplicative inverse:

$$z_1-z_2=z_1+(-z_2), \tag{1.1.22}$$

$$\frac{z_1}{z_2}=z_1z_2^{-1}, \quad (z_2\neq 0). \tag{1.1.23}$$

So if $z_1=(x_1,y_1)$ and $z_2=(x_2,y_2)$, then

$$z_1-z_2=(x_1-x_2,y_1-y_2)=(x_1-x_2)+i(y_1-y_2), \tag{1.1.24}$$

and

$$\frac{z_1}{z_2}=\frac{x_1x_2+y_1y_2}{x_2^2+y_2^2}+i\frac{y_1x_2-x_1y_2}{x_2^2+y_2^2}, \quad (z_2\neq 0). \tag{1.1.25}$$

According to the definition of complex conjugate, the operations about the complex conjugate have the following properties:

(1) $\overline{z_1+z_2}=\overline{z_1}+\overline{z_2}$, $\overline{z_1-z_2}=\overline{z_1}-\overline{z_2}$, $\overline{z_1z_2}=\overline{z_1}\,\overline{z_2}$, $\overline{\left(\frac{z_1}{z_2}\right)}=\frac{\overline{z_1}}{\overline{z_2}}$, $(z_2\neq 0)$;

(2) $\operatorname{Re} z=\frac{z+\overline{z}}{2}$, $\operatorname{Im} z=\frac{z-\overline{z}}{2i}$;

(3) $z\overline{z}=|z|^2$.

It suggests the method for determining a quotient $\frac{z_1}{z_2}$, which is based on multiplying both the numerator and the denominator of $\frac{z_1}{z_2}$ by $\overline{z_2}$, so that the denominator becomes the real number $|z_2|^2$. That is

$$\frac{z_1}{z_2}=\frac{z_1\overline{z_2}}{|z_2|^2}. \tag{1.1.26}$$

According to the geometric meaning of a complex number, the sum and difference of two complex numbers are in accordance with the sum and difference of two vectors. Then we turn now to the *triangle inequality*, which provides an upper bound for the modulus of the sum of two complex numbers z_1 and z_2:

$$|z_1+z_2|\leqslant|z_1|+|z_2|. \tag{1.1.27}$$

This important inequality is geometrically evident in Fig. 1.1.4, since it is merely a statement that the length of one side of a triangle is less than or equal to the sum of the

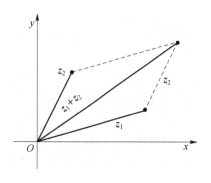

Fig. 1.1.4

lengths of the other two sides.

An immediate consequence of the triangle inequality is the fact that
$$|z_1+z_2| \geqslant ||z_1|-|z_2||. \qquad (1.1.28)$$
To derive inequality (1.1.28), we write
$$|z_1|=|(z_1+z_2)+(-z_2)| \leqslant |z_1+z_2|+|-z_2|,$$
which means that
$$|z_1+z_2| \geqslant |z_1|-|z_2|. \qquad (1.1.29)$$
This is inequality (1.1.28) when $|z_1| \geqslant |z_2|$. If $|z_1|<|z_2|$, we need only interchange z_1 and z_2 in inequality (1.1.29) to get
$$|z_1+z_2| \geqslant -(|z_1|-|z_2|),$$
which is the desired result. Inequality (1.1.28) tells us, of course, that the length of one side of a triangle is greater than or equal to the difference of the lengths of the other two sides.

Because $|-z_2|=|z_2|$, one can replace z_2 by $-z_2$ in inequalities (1.1.27) and (1.1.28) to summarize the result in a particularly useful form:
$$||z_1|-|z_2|| \leqslant |z_1 \pm z_2| \leqslant |z_1|+|z_2|. \qquad (1.1.30)$$

The triangle inequality (1.1.27) can be generalized by means of mathematical induction to sums involving any finite number of terms:
$$|z_1+z_2+\cdots+z_n| \leqslant |z_1|+|z_2|+\cdots+|z_n|, \quad (n=2,3,\cdots). \qquad (1.1.31)$$

To give details of the induction proof here, we note that when $n=2$, inequality (1.1.31) is just inequality (1.1.27). Furthermore, if inequality (1.1.31) is assumed to be valid when $n=m$, it must also hold when $n=m+1$ since, by inequality (1.1.27),
$$|(z_1+z_2+\cdots+z_m)+z_{m+1}| \leqslant |z_1+z_2+\cdots+z_m|+|z_{m+1}| \leqslant$$
$$(|z_1|+|z_2|+\cdots+|z_m|)+|z_{m+1}|.$$

With the aid of the exponential form of a complex number, we can rewrite the product of two complex numbers. If $z_1=r_1 e^{i\theta_1}$ and $z_2=r_2 e^{i\theta_2}$, the product $z_1 z_2$ has exponential form
$$z_1 z_2=r_1 r_2 e^{i\theta_1} e^{i\theta_2}=r_1 r_2 e^{i(\theta_1+\theta_2)}. \qquad (1.1.32)$$
Moreover,
$$\frac{z_1}{z_2}=\frac{r_1}{r_2}\frac{e^{i\theta_1}-e^{i\theta_2}}{e^{i\theta_2}-e^{i\theta_2}}=\frac{r_1}{r_2}\frac{e^{i(\theta_1-\theta_2)}}{e^{i0}}=\frac{r_1}{r_2}e^{i(\theta_1-\theta_2)}. \qquad (1.1.33)$$

Because $1=1e^{i0}$, it follows from expression (1.1.33) that the inverse of any nonzero complex number $z=re^{i\theta}$ is
$$z^{-1}=\frac{1}{z}=\frac{1}{r}e^{-i\theta}. \qquad (1.1.34)$$

Expressions (1.1.32), (1.1.33) and (1.1.34) are, of course, easily remembered by applying the usual algebraic rules for real numbers and e^x.

Expression (1.1.32) yields an important identity involving arguments:
$$\arg(z_1 z_2)=\arg z_1 + \arg z_2. \qquad (1.1.35)$$

It is to be interpreted as saying that if values of two of these three (multiple-valued) arguments are specified, then there is a value of the third such that the equation holds.

Example 1.1.4 When $z_1 = -1$ and $z_2 = i$,
$$\mathrm{Arg}(z_1 z_2) = \mathrm{Arg}(-i) = -\frac{\pi}{2},$$
but
$$\mathrm{Arg}\, z_1 + \mathrm{Arg}\, z_2 = \pi + \frac{\pi}{2} = \frac{3\pi}{2}.$$

If, however, we take the values of arg z_1 and arg z_2 just used and select the value
$$\mathrm{Arg}(z_1 z_2) + 2\pi = -\frac{\pi}{2} + 2\pi = \frac{3\pi}{2}$$
of $\arg(z_1 z_2)$, we find that equation (1.1.35) is satisfied.

Statement (1.1.35) tells us that
$$\arg\left(\frac{z_1}{z_2}\right) = \arg(z_1 z_2^{-1}) = \arg z_1 + \arg(z_2^{-1}) = \arg z_1 - \arg z_2. \tag{1.1.36}$$

Example 1.1.5 In order to find the principal argument Arg z when
$$z = \frac{-2}{1+\sqrt{3}i},$$
observe that
$$\arg z = \arg(-2) - \arg(1+\sqrt{3}i).$$

Since
$$\mathrm{Arg}(-2) = \pi \quad \text{and} \quad \mathrm{Arg}(1+\sqrt{3}i) = \frac{\pi}{3},$$
one value of arg z is $\frac{2\pi}{3}$, and because $\frac{2\pi}{3}$ is between $-\pi$ and π, we find that Arg $z = \frac{2\pi}{3}$.

Another important result that can be obtained formally by applying rules for real numbers to $z = re^{i\theta}$ is
$$z^n = r^n e^{in\theta}, \quad (n=0, \pm 1, \pm 2, \cdots). \tag{1.1.37}$$

Proof It is easily verified for positive values of n by mathematical induction. To be specific, we first note that it becomes $z = re^{i\theta}$ when $n = 1$. Next, we assume that it is valid when $n = m$, where m is any positive integer. In view of expression (1.1.32) for the product of two nonzero complex numbers in exponential form, it is then valid for $n = m+1$:
$$z^{m+1} = zz^m = re^{i\theta} r^m e^{im\theta} = r^{m+1} e^{i(m+1)\theta}.$$

Expression (1.1.37) is thus verified when n is a positive integer. It also holds when $n = 0$, with the convention that $z^0 = 1$. If $n = -1, -2, \cdots$, on the other hand, we define z^n in terms of the multiplicative inverse of z by writing
$$z^n = (z^{-1})^m \quad \text{where} \quad m = -n = 1, 2, \cdots.$$

Then, since expression (1.1.37) is valid for positive integer powers, it follows from the exponential form (1.1.34) of z^{-1} that
$$z^n = \left[\frac{1}{r} e^{i(-\theta)}\right]^m = \left(\frac{1}{r}\right)^m e^{im(-\theta)} = \left(\frac{1}{r}\right)^{-n} e^{i(-n)(-\theta)}$$
$$= r^n e^{in\theta}, \quad (n = -1, -2, \cdots).$$

Expression (1.1.37) is now established for all integer powers.

Observe that if $r=1$, expression (1.1.37) becomes
$$(e^{i\theta})^n = e^{in\theta}, \quad (n=0, \pm 1, \pm 2, \cdots). \tag{1.1.38}$$
When written in the form
$$(\cos\theta + i\sin\theta)^n = \cos n\theta + i\sin n\theta \quad (n=0, \pm 1, \pm 2, \cdots), \tag{1.1.39}$$
it is known as **de Moivre's formula**.

Expression (1.1.37) is useful in finding powers of complex numbers even when they are given in rectangular form and the result is desired in that form.

Example 1.1.6 Put $(\sqrt{3}+i)^7$ in rectangular form.

Solution We can write
$$(\sqrt{3}+i)^7 = (2e^{i\frac{\pi}{6}})^7 = 2^7 e^{i\frac{7\pi}{6}} = (2^6 e^{i\pi})(2e^{i\frac{\pi}{6}}) = -64(\sqrt{3}+i).$$

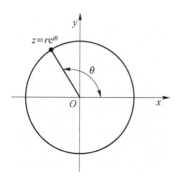

Fig. 1.1.5

Consider now a point $z=re^{i\theta}$, lying on a circle centered at the origin with radius r (Fig. 1.1.5). As θ is increased, z moves around the circle in the counterclockwise direction. In particular, when θ is increased by 2π, we arrive at the original point, and the same is true when θ is decreased by 2π. It is, therefore, evident from Fig. 1.1.5 that two nonzero complex numbers
$$z_1 = r_1 e^{i\theta_1} \quad \text{and} \quad z_2 = r_2 e^{i\theta_2}$$
are equal if and only if
$$r_1 = r_2 \quad \text{and} \quad \theta_1 = \theta_2 + 2k\pi,$$
with k being some integer ($k=0, \pm 1, \pm 2, \cdots$).

This observation, together with the expression $z^n = r^n e^{in\theta}$ for integer powers of complex numbers $z=re^{i\theta}$, is useful in finding the **nth roots** of any nonzero complex number $z_0 = r_0 e^{i\theta_0}$, where n has one of the values $n=2, 3, \cdots$.

The method starts with the fact that an nth root of z_0 is a nonzero number $z=re^{i\theta}$ such that $z^n = z_0$, or
$$r^n e^{in\theta} = r_0 e^{i\theta_0}.$$
According to the statement above,
$$r^n = r_0 \quad \text{and} \quad n\theta = \theta_0 + 2k\pi,$$
where k is any integer ($k=0, \pm 1, \pm 2, \cdots$). So $r = \sqrt[n]{r_0}$, where this radical denotes the unique positive nth root of the positive real number r_0, and
$$\theta = \frac{\theta_0 + 2k\pi}{n} = \frac{\theta_0}{n} + \frac{2k\pi}{n}, \quad (k=0, \pm 1, \pm 2, \cdots).$$

Consequently, the complex numbers
$$z = \sqrt[n]{r_0} \exp\left[i\left(\frac{\theta_0}{n} + \frac{2k\pi}{n}\right)\right], \quad (k=0, \pm 1, \pm 2, \cdots) \tag{1.1.40}$$

are the **nth roots** of z_0. We are able to see immediately from this exponential form of the roots that they all lie on the circle $|z|=\sqrt[n]{r_0}$ about the origin and are equally spaced every $\dfrac{2\pi}{n}$ radians, starting with argument $\dfrac{\theta_0}{n}$. Evidently, then, all of the distinct roots are obtained when $k=0, 1, 2, \cdots, n-1$, and no further roots arise with other values of k. We let $c_k (k=0,1,2,\cdots,n-1)$ denote these distinct roots and write

$$c_k = \sqrt[n]{r_0}\exp\left[i\left(\dfrac{\theta_0}{n}+\dfrac{2k\pi}{n}\right)\right], \quad (k=0, 1, 2, \cdots, n-1). \tag{1.1.41}$$

(See Fig. 1.1.6)

The number $\sqrt[n]{r_0}$ is the length of each of the radius vectors representing the nth roots. The first root c_0 has argument $\dfrac{\theta_0}{n}$; and the two roots when $n=2$ lie at the opposite ends of a diameter of the circle $|z|=\sqrt[n]{r_0}$, the second root being $-c_0$. When $n\geqslant 3$, the roots lie at the vertices of a regular polygon of n sides inscribed in that circle.

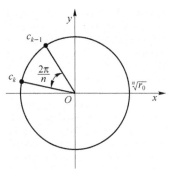

Fig. 1.1.6

We shall let $z_0^{\frac{1}{n}}$ denote the set of nth roots of z_0. If, in particular, z_0 is a positive real number r_0, the symbol $r_0^{\frac{1}{n}}$ denotes the entire set of roots; and the symbol $\sqrt[n]{r_0}$ in expression (1.1.41) is reserved for the one positive root. When the value of θ_0 that is used in expression (1.1.41) is the principal value of $\arg z_0 (-\pi<\theta_0\leqslant\pi)$, the number c_0 is referred to as the **principal root**. Thus when z_0 is a positive real number r_0, its principal root is $\sqrt[n]{r_0}$.

Finally, a convenient way to remember expression (1.1.41) is to write z_0 in its most general exponential form

$$z_0 = r_0 e^{i(\theta_0+2k\pi)}, \quad (k=0, \pm 1, \pm 2, \cdots) \tag{1.1.42}$$

and to formally apply laws of fractional exponents involving real numbers, keeping in mind that there are precisely n roots:

$$z_0^{\frac{1}{n}} = [r_0 e^{i(\theta_0+2k\pi)}]^{\frac{1}{n}} = \sqrt[n]{r_0}\exp\left[\dfrac{i(\theta_0+2k\pi)}{n}\right] \tag{1.1.43}$$

$$= \sqrt[n]{r_0}\exp\left[i\left(\dfrac{\theta_0}{n}+\dfrac{2k\pi}{n}\right)\right], \quad (k=0, 1, 2, \cdots, n-1)$$

Example 1.1.7 Determine the nth roots of unity.

Solution We can write

$$1 = 1\exp[i(0+2k\pi)], \quad (k=0, \pm 1, \pm 2, \cdots)$$

and find that

$$1^{\frac{1}{n}} = \sqrt[n]{1}\exp\left[i\left(\dfrac{0}{n}+\dfrac{2k\pi}{n}\right)\right] = \exp\left[i\left(\dfrac{2k\pi}{n}\right)\right], \quad (k=0, 1, 2, \cdots, n-1).$$

When $n=2$, these roots are, of course, ± 1. When $n\geqslant 3$, the regular polygon at whose vertices the roots lie is inscribed in the unit circle $|z|=1$, with one vertex corresponding to the principal root $z=1, (k=0)$.

If we write
$$\omega_n = \exp\left(i\frac{2\pi}{n}\right),$$
it follows form property of $e^{i\theta}$ that
$$\omega_n^k = \exp\left(i\frac{2k\pi}{n}\right), \quad (k=0, 1, 2, \cdots, n-1).$$
Hence the distinct nth roots of unity just found are simply
$$1, \omega_n, \omega_n^2, \cdots, \omega_n^{n-1}.$$
See Fig. 1.1.7, where the cases $n=3$, 4 and 6 are illustrated. Note that $\omega_n^n = 1$.

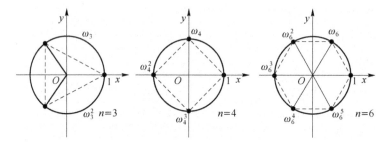

Fig. 1.1.7

Finally, it is worthwhile observing that if c is any particular nth root of a nonzero complex number z_0, the set of nth roots can be put in the form
$$c, c\omega_n, c\omega_n^2, \cdots, c\omega_n^{n-1}.$$
This is because multiplication of any nonzero complex number by ω_n increases the argument of that number by $\frac{2\pi}{n}$, while leaving its modulus unchanged.

Example 1.1.8 To evaluate $(-8i)^{\frac{1}{3}}$.

Solution We can write
$$-8i = 8\exp\left[i\left(-\frac{\pi}{2}+2k\pi\right)\right], \quad (k=0, \pm 1, \pm 2, \cdots)$$
to see that the desired roots are
$$c_k = 2\exp\left[i\left(-\frac{\pi}{6}+\frac{2k\pi}{3}\right)\right], \quad (k=0, 1, 2).$$
They lie at the vertices of an equilateral triangle, inscribed in the circle $|z|=2$, and are equally spaced around that circle every $\frac{2\pi}{3}$ radians, starting with the principal root
$$c_0 = 2\exp\left[i\left(-\frac{\pi}{6}\right)\right] = 2\left(\cos\frac{\pi}{6} - i\sin\frac{\pi}{6}\right) = \sqrt{3} - i.$$
Without any further calculations, it is then evident that $c_1 = 2i$; and, since c_2 is symmetric to c_0 with respect to the imaginary axis, we know that $c_2 = -\sqrt{3} - i$.

These roots can, of course, be written as
$$c_0, c_0\omega_3, c_0\omega_3^2 \quad \text{where} \quad \omega_3 = \exp\left(i\frac{2\pi}{3}\right).$$

1.1.3 Regions in the complex plane

In this section, we are concerned with sets of complex numbers, or points in the z plane, and their closeness to one another.

Definition 1.1.7 An **ε neighborhood** of a given point z_0 is defined as a set

$$|z-z_0|<\varepsilon, \tag{1.1.44}$$

which consists of all points z lying inside but not on a circle centered at z_0 and with a specified positive radius ε. When the value of ε is understood or is immaterial in the discussion, the set (1.1.44) is often referred to as just a neighborhood. Occasionally, it is convenient to speak of *a deleted neighborhood*

$$0<|z-z_0|<\varepsilon \tag{1.1.45}$$

consisting of all points z in an ε neighborhood of z_0 except for the point z_0 itself.

Definition 1.1.8 A point z_0 is said to be an ***interior point*** of a set S whenever there is some neighborhood of z_0 that contains only points of S; it is called an ***exterior point*** of S when there exists a neighborhood of it containing no points of S. A ***boundary point*** is, therefore, a point all of whose neighborhoods contain points in S and points not in S.

The circle $|z|=1$, for instance, is the boundary of each of the sets

$$|z|<1 \text{ and } |z|\leqslant 1. \tag{1.1.46}$$

Definition 1.1.9 A set is ***open*** if each of its points is an interior point. A set is ***closed*** if it contains all of its boundary points, and the ***closure*** of a set S is the closed set consisting of all points in S together with the boundary of S.

According to the definition 1.1.9, the first of the sets (1.1.46) is open and that the second is its closure.

Some sets are, of course, neither open nor closed. For a set to be not open, there must be a boundary point that is contained in the set; and if a set is not closed, there exists a boundary point not contained in the set. Observe that the punctured disk $0<|z|\leqslant 1$ is neither open nor closed. The set of all complex numbers is, on the other hand, both open and closed since it has no boundary points.

Definition 1.1.10 An open set S is ***connected*** if each pair of points z_1 and z_2 in it can be joined by a polygonal line, consisting of a finite number of line segments joined end to end, that lies entirely in S.

For example, the open set $|z|<1$ is connected. The annulus $1<|z|<2$ is, of course, open and it is also connected.

Definition 1.1.11 An open set that is connected is called a ***domain***. A domain together with some, none, or all of its boundary points is referred to as a ***region***.

Definition 1.1.12 A set S is ***bounded*** if every point of S lies inside some circle $|z|=R$; otherwise, it is ***unbounded***.

For instance, both of the sets (1.1.46) are bounded regions, and the half plane Re $|z|\geqslant 0$ is unbounded.

Definition 1.1.13 A point z_0 is said to be an ***accumulation point*** of a set S if each deleted neighborhood of z_0 contains at least one point of S.

It follows that if a set S is closed, then it contains each of its accumulation points. For if an accumulation point z_0 was not in S, it would be a boundary point of S; but this contradicts the fact that a closed set contains all of its boundary points. It is left as an exercise to show that the converse is, in fact, true. Thus, a set is closed if and only if it contains all of its accumulation points.

Evidently, a point z_0 is not an accumulation point of a set S whenever there exists some deleted neighborhood of z_0 that does not contain points of S.

Exercises 1.1

1. Show that

   (a) $\text{Re}(iz) = -\text{Im } z$; (b) $\text{Im}(iz) = \text{Re } z$.

2. Reduce each of these quantities to a real number:

   (a) $\dfrac{1+2i}{3-4i} + \dfrac{2-i}{5i}$; (b) $\dfrac{5i}{(1-i)(2-i)(3-i)}$; (c) $(1-i)^4$.

3. Use mathematical induction to verify the binomial formula

$$(z_1 + z_2)^n = \sum_{k=0}^{n} \binom{n}{k} z_1^k z_2^{n-k}, \quad (n=1, 2, \cdots).$$

4. By factoring $z^4 - 4z^2 + 3$ into two quadratic factors and then using the triangle inequality, show that if z lies on the circle $|z|=2$, then

$$\left| \frac{1}{z^4 - 4z^2 + 3} \right| \leqslant \frac{1}{3}.$$

5. Prove that

   (a) z is real if and only if $\bar{z} = z$;

   (b) z is either real or pure imaginary if and only if $\overline{z^2} = z^2$.

6. Follow the steps below to give an algebraic derivation of the triangle inequality

$$|z_1 + z_2| \leqslant |\overline{z_1}| + |\overline{z_2}|.$$

   (a) Show that $|z_1+z_2|^2 = (z_1+z_2)(\overline{z_1}+\overline{z_2}) = z_1\overline{z_1} + (z_1\overline{z_2} + \overline{z_1}z_2) + z_2\overline{z_2}$.

   (b) Point out why $z_1\overline{z_2} + \overline{z_1}z_2 = 2\text{Re}(z_1\overline{z_2}) \leqslant 2|z_1||z_2|$.

   (c) Use the results in parts (a) and (b) to obtain the inequality $|z_1+z_2|^2 \leqslant (|z_1|+|z_2|)^2$, and note how the triangle inequality follows.

7. Find the principal argument Arg z when

   (a) $z = \dfrac{i}{-2-2i}$; (b) $z = (\sqrt{3}-i)^6$; (c) $z = 1 - \sqrt{3}i$.

8. Show that

   (a) $i(1-\sqrt{3}i)(\sqrt{3}+i) = 2(1+\sqrt{3}i)$; (b) $\dfrac{5i}{2+i} = 1 + 2i$;

(c) $(-1+i)^7 = -8(1+i)$; (d) $(1+\sqrt{3}i)^{-10} = 2^{-11}(-1+\sqrt{3}i)$.

9. Establish the identity
$$1+z+z^2+\cdots+z^n = \frac{1-z^{n+1}}{1-z}, \quad (z\neq 1)$$
and then use it to derive Lagrange's trigonometric identity:
$$1+\cos\theta+\cos 2\theta+\cdots+\cos n\theta = \frac{1}{2}+\frac{\sin\left[(2n+1)\dfrac{\theta}{2}\right]}{2\sin\left(\dfrac{\theta}{2}\right)}, \quad (0<\theta\leqslant 2\pi).$$

Suggestion As for the first identity, write $S=1+z+z^2+\cdots+z^n$ and consider the difference $S-zS$. To derive the second identity, write $z=e^{i\theta}$ in the first one.

10. Evaluate the following expressions
    (a) $\sqrt{2i}$; (b) $\sqrt[3]{i}$; (c) $\sqrt[4]{-1}$; (d) $\sqrt{1+i}$.

11. Write down the trigonometric and exponential expressions of the following complex numbers:
    (a) $1+i$; (b) i; (c) -2; (d) $1-\cos\theta+i\sin\theta$, $(0\leqslant\theta\leqslant\pi)$.

12. Let n be a natural number and $x_n+iy_n=(1+i\sqrt{3})^n$, show that
$$x_{n-1}y_n - x_n y_{n-1} = 4^{n-1}\sqrt{3}.$$

13. Show that if c is any nth root of unity other than unity itself, then
$$1+c+c^2+\cdots+c^{n-1}=0.$$

14. Sketch the following sets and determine which are domains:
    (a) $|z-2+i|\leqslant 1$; (b) $|2z+3|>4$; (c) $\text{Im } z>1$; (d) $\text{Im } z=1$;
    (e) $0\leqslant\arg z\leqslant\dfrac{\pi}{4}$, $(z\neq 0)$; (f) $|z-4|\geqslant|z|$.

15. Which sets in Exercise 14 are neither open nor closed?

16. Which sets in Exercise 14 are bounded?

17. Prove that if a set contains each of its accumulation points, then it must be a closed set.

18. Show that any point z_0 of a domain is an accumulation point of that domain.

1.2 Functions of a complex variable

1.2.1 Definition of function of a complex variable

Definition 1.2.1 Let S be a set of complex numbers. A *function* f defined on S is a rule that assigns to each z in S a complex number w. The number w is called the *value* of f at z and is defined by $f(z)$; that is, $w=f(z)$. The set S is called the *domain of definition* of f.

It must be emphasized that both a domain of definition and a rule are needed in order for a function to be well defined. When the domain of definition is not mentioned, we agree that the largest possible set is to be taken. Also, it is not always convenient to use notation that distinguishes between a given function and its values.

Suppose that $w=u+iv$ is the value of a function f at $z=x+iy$, so that
$$u+iv=f(x+iy).$$
Each of the real numbers u and v depends on the real variables x and y, and it follows that $f(z)$ can be expressed in terms of a pair of real-valued functions of the real variables x and y:
$$f(z)=u(x,y)+iv(x,y). \tag{1.2.1}$$
If the polar coordinates r and θ, instead of x and y, are used, then
$$u+iv=f(re^{i\theta}),$$
where $w=u+iv$ and $z=re^{i\theta}$. In that case, we may write
$$f(z)=u(r,\theta)+iv(r,\theta). \tag{1.2.2}$$

Example 1.2.1 If $f(z)=z^2$, then
$$f(x+iy)=(x+iy)^2=x^2-y^2+i2xy.$$
Hence
$$u(x,y)=x^2-y^2 \quad \text{and} \quad v(x,y)=2xy.$$
When polar coordinates are used,
$$f(re^{i\theta})=(re^{i\theta})^2=r^2 e^{i2\theta}=r^2\cos 2\theta+ir^2\sin 2\theta.$$
Consequently,
$$u(r,\theta)=r^2\cos 2\theta \quad \text{and} \quad v(r,\theta)=r^2\sin 2\theta.$$

If, in either of equations (1.2.1) and (1.2.2), the function v always has value zero, then the value of f is always real. That is, f is a real-valued function of a complex variable.

Example 1.2.2 A real-valued function that is used to illustrate some important concepts later in this chapter is
$$f(z)=|z|^2=x^2+y^2+i0.$$

If n is zero or a positive integer and if $a_0, a_1, a_2, \cdots, a_n$ are complex constants, where $a_n \neq 0$, the function
$$P(z)=a_0+a_1 z+a_2 z^2+\cdots+a_n z^n$$
is a ***polynomial of degree*** n. Note that the sum here has a finite number of terms and that the domain of definition is the entire z plane. Quotients $\dfrac{P(z)}{Q(z)}$ of polynomials are called rational functions and are defined at each point z where $Q(z)\neq 0$. Polynomials and rational functions constitute elementary, but important, classes of functions of a complex variable.

Example 1.2.3 Let z denote any nonzero complex number. We know that $z^{\frac{1}{2}}$ has two values
$$z^{\frac{1}{2}}=\pm\sqrt{r}\exp\left(i\frac{\Theta}{2}\right),$$
where $r=|z|$ and $\Theta(-\pi<\Theta\leqslant\pi)$ is the principal value of arg z. But, if we choose only the

positive value of $\pm\sqrt{r}$ and write

$$f(z)=\sqrt{r}\exp\left(i\frac{\Theta}{2}\right), \quad (r>0, -\pi<\Theta\leqslant\pi), \tag{1.2.3}$$

the (single-valued) function (1.2.3) is well defined on the set of nonzero numbers in the z plane. Since zero is the only square root of zero, we also write $f(0)=0$. The function f is then well defined on the entire plane.

1.2.2 Complex mappings

Properties of a real-valued function of a real variable are often exhibited by the graph of the function. But when $w=f(z)$, where z and w are complex, no such convenient graphical representation of the function f is available because each of the number z and w is located in a plane rather than on a line. One can, however, display some information about the function by indicating pairs of corresponding points $z=(x, y)$ and $w=(u, v)$. To do this, it is generally simpler to draw the z and w planes separately.

When a function f is thought of in this way, it is often referred to as a ***mapping***, or ***transformation***. The ***image*** of a point z in the domain of definition S is the point $w=f(z)$, and the set of images of all points in a set T that is contained in S is called ***range*** of f. The ***inverse image*** of a point w is the set of all points z in the domain of definition of f that have w as their images. The inverse image of a point may contain just one point, many points, or none at all. The last case occurs, of course, when w is not in the range of f.

Terms such as ***translation***, ***rotation***, and ***reflection*** are used to convey dominant geometric characteristics of certain mappings. In such cases, it is sometimes convenient to consider the z and w planes to be the same. For example, the mapping

$$w=z+1=(x+1)+iy,$$

where $z=x+iy$, can be thought of as a translation of each point z one unit to the right. Since $i=e^{i\frac{\pi}{2}}$, the mapping

$$w=iz=r\exp\left[i\left(\theta+\frac{\pi}{2}\right)\right],$$

where $z=re^{i\theta}$, rotates the radius vector for each nonzero point z through a right angle about the origin in the counterclockwise direction; and the mapping

$$w=\bar{z}=x-iy$$

transforms each point $z=x+iy$ into its reflection in the real axis.

More information is usually exhibited by sketching images of curves and regions than by simply indicating images of individual points. In the following three examples, we illustrate this with the transformation $w=z^2$. We begin by finding the images of some ***curves*** in the z plane.

Example 1.2.4 According to Example 1.2.2, the mapping $w=z^2$ can be thought of as the transformation

$$u=x^2-y^2, \quad v=2xy \tag{1.2.4}$$

from the xy plane into the uv plane. This form of the mapping is especially useful in finding the images of certain hyperbolas.

It is easy to show, for instance, that each branch of a hyperbola
$$x^2-y^2=c_1, \quad (c_1>0) \tag{1.2.5}$$
is mapped in a one to one manner onto the vertical line $u=c_1$. We start by noting from the first of equations (1.2.4) that $u=c_1$ when (x,y) is a point lying on either branch. When, in particular, it lies on the right-hand branch, the second equation of (1.2.4) tells us that $v=2y\sqrt{y^2+c_1}$. Thus the image of the right-hand branch can be expressed parametrically as
$$u=c_1, \quad v=2y\sqrt{y^2+c_1}, \quad (-\infty<y<\infty);$$
and it is evident that the image of a point (x,y) on that branch moves upward along the entire line as (x,y) traces out the branch in the upward direction (Fig. 1.2.1). Likewise, since the pair of equations
$$u=c_1, \quad v=-2y\sqrt{y^2+c_1}, \quad (-\infty<y<\infty)$$
furnishes a parametric representation for the image of the left-hand of the hyperbola, the image of a point going downward along the entire left-hand branch is seen to move up the entire line $u=c_1$.

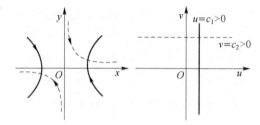

Fig. 1.2.1

On the other hand, each branch of a hyperbola
$$2xy=c_2, \quad (c_2>0) \tag{1.2.6}$$
is transformed into the line $v=c_2$, as indicated in Fig. 1.2.1. To verify this, we note from the second equation of (1.2.4) that $v=c_2$ when (x,y) is a point on either branch. Suppose that $v=c_2$ is on the branch lying in the first quadrant. Then, since $y=\dfrac{c_2}{2x}$, the first equation of (1.2.4) reveals that the branch's image has parametric representation
$$u=x^2-\frac{c_2^2}{4x^2}, \quad v=c_2, \quad (0<x<\infty).$$

Observe that
$$\lim_{\substack{x\to 0\\x>0}} u=-\infty \quad \text{and} \quad \lim_{x\to\infty} u=\infty.$$

Since u depends continuously on x, then, it is clear that as (x,y) travels down the entire upper branch of hyperbola (1.2.6), its image moves to the right along the entire horizontal line $v=c_2$. In as much as the image of the lower branch has parametric representation
$$u=\frac{c_2^2}{4y^2}-y^2, \quad v=c_2, \quad (-\infty<y<0)$$

and since
$$\lim_{\substack{y\to -\infty}} u = -\infty \quad \text{and} \quad \lim_{\substack{y\to 0 \\ y<0}} u = \infty,$$
it follows that the image of a point moving upward along the entire lower branch also travels to the right along the entire line $v=c_2$ (see Fig. 1.2.1).

Example 1.2.5 The domain $x>0$, $y>0$, $xy<1$ consists of all points lying on the upper branches of hyperbolas from the family $2xy=c$, where $0<c<2$ (Fig. 1.2.2). We know from Example 1.2.4 that as a point travels downward along the entirety of such a branch, its image under the transformation $w=z^2$ moves to the right along the entire line $v=c$. Since, for all values of c between 0 and 2, these upper branches fill out the domain $x>0$, $y>0$, $xy<1$, that domain is mapped onto horizontal strip $0<v<2$.

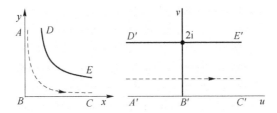

Fig. 1.2.2

In view of equations (1.2.4), the image of a point $(0,y)$ in the z plane is $(-y^2,0)$. Hence as $(0,y)$ travels downward to the origin along the y axis, its image moves to the right negative u axis and reaches the origin in the w plane. Then, since the image of a point $(x,0)$ is $(x^2,0)$, that image moves to the right from the origin along the u axis as $(x,0)$ moves to the right from the origin along the x axis. The image of the upper branch of the hyperbola $xy=1$ is, of course, the horizontal line $v=2$. Evidently, then, the closed region $x\geqslant 0$, $y\geqslant 0$, $xy\leqslant 1$ is mapped onto the closed strip $0\leqslant v\leqslant 2$, as indicated in Fig. 1.2.2.

Our last example here illustrates how polar coordinates can be useful in analyzing certain mappings.

Example 1.2.6 The mapping $w=z^2$ becomes
$$w = r^2 e^{i2\theta} \tag{1.2.7}$$
when $z=re^{i\theta}$. Evidently, then, the image $w=\rho e^{i\varphi}$ of any nonzero point z is found by squaring the modulus $r=|z|$ and doubling the value θ of arg z that is used
$$\rho = r^2 \quad \text{and} \quad \varphi = 2\theta. \tag{1.2.8}$$
Observe that point $z=r_0 e^{i\theta}$ on a circle $r=r_0$ are transformed into points $w=r_0^2 e^{i2\theta}$ on the circle $\rho=r_0^2$. As a point on the first circle moves counterclockwise from the positive real axis to the positive imaginary axis, its image on the second circle moves counterclockwise from the positive real axis to the negative real axis (see Fig. 1.2.3). So, as all possible positive values of r_0 are chosen, the corresponding arcs in the z and w planes fill out the first quadrant and the upper half plane, respectively. The transformation $w=z^2$ is, then, a one to one mapping of the first quadrant $r\geqslant 0$, $0\leqslant \theta \leqslant \dfrac{\pi}{2}$ in the z plane onto the upper half $\rho \geqslant 0$,

$0 \leqslant \varphi \leqslant \pi$ of the w plane, as indicated in Fig. 1.2.3. The point $z=0$ is, of course, mapped onto the point $w=0$.

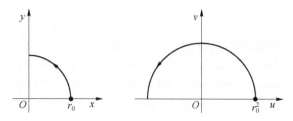

Fig. 1.2.3

The transformation $w=z^2$ also maps the upper half plane $r \geqslant 0$, $0 \leqslant \theta \leqslant \pi$ onto the entire w plane. However, in this case, the transformation is not one to one since both the positive and negative real axes in the z plane are mapped onto the positive real axis in the w plane.

When n is a positive integer greater than 2, various mapping properties of the transformation $w=z^n$, or $w=r^n e^{in\theta}$, are similar to those of $w=z^2$. Such a transformation maps the entire z plane onto the entire w plane, where each nonzero point in the w plane is the image of n distinct points in the z plane. The circle $r=r_0$ is mapped onto the circle $\rho=r_0^n$; and the sector $r \leqslant r_0$, $0 \leqslant \theta \leqslant \frac{2\pi}{n}$ is mapped onto the disk $\rho \leqslant r_0^n$, but not in a one to one manner.

Exercises 1.2

1. For each of the functions below, describe the domain of definition that is understood:

(a) $f(z)=\dfrac{1}{z^2+1}$;   (b) $f(z)=\text{Arg}\left(\dfrac{1}{z}\right)$;

(c) $f(z)=\dfrac{z}{z+\bar{z}}$;   (d) $f(z)=\dfrac{1}{1-|z|^2}$.

2. Write the function $f(z)=z^3+z+1$ in the form $f(z)=u(x,y)+iv(x,y)$.

3. Suppose that $f(z) = x^2 - y^2 - 2y + i(2x - 2xy)$, where $z = x + iy$. Use the expressions

$$x=\frac{z+\bar{z}}{2} \quad \text{and} \quad y=\frac{z-\bar{z}}{2i},$$

to write $f(z)$ in terms of z, and simplify the result.

4. Write the function

$$f(z)=z+\frac{1}{z}, \quad (z \neq 0)$$

in the form $f(z)=u(r,\theta)+iv(r,\theta)$.

5. Find and sketch, showing corresponding orientations, the images of the hyperbolas

$$x^2-y^2=c_1, \quad (c_1<0) \quad \text{and} \quad 2xy=c_2, \quad (c_2<0)$$

under the transformation $w=z^2$.

6. Show that the lines $ay=x$ ($a\neq 0$) are mapped onto the spirals $\rho=\exp(a\varphi)$ under the transformation $w=\exp z$, where $w=\rho\exp(i\varphi)$.

1.3 Limit and continuity of a complex function

1.3.1 Limit of a complex function

Definition 1.3.1 Suppose that w_0 is a complex number, and a function f is defined at all points z in some deleted neighborhood of z_0. If for each positive number ε, there is a positive number δ such that
$$|f(z)-w_0|<\varepsilon \quad \text{whenever} \quad 0<|z-z_0|<\delta, \tag{1.3.1}$$
then the ***limit*** of $f(z)$ as z approaches z_0 is w_0, and it can be denoted as
$$\lim_{z\to z_0} f(z)=w_0. \tag{1.3.2}$$

Geometrically, this definition says that for each ε neighborhood $|w-w_0|<\varepsilon$ of w_0, there is a deleted δ neighborhood $0<|z-z_0|<\delta$ of z_0 such that every point z in it has an image w lying in the ε neighborhood (Fig. 1.3.1). Note that even though all points in the deleted neighborhood $0<|z-z_0|<\delta$ are to be considered, their images need not fill up the entire neighborhood $|w-w_0|<\varepsilon$. If f has the constant value w_0, for instance, the image of z is always the center of that neighborhood. Note, too, that once a δ has been found, it can be replaced by any smaller positive number, such as $\dfrac{\delta}{2}$.

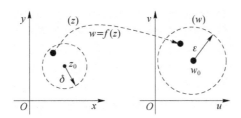

Fig. 1.3.1

Note

1. It is easy to show that **when a limit of a function $f(z)$ exists at a point z_0, it is unique**.

Proof Suppose that
$$\lim_{z\to z_0} f(z)=w_0 \quad \text{and} \quad \lim_{z\to z_0} f(z)=w_1.$$
Then, for each positive number ε, there is a positive number δ_0 and δ_1 such that
$$|f(z)-w_0|<\varepsilon \quad \text{whenever} \quad 0<|z-z_0|<\delta_0$$
and
$$|f(z)-w_1|<\varepsilon \quad \text{whenever} \quad 0<|z-z_0|<\delta_1.$$

So if $0 < |z-z_0| < \delta$, where δ is any positive number that is smaller than δ_0 and δ_1, we find that
$$|w_1 - w_0| = |[f(z) - w_0] - [f(z) - w_1]| \leqslant |f(z) - w_0| + |[f(z) - w_1]| < \varepsilon + \varepsilon = 2\varepsilon.$$
But $|w_1 - w_0|$ is a nonnegative constant, and ε can be chosen arbitrarily small.

Hence $w_1 - w_0 = 0$, or $w_1 = w_0$.

2. Definition (1.3.1) requires that f is defined at all points in some deleted neighborhood of z_0. Such a deleted neighborhood, of course, always exists when z_0 is an interior point of a region on which f is defined. We can extend the definition of limit to the case in which z_0 is a boundary point of the region by agreeing that the first of inequalities (1.3.1) need be satisfied by only those points z that lie in both the region and the deleted neighborhood.

Example 1.3.1 Show that if $f(z) = \dfrac{i\bar{z}}{2}$ in the open disk $|z| < 1$, then
$$\lim_{z \to 1} f(z) = \dfrac{i}{2}.$$

Proof Observe that when z is in the disk $|z| < 1$,
$$\left| f(z) - \dfrac{i}{2} \right| = \left| \dfrac{i\bar{z}}{2} - \dfrac{i}{2} \right| = \dfrac{|z-1|}{2}.$$
Hence, for any such z and each positive number ε,
$$\left| f(z) - \dfrac{i}{2} \right| < \varepsilon \quad \text{whenever} \quad 0 < |z-1| < 2\varepsilon.$$
Thus condition (1.3.1) is satisfied by points in the region $|z| < 1$ when δ is equal to 2ε or any smaller positive number.

3. If limit (1.3.2) exists, the symbol $z \to z_0$ implies that z is allowed to approach z_0 in an arbitrary manner, not just from some particular direction. The next example emphasizes this.

Example 1.3.2 Show that if $f(z) = \dfrac{z}{\bar{z}}$, the limit $\lim_{z \to 0} f(z)$ does not exist.

Proof If it did exist, it could be found by letting the point $z = (x, y)$ approach the origin in any manner. But when $z = (x, 0)$ is a nonzero point on the real axis,
$$f(z) = \dfrac{x + i0}{x - i0} = 1;$$
and when $z = (0, y)$ is a nonzero point on the imaginary axis,
$$f(z) = \dfrac{0 + iy}{0 - iy} = -1.$$
Thus, by letting z approach the origin along the real axis, we would find that the desired limit is 1. An approach along the imaginary axis would, on the other hand, yield the limit -1. Since a limit is unique, we must conclude that the limit does not exist.

While definition (1.3.1) provides a means of testing whether a given point w_0 is a limit, it does not directly provide a method for determining that limit. Theorems on limits, presented in the next section, will enable us to actually find many limits.

By the definition of limit, we can derive the following theorems.

Theorem 1.3.1 Suppose that
$$f(z)=u(x, y)+iv(x, y), \quad (z=x+iy)$$
and
$$z_0=x_0+iy_0, \quad w_0=u_0+iv_0,$$
then
$$\lim_{z \to z_0} f(z)=w_0 \tag{1.3.3}$$
if and only if
$$\lim_{(x,y) \to (x_0,y_0)} u(x,y)=u_0 \quad \text{and} \quad \lim_{(x,y) \to (x_0,y_0)} v(x,y)=v_0. \tag{1.3.4}$$

Proof To prove the theorem, we first assume that limits (1.3.4) hold and obtain limit (1.3.3). Limits (1.3.4) tell us that for each positive number ε, there exist positive numbers δ_1 and δ_2 such that
$$|u-u_0|<\frac{\varepsilon}{2} \quad \text{whenever} \quad 0<\sqrt{(x-x_0)^2+(y-y_0)^2}<\delta_1 \tag{1.3.5}$$
and
$$|v-v_0|<\frac{\varepsilon}{2} \quad \text{whenever} \quad 0<\sqrt{(x-x_0)^2+(y-y_0)^2}<\delta_2. \tag{1.3.6}$$

Let δ be any positive number smaller than δ_1 and δ_2. Since
$$|(u+iv)-(u_0+iv_0)|=|(u-u_0)+i(v-v_0)| \leqslant |u-u_0|+|v-v_0|$$
and
$$\sqrt{(x-x_0)^2+(y-y_0)^2}=|(x-x_0)+i(y-y_0)|=|(x+iy)-(x_0+iy_0)|,$$
it follows from statements (1.3.5) and (1.3.6) that
$$|(u+iv)-(u_0+iv_0)|<\frac{\varepsilon}{2}+\frac{\varepsilon}{2}=\varepsilon$$
whenever
$$0<|(x+iy)-(x_0+iy_0)|<\delta.$$
That is, limit (1.3.3) holds.

Let us now start with the assumption that limit (1.3.3) holds. With that assumption, we know that for each positive number ε, there is a positive number δ such that
$$|(u+iv)-(u_0+iv_0)|<\varepsilon \tag{1.3.7}$$
whenever
$$0<|(x+iy)-(x_0+iy_0)|<\delta. \tag{1.3.8}$$
But
$$|u-u_0| \leqslant |(u-u_0)+i(v-v_0)|=|(u+iv)-(u_0+iv_0)|,$$
$$|v-v_0| \leqslant |(u-u_0)+i(v-v_0)|=|(u+iv)-(u_0+iv_0)|,$$
and
$$|(x+iy)-(x_0+iy_0)|=|(x-x_0)+i(y-y_0)|=\sqrt{(x-x_0)^2+(y-y_0)^2}.$$
Hence it follows from inequalities (1.3.7) and (1.3.8) that
$$|u-u_0|<\varepsilon \quad \text{and} \quad |v-v_0|<\varepsilon$$
whenever

$$0<\sqrt{(x-x_0)^2+(y-y_0)^2}<\delta.$$

This establishes limits (1.3.4), and the proof of the theorem is completed.

Theorem 1.3.2 Suppose that
$$\lim_{z\to z_0}f(z)=w_0 \quad \text{and} \quad \lim_{z\to z_0}F(z)=W_0, \tag{1.3.9}$$
then
$$\lim_{z\to z_0}[f(z)+F(z)]=w_0+W_0, \tag{1.3.10}$$
$$\lim_{z\to z_0}[f(z)F(z)]=w_0W_0; \tag{1.3.11}$$

And if $W_0\neq 0$,
$$\lim_{z\to z_0}\frac{f(z)}{F(z)}=\frac{w_0}{W_0}. \tag{1.3.12}$$

This important theorem can be proved directly by using the definition of the limit of a function of a complex variable. But, with the aid of Theorem 1.3.1, it follows almost immediately from theorems on limits of real-valued functions of two real variables.

It is sometimes convenient to include with the complex plane the **point at infinity**, denoted by ∞, and to use limits involving it. The complex plane together with this point is called the **extended complex plane**. To visualize the point at infinity, one can think of the complex plane as passing through the equator of a unit sphere centered at the origin (Fig. 1.3.2). To each point z in the plane there corresponds exactly one point P on the surface of the sphere. The point P is the point where the line through z and the north pole N intersects the sphere. In such manner, to each point P on the surface of the sphere, other than the north pole N, there corresponds exactly one point z in the plane. By letting the point N of the sphere correspond to the point at infinity, we obtain a one to one correspondence between the points of the sphere and the points of the extended complex plane. The sphere is known as the **Riemann sphere**, and the correspondence is called a stereographic projection.

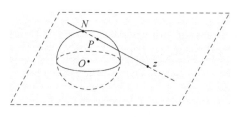

Fig. 1.3.2

Observe that the exterior of the unit circle centered at the origin in the complex plane corresponds to the upper hemisphere with the equator and the point N deleted. Moreover, for each small positive number ε, those points in the complex plane exterior to the circle $|z|=\dfrac{1}{\varepsilon}$ correspond to points on the sphere close to N. We thus call the $|z|>\dfrac{1}{\varepsilon}$ an ε neighborhood, or neighborhood, of ∞.

Let us agree that in referring to a point z, we mean a point in the finite plane.

Hereafter, when the point at infinity is to be considered, it will be specifically mentioned.

A meaning is now readily given to the statement
$$\lim_{z \to z_0} f(z) = w_0$$
when either z_0 or w_0, or possibly each of these numbers, is replaced by the point at infinity. In the definition of limit, we simply replace the appropriate neighborhoods of z_0 and w_0 by neighborhoods of ∞. The proof of the following theorem illustrates how this is done.

Theorem 1.3.3 If z_0 and w_0 are points in z and w plane, then

$$\lim_{z \to z_0} f(z) = \infty \quad \text{if and only if} \quad \lim_{z \to z_0} \frac{1}{f(z)} = 0 \tag{1.3.13}$$

and

$$\lim_{z \to \infty} f(z) = w_0 \quad \text{if and only if} \quad \lim_{z \to 0} f\left(\frac{1}{z}\right) = w_0, \tag{1.3.14}$$

$$\lim_{z \to \infty} f(z) = \infty \quad \text{if and only if} \quad \lim_{z \to 0} \frac{1}{f\left(\frac{1}{z}\right)} = 0. \tag{1.3.15}$$

Proof We start the proof by noting that the first of limits (1.3.13) means that for each positive number ε, there is a positive number δ such that

$$|f(z)| > \frac{1}{\varepsilon} \quad \text{whenever} \quad 0 < |z - z_0| < \delta. \tag{1.3.16}$$

That is, the point $w = f(z)$ lies in the ε neighborhood $|w| > \frac{1}{\varepsilon}$ of ∞ whenever z lies in the deleted neighborhood $0 < |z - z_0| < \delta$ of z_0. Since statement (1.3.16) can be written as

$$\left| \frac{1}{f(z)} - 0 \right| < \varepsilon \quad \text{whenever} \quad 0 < |z - z_0| < \delta,$$

the second of limits (1.3.13) follows.

The first of limits (1.3.14) means that for each positive number ε, a positive number δ exists such that

$$|f(z) - w_0| < \varepsilon \quad \text{whenever} \quad |z| > \frac{1}{\delta}. \tag{1.3.17}$$

Replacing z by $\frac{1}{z}$ in statement (1.3.17) and then writing the result as

$$\left| f\left(\frac{1}{z}\right) - w_0 \right| < \varepsilon \quad \text{whenever} \quad 0 < |z - 0| < \delta,$$

we arrive at the second of limits (1.3.14).

Finally, the first of limits (1.3.15) is to be interpreted as saying that for each positive number ε, there is a positive number δ such that

$$|f(z)| > \frac{1}{\varepsilon} \quad \text{whenever} \quad |z| > \frac{1}{\delta}. \tag{1.3.18}$$

When z is replaced by $\frac{1}{z}$, this statement can be put in the form

$$\left| \frac{1}{f\left(\frac{1}{z}\right)} - 0 \right| < \varepsilon \quad \text{whenever} \quad 0 < |z - 0| < \delta;$$

and this gives us the second of limits (1.3.15).

Example 1.3.3 Observe that

$$\lim_{z\to -1}\frac{iz+3}{z+1}=\infty \quad \text{since} \quad \lim_{z\to -1}\frac{z+1}{iz+3}=0$$

and

$$\lim_{z\to\infty}\frac{2z+i}{z+1}=2 \quad \text{since} \quad \lim_{z\to 0}\frac{\left(\frac{2}{z}\right)+i}{\left(\frac{1}{z}\right)+1}=\lim_{z\to 0}\frac{2+iz}{1+z}=2.$$

Furthermore,

$$\lim_{z\to\infty}\frac{2z^3-1}{z^2+1}=\infty \quad \text{since} \quad \lim_{z\to 0}\frac{\left(\frac{1}{z^2}\right)+1}{\left(\frac{2}{z^3}\right)-1}=\lim_{z\to 0}\frac{z+z^3}{2-z^3}=0.$$

1.3.2 Continuity of a complex function

Definition 1.3.2 A function f is **continuous** at a point z_0 if all of the following conditions are satisfied:

(i) $\lim_{z\to z_0} f(z)$ exists, (1.3.19)

(ii) $f(z_0)$ exists, (1.3.20)

(iii) $\lim_{z\to z_0} f(z) = f(z_0)$. (1.3.21)

Observe that statement (1.3.21) actually contains statements (1.3.19) and (1.3.20), since the existence of the quantity on each side of the equation there is needed. Statement (1.3.21) says, of course, that for each positive number ε, there is a positive number δ such that

$$|f(z)-f(z_0)|<\varepsilon \quad \text{whenever} \quad |z-z_0|<\delta. \tag{1.3.22}$$

A function of a complex variable is said to **be continuous in a region** R if it is continuous at each point in R.

If two functions are continuous at a point, their sum and product are also continuous at that point; their quotient is continuous at any such point if the denominator is not zero here. These observations are direct consequences of Theorem 1.3.2.

We turn now to two expected properties of continuous functions whose verifications are not so immediate. Our proofs depend on definition (1.3.22) of continuity, and we present the results as theorems.

Theorem 1.3.4 The composition of two continuous function is continuous.

Proof Let $w=f(z)$ be a function that is defined for all z in a neighborhood $|z-z_0|<\delta$ of a point z_0, and we let $W=g(w)$ be a function whose domain of definition contains the image of that neighborhood under f. The composition $W=g[f(z)]$ is continuous at z_0 and that g is continuous at the point $f(z_0)$ in the w plane. In view of the continuity of g at $f(z_0)$, there is, for each positive number ε, there is a positive number γ such that

$$|g[f(z)]-g[f(z_0)]|<\varepsilon \quad \text{whenever} \quad |f(z)-f(z_0)|<\gamma.$$

But the continuity of f at z_0 ensures that the neighborhood $|z-z_0|<\delta$ can be made small enough that the second of these inequalities holds. The continuity of the composition $g[f(z)]$ is, therefore, established.

Theorem 1.3.5 If $f(z)$ is continuous at z_0 and $f(z_0)\neq 0$, then $f(z)\neq 0$ throughout some neighborhood of that point.

Proof Assuming that $f(z)$ is, in fact, continuous and nonzero at z_0, we can prove Theorem 1.3.5 by assigning the positive value $\dfrac{|f(z_0)|}{2}$ to the number ε in statement (1.3.22). This tells us that there is a positive number δ such that

$$|f(z)-f(z_0)|<\frac{|f(z_0)|}{2} \quad \text{whenever} \quad |z-z_0|<\delta.$$

So if there is a point z in the neighborhood $|z-z_0|<\delta$ at which $f(z)=0$, we have the contradiction

$$|f(z_0)|<\frac{|f(z_0)|}{2};$$

and the theorem is proved.

The continuity of a function

$$f(z)=u(x,y)+iv(x,y) \tag{1.3.23}$$

is closely related to the continuity of its component functions $u(x,y)$ and (x,y). We note, for instance, how it follows from Theorem 1.3.1 that the function (1.3.23) is continuous at a point $z_0=(x_0,y_0)$ if and only if its component functions are continuous there. Our proof of the next theorem illustrates the use of this statement. The theorem is extremely important and will be used often in later chapters, especially in applications. Before stating the theorem, we recall from Section 1.1.3 that a region R is closed if it contains all of its boundary points and that it is bounded if it lies inside some circle centered at the origin.

Theorem 1.3.6 If a function f is continuous throughout a region R that is both closed and bounded, there exists a nonnegative real number M such that

$$|f(z)|\leqslant M \tag{1.3.24}$$

for all points z in R, where equality holds for at least one such z.

Proof To prove this, we assume that the function f in equation (1.3.23) is continuous and note that how it follows that the function

$$\sqrt{[u(x,y)]^2+[v(x,y)]^2}$$

is continuous throughout R and thus reaches a maximum value M somewhere in R. Inequality (1.3.24) thus holds, and we say that f is **bounded** on R.

Exercises 1.3

1. Use definition of limit to prove that

(a) $\lim\limits_{z\to z_0}\text{Re } z=\text{Re } z_0$; (b) $\lim\limits_{z\to z_0}\overline{z}=\overline{z_0}$; (c) $\lim\limits_{z\to 0}\dfrac{(\overline{z})^2}{z}=0$.

2. Let a, b and c denote complex constants. Then use definition of limit to show that
(a) $\lim\limits_{z\to z_0}(az+b)=az_0+b$; (b) $\lim\limits_{z\to z_0}(z^2+c)=z_0^2+c$;
(c) $\lim\limits_{z\to 1-i}[x+i(2x+y)]=1+i$, $(z=x+iy)$.

3. Let n be a positive integer and let $P(z)$ and $Q(z)$ be polynomials, where $Q(z_0)\neq 0$, to find
(a) $\lim\limits_{z\to z_0}\dfrac{1}{z^n}$, $(z_0\neq 0)$; (b) $\lim\limits_{z\to i}\dfrac{iz^3-1}{z+i}$; (c) $\lim\limits_{z\to z_0}\dfrac{P(z)}{Q(z)}$.

4. Show that the limit of the function
$$f(z)=\left(\dfrac{\bar z}{z}\right)^2$$
as z tends to 0 does not exist.

5. Use definition of limit to prove that
$$\text{if } \lim\limits_{z\to z_0}f(z)=w_0, \text{ then } \lim\limits_{z\to z_0}|f(z)|=|w_0|.$$

6. Show that
(a) $\lim\limits_{z\to\infty}\dfrac{4z^2}{(z-1)^2}=4$; (b) $\lim\limits_{z\to 1}\dfrac{1}{(z-1)^3}=\infty$; (c) $\lim\limits_{z\to\infty}\dfrac{z^2+1}{z-1}=\infty$.

7. Show that when
$$T(z)=\dfrac{az+b}{cz+d}, \quad (ad-bc\neq 0),$$
then
(a) $\lim\limits_{z\to\infty}T(z)=\infty$ if $c=0$;

(b) $\lim\limits_{z\to\infty}T(z)=\dfrac{a}{c}$ and $\lim\limits_{z\to -\frac{d}{c}}T(z)=\infty$ if $c\neq 0$.

Chapter 2　Analytic Functions

In this chapter, we are to consider the differential calculus of functions of a complex variable. We start this chapter by defining the derivative of a function of a complex variable, followed by necessary and/or sufficient conditions on the differentiability of such a function. We then come to the definition of the analytic function of a complex variable. Finally we will extend various elementary functions studied in calculus to corresponding functions of a complex variable. We will stress the similarities, and more importantly, differences of the function of a complex variable with that of a real variable.

2.1　Derivatives of complex functions

2.1.1　Derivatives

Definition 2.1.1　Let $f(z)$ be a function defined in the domain D, which contains a neighborhood $|z-z_0|<\varepsilon$ of a point z_0. It is said to be differentiable at z_0, if the limit

$$\lim_{z \to z_0} \frac{f(z)-f(z_0)}{z-z_0} \tag{2.1.1}$$

exists. The limit, called the **derivative** of $f(z)$ at z_0, is denoted by $f'(z_0)$.

Writing $\Delta z = z - z_0$ in the definition above, we get the equivalent form

$$f'(z_0) = \lim_{\Delta z \to 0} \frac{f(z_0 + \Delta z) - f(z_0)}{\Delta z}. \tag{2.1.2}$$

Introduce $\Delta w = f(z_0 + \Delta z) - f(z_0)$, which denotes the change in the value of $w = f(z)$ corresponding to the change Δz in the point z_0. If we write $\dfrac{\mathrm{d}w}{\mathrm{d}z}$ for $f'(z_0)$, we can also write the definition as

$$\frac{\mathrm{d}w}{\mathrm{d}z} = \lim_{\Delta z \to 0} \frac{\Delta w}{\Delta z}. \tag{2.1.3}$$

Example 2.1.1　Consider the function $f(z) = z^2$. At any point z,

$$\lim_{\Delta z \to 0} \frac{\Delta w}{\Delta z} = \lim_{\Delta z \to 0} \frac{(z + \Delta z)^2 - z^2}{\Delta z} = \lim_{\Delta z \to 0} (2z + \Delta z) = 2z.$$

So $f'(z)=2z$ at any point z.

Example 2.1.2 Consider the function $f(z)=|z|^2$. At any point z,
$$\frac{\Delta w}{\Delta z}=\frac{|z+\Delta z|^2-|z|^2}{\Delta z}=\frac{(z+\Delta z)(\bar{z}+\overline{\Delta z})-z\bar{z}}{\Delta z}=\bar{z}+\overline{\Delta z}+z\frac{\overline{\Delta z}}{\Delta z}.$$

If the limit of $\frac{\Delta w}{\Delta z}$ exists, it can be found by letting the point $\Delta z=(\Delta x,\Delta y)$ approach the origin $(0,0)$ in the complex plane in any manner. Particularly, as Δz approaches the origin $(0,0)$ horizontally through the points $(\Delta x,0)$ on the real axis, or equivalently, $\Delta z=\Delta x$, then $\overline{\Delta z}=\Delta x$, and
$$\frac{\Delta w}{\Delta z}=\bar{z}+\overline{\Delta z}+z\frac{\overline{\Delta z}}{\Delta z}=\bar{z}+\Delta x+z.$$

Let $\Delta x \to 0$, then the limit should be $\bar{z}+z$, if it exists. On the other hand, when Δz approaches the origin $(0,0)$ vertically through the points $(0,\Delta y)$ on the imaginary axis, then $\Delta z=\mathrm{i}\Delta y$, $\overline{\Delta z}=-\mathrm{i}\Delta y$, and
$$\frac{\Delta w}{\Delta z}=\bar{z}+\overline{\Delta z}+z\frac{\overline{\Delta z}}{\Delta z}=\bar{z}-\mathrm{i}\Delta y-z.$$

Let $\Delta y \to 0$, then the limit should be $\bar{z}-z$, if it exists. Hence if the limit of $\frac{\Delta w}{\Delta z}$ exists as Δz tends to zero, the uniqueness of limits tells us that
$$\bar{z}+z=\bar{z}-z,$$
or $z=0$. Thus $f'(z)$ cannot exist for any $z\neq 0$.

When $z=0$, the expression for $\frac{\Delta w}{\Delta z}$ becomes
$$\frac{\Delta w}{\Delta z}=\overline{\Delta z},$$
which obviously tends to zero as Δz approaches zero. So in all, $f'(z)$ only exists at $z=0$, and $f'(0)=0$.

Example 2.1.2 shows that a function can be differentiable at a certain point, but nowhere else in any neighborhood of that point. Note that for $f(z)=|z|^2$, we have
$$u(x,y)=x^2+y^2, \quad v(x,y)=0.$$
So Example 2.1.2 also shows that the real and imaginary components of a function of a complex variable can have continuous partial derivatives of all orders at a point, yet the function fails to be differentiable there. Recall that $f(z)$ is continuous at a point if and only if $u(x,y)$ and $v(x,y)$ are continuous at the point.

The function $f(z)=|z|^2$ is continuous at each point in the complex plane. So Example 2.1.2 also shows that the continuity of a function at a point does not imply the existence of a derivative there. However, it is true that the existence of the derivative of a function at a point does imply the continuity of the function at that point. To see this, assume that $f'(z_0)$ exists, then
$$\lim_{z \to z_0}(f(z)-f(z_0))=\lim_{z \to z_0}\frac{f(z)-f(z_0)}{z-z_0}\lim_{z \to z_0}(z-z_0)=f'(z_0)\cdot 0=0,$$
which implies that $f(z)$ is continuous at z_0.

Geometric interpretations of derivatives of functions of a complex variable are not so intuitionistic as they are for derivatives of functions of a real variable.

The definition of derivative of a function of a complex variable is identical in form to that of a function of a real variable. In fact, the basic differentiation formulas given below without proof can be derived by essentially the same steps as those used in calculus.

2.1.2 Some properties of derivatives

Let c be a complex constant, and assume that $f(z)$ is differentiable at a point z. It is easy to show that

$$\frac{\mathrm{d}}{\mathrm{d}z}c=0, \quad \frac{\mathrm{d}}{\mathrm{d}z}z=1, \quad \frac{\mathrm{d}}{\mathrm{d}z}(cf(z))=cf'(z). \tag{2.1.4}$$

If n is a positive integer, it holds that

$$\frac{\mathrm{d}}{\mathrm{d}z}z^n = nz^{n-1}. \tag{2.1.5}$$

This formula remains valid when n is a negative integer, provided that $z \neq 0$.

If two functions $f(z)$ and $g(z)$ are both differentiable at a point z, then

$$\frac{\mathrm{d}}{\mathrm{d}z}[f(z) \pm g(z)] = f'(z) \pm g'(z), \tag{2.1.6}$$

$$\frac{\mathrm{d}}{\mathrm{d}z}[f(z)g(z)] = f'(z)g(z) + f(z)g'(z), \tag{2.1.7}$$

and, when $g(z) \neq 0$,

$$\frac{\mathrm{d}}{\mathrm{d}z}\left[\frac{f(z)}{g(z)}\right] = \frac{f'(z)g(z) - f(z)g'(z)}{[g(z)]^2}. \tag{2.1.8}$$

There is also a chain rule for differentiating composite functions. Suppose that f is differentiable at a point z, and g is differentiable at the point $f(z)$. Then the function $F(z) = g[f(z)]$ is differentiable at z, and

$$F'(z) = g'[f(z)]f'(z). \tag{2.1.9}$$

If we write $w = f(z)$ and $W = g(w)$, so that $W = F(z)$, then the chain rule can be written as

$$\frac{\mathrm{d}W}{\mathrm{d}z} = \frac{\mathrm{d}W}{\mathrm{d}w}\frac{\mathrm{d}w}{\mathrm{d}z}. \tag{2.1.10}$$

Example 2.1.3 To find the derivative of $(3z^2 + 2\mathrm{i})^3$, write $w = 3z^2 + 2\mathrm{i}$ and $W = w^3$. Then

$$\frac{\mathrm{d}}{\mathrm{d}z}(3z^2 + 2\mathrm{i})^3 = 3w^2 \cdot 6z = 18z\,(3z^2 + 2\mathrm{i})^2.$$

2.1.3 A necessary condition on differentiability

The definition of the derivative serves as a necessary and sufficient condition on the existence of the derivative of a function of a complex variable at a certain point. But it may not be easy to determine whether the limit exists in some cases. We are now to propose some other necessary and/or sufficient conditions. We will first obtain a pair of equations that the first-order partial derivatives of the component functions u and v of a function

$$f(z)=u(x,y)+iv(x,y)$$

must satisfy when the derivative of f exists at a point $z_0=x_0+iy_0$. Write $\Delta z=\Delta x+i\Delta y$, and

$$\Delta w = f(z_0+\Delta z)-f(z_0)$$
$$=[u(x_0+\Delta x,y_0+\Delta y)-u(x_0,y_0)]+i[v(x_0+\Delta x,y_0+\Delta y)-v(x_0,y_0)].$$

Since $f(z)=u(x,y)+iv(x,y)$ is differentiable at $z_0=x_0+iy_0$, so

$$f'(z_0)=\lim_{\Delta z\to 0}\frac{\Delta w}{\Delta z}=\lim_{(\Delta x,\Delta y)\to(0,0)}\operatorname{Re}\frac{\Delta w}{\Delta z}+i\lim_{(\Delta x,\Delta y)\to(0,0)}\operatorname{Im}\frac{\Delta w}{\Delta z}.$$

Keep in mind that the limit should exist as $(\Delta x,\Delta y)$ approaches the origin $(0,0)$ in any manner. In particular, if we let $(\Delta x,\Delta y)$ tend to $(0,0)$ horizontally through the points $(\Delta x,0)$ on the real axis, or equivalently, $\Delta y=0$, then the quotient $\frac{\Delta w}{\Delta z}$ becomes

$$\frac{\Delta w}{\Delta z}=\frac{u(x_0+\Delta x,\ y_0)-u(x_0,y_0)}{\Delta x}+i\frac{v(x_0+\Delta x,y_0)-v(x_0,y_0)}{\Delta x}.$$

Thus,

$$\lim_{(\Delta x,\Delta y)\to(0,0)}\operatorname{Re}\frac{\Delta w}{\Delta z}=\lim_{\Delta x\to 0}\frac{u(x_0+\Delta x,y_0)-u(x_0,y_0)}{\Delta x}=u_x(x_0,y_0),$$

and

$$\lim_{(\Delta x,\Delta y)\to(0,0)}\operatorname{Im}\frac{\Delta w}{\Delta z}=\lim_{\Delta x\to 0}\frac{v(x_0+\Delta x,y_0)-v(x_0,\ y_0)}{\Delta x}=v_x(x_0,y_0),$$

where $u_x(x_0,y_0)$ and $v_x(x_0,y_0)$ denote, respectively, the first-order partial derivatives with respect to x of the functions u and v at (x_0,y_0). So we have

$$f'(z_0)=u_x(x_0,y_0)+iv_x(x_0,y_0). \tag{2.1.11}$$

On the other hand, if we let $(\Delta x,\Delta y)$ approach $(0,0)$ vertically through the points $(0,\Delta y)$ on the imaginary axis, that is, $\Delta x=0$, then we have

$$\frac{\Delta w}{\Delta z}=\frac{u(\Delta x_0,y_0+\Delta y)-u(x_0,y_0)}{i\Delta y}+i\frac{v(\Delta x_0,y_0+\Delta y)-v(x_0,y_0)}{i\Delta y}$$
$$=\frac{v(\Delta x_0,y_0+\Delta y)-v(x_0,y_0)}{\Delta y}-i\frac{u(\Delta x_0,y_0+\Delta y)-u(x_0,y_0)}{\Delta y}.$$

Hence

$$\lim_{(\Delta x,\Delta y)\to(0,0)}\operatorname{Re}\frac{\Delta w}{\Delta z}=\lim_{\Delta y\to 0}\frac{v(x_0,y_0+\Delta y)-v(x_0,y_0)}{\Delta y}=v_y(x_0,y_0),$$

and

$$\lim_{(\Delta x,\Delta y)\to(0,0)}\operatorname{Im}\frac{\Delta w}{\Delta z}=\lim_{\Delta y\to 0}-\frac{u(x_0,y_0+\Delta y)-u(x_0,y_0)}{\Delta y}=-u_y(x_0,y_0).$$

In this case, we get another formula

$$f'(z_0)=v_y(x_0,y_0)-iu_y(x_0,y_0). \tag{2.1.12}$$

We have obtained two formulas for $f'(z_0)$ as

$$f'(z_0)=u_x(x_0,y_0)+iv_x(x_0,y_0)=v_y(x_0,y_0)-iu_y(x_0,y_0).$$

So it must hold that

$$u_x(x_0,y_0)=v_y(x_0,y_0)\quad\text{and}\quad u_y(x_0,y_0)=-v_x(x_0,y_0). \tag{2.1.13}$$

These equations are known as the **Cauchy-Riemann equations**, so named in honor of the French mathematician Cauchy A. L. (1789—1857), who discovered and used them, and in honor of the German mathematician Riemann G. F. B. (1826—1866), who made them fundamental in his development of the theory of functions of a complex variable.

We summarize the above results in the following theorem.

Theorem 2.1.1 Suppose that $f(z)=u(x,y)+iv(x,y)$ is differentiable at a point $z_0 = x_0+iy_0$. Then $u(x,y)$ and $v(x,y)$ must satisfy the Cauchy-Riemann equations
$$u_x(x_0,y_0)=v_y(x_0,y_0), \quad u_y(x_0,y_0)=-v_x(x_0,y_0).$$
And $f'(z_0)$ can be written as
$$f'(z_0)=u_x(x_0,y_0)+iv_x(x_0,y_0)=v_y(x_0,y_0)-iu_y(x_0,y_0). \qquad (2.1.14)$$

Example 2.1.4 We have already shown that the function
$$f(z)=z^2=x^2-y^2+i2xy$$
is differentiable everywhere, and $f'(z)=2z$. To verify that the Cauchy-Riemann equations are satisfied everywhere, write
$$u(x,y)=x^2-y^2, \quad v(x,y)=2xy.$$
Thus
$$u_x=2x=v_y, \quad u_y=-2y=-v_x.$$
Furthermore, from Theorem 2.1.1 we know that
$$f'(z)=u_x+iv_x=2x+i2y=2z.$$

The Cauchy-Riemann equations are necessary conditions for the existence of the derivative of a function at a point. They can often be used to locate points at which a function does not have a derivative.

Example 2.1.5 For the function $f(z)=|z|^2$, we have
$$u(x,y)=x^2+y^2, \quad v(x,y)=0.$$
If the Cauchy-Riemann equations are to be satisfied at some point (x,y), $u_x=v_y$ gives $2x=0$, while $u_y=-v_x$ gives $2y=0$. These show that the Cauchy-Riemann equations are only satisfied at the point $(0,0)$. So Theorem 2.1.1 tells us that $f'(z)$ does not exist at any nonzero point. This coincides with the result in Example 2.1.2. Note that the theorem does not ensure that $f'(0)$ exists, since satisfying the Cauchy-Riemann is just a necessary condition for differentiability. The following example illustrates this.

Example 2.1.6 Consider the function
$$f(z)=\begin{cases} \dfrac{\overline{z}^2}{z}, & z\neq 0, \\ 0, & z=0. \end{cases}$$
Direct calculations show that
$$u(x,y)=\begin{cases} \dfrac{x^3-3xy^2}{x^2+y^2}, & (x,y)\neq(0,0), \\ 0, & (x,y)=(0,0), \end{cases}$$
$$v(x,y)=\begin{cases} \dfrac{y^3-3x^2y}{x^2+y^2}, & (x,y)\neq(0,0), \\ 0, & (x,y)=(0,0). \end{cases}$$
Their partial derivatives at $(0,0)$ can be calculated as

$$u_x(0,0)=\lim_{\Delta x\to 0}\frac{u(\Delta x,0)-u(0,0)}{\Delta x}=\lim_{\Delta x\to 0}\frac{\Delta x}{\Delta x}=1,$$

$$u_y(0,0)=\lim_{\Delta y\to 0}\frac{u(0,\Delta y)-u(0,0)}{\Delta y}=\lim_{\Delta y\to 0}\frac{0}{\Delta y}=0,$$

$$v_x(0,0)=\lim_{\Delta x\to 0}\frac{v(\Delta x,0)-v(0,0)}{\Delta x}=\lim_{\Delta x\to 0}\frac{0}{\Delta x}=0,$$

$$v_y(0,0)=\lim_{\Delta y\to 0}\frac{v(0,\Delta y)-v(0,0)}{\Delta y}=\lim_{\Delta y\to 0}\frac{\Delta y}{\Delta y}=1.$$

So obviously,
$$u_x(0,0)=1=v_y(0,0),\quad u_y(0,0)=0=-v_x(0,0),$$
that is, the Cauchy-Riemann equations are satisfied at the point $z=0$.

However, at the point $z=0$,
$$\frac{\Delta w}{\Delta z}=\frac{f(\Delta z)-f(0)}{\Delta z}=\frac{\overline{\Delta z}^2}{\Delta z^2}=\left(\frac{\overline{\Delta z}}{\Delta z}\right)^2.$$

If we let $(\Delta x,\Delta y)$ approach $(0,0)$ along the line $\Delta y=k\Delta x$, that is, $\Delta z=\Delta x+ik\Delta x$, then
$$\lim_{\Delta x\to 0}\frac{\Delta w}{\Delta z}=\lim_{\Delta x\to 0}\left(\frac{\overline{\Delta z}}{\Delta z}\right)^2=\lim_{\Delta x\to 0}\left(\frac{\Delta x(1-ik)}{\Delta x(1+ik)}\right)^2=\left(\frac{1-ik}{1+ik}\right)^2,$$
which depends on the slope k of the line. This implies that $f'(0)$ does not exist.

2.1.4 Sufficient conditions on differentiability

We present here a sufficient condition to ensure the existence of the derivative of a function at some point in the following theorem without proof.

Theorem 2.1.2 Suppose that the function $f(z)=u(x,y)+iv(x,y)$ is defined throughout some neighborhood of a point $z_0=x_0+iy_0$, and suppose that u_x, u_y, v_x, v_y exist everywhere in that neighborhood. If u_x, u_y, v_x, v_y are continuous at (x_0,y_0), and satisfy the Cauchy-Riemann equations
$$u_x(x_0,y_0)=v_y(x_0,y_0),\quad u_y(x_0,y_0)=-v_x(x_0,y_0),$$
then $f'(z_0)$ exists, and
$$f'(z_0)=u_x(x_0,y_0)+iv_x(x_0,y_0).$$

Example 2.1.7 Consider the exponential function of a complex variable defined by
$$f(z)=e^z=e^{x+iy}=e^x e^{iy}=e^x\cos y+ie^x\sin y.$$
For more details on the exponential function of a complex variable, please refer to Section 2.3 in this chapter. Obviously,
$$u(x,y)=e^x\cos y,\quad v(x,y)=e^x\sin y.$$
Since $u_x=e^x\cos y=v_y$ and $u_y=-e^x\sin y=-v_x$ everywhere and these derivatives are everywhere continuous, the conditions in Theorem 2.1.2 are satisfied. Thus $f'(z)$ exists everywhere, and
$$f'(z)=u_x+iv_x=e^x\cos y+ie^x\sin y=f(z).$$

Example 2.1.8 Let us take another look at the function $f(z)=|z|^2$, where $u(x,y)=x^2+y^2$, $v(x,y)=0$. We have already shown in Example 2.1.5 that this function is not

analytic at any nonzero point. Since
$$u_x = 2x, \quad u_y = 2y, \quad v_x = v_y = 0,$$
the Cauchy-Riemann equations are satisfied at $z = 0$, and these partial derivatives are continuous at $z=0$. So Theorem 2.1.2 tells us that $f(z)$ has a derivative at $z=0$, and
$$f'(0) = 0 + i0 = 0.$$
The derivative does not exist at any nonzero point. These coincide with the results in Example 2.1.2.

It is sometimes more convenient to use the polar coordinate when $z_0 \neq 0$. We are to restate Theorem 2.1.2 in polar coordinates using the coordinate transformation
$$x = r\cos\theta, \quad y = r\sin\theta. \tag{2.1.15}$$
With the polar coordinate, the function becomes
$$f(z) = f(re^{i\theta}) = u(r,\theta) + iv(r,\theta).$$
Suppose that the hypothesis in Theorem 2.1.2 are satisfied. Then by the chain rule, we have
$$\frac{\partial u}{\partial r} = \frac{\partial u}{\partial x}\frac{\partial x}{\partial r} + \frac{\partial u}{\partial y}\frac{\partial y}{\partial r}, \quad \frac{\partial u}{\partial \theta} = \frac{\partial u}{\partial x}\frac{\partial x}{\partial \theta} + \frac{\partial u}{\partial y}\frac{\partial y}{\partial \theta},$$
$$\frac{\partial v}{\partial r} = \frac{\partial v}{\partial x}\frac{\partial x}{\partial r} + \frac{\partial v}{\partial y}\frac{\partial y}{\partial r}, \quad \frac{\partial v}{\partial \theta} = \frac{\partial v}{\partial x}\frac{\partial x}{\partial \theta} + \frac{\partial v}{\partial y}\frac{\partial y}{\partial \theta}.$$
More precisely,
$$u_r = u_x \cos\theta + u_y \sin\theta = u_y \sin\theta + u_x \cos\theta,$$
$$u_\theta = -u_x r\sin\theta + u_y r\cos\theta = r(-u_x \sin\theta + u_y \cos\theta),$$
$$v_r = v_x \cos\theta + v_y \sin\theta = v_y \sin\theta + v_x \cos\theta,$$
$$v_\theta = -v_x r\sin\theta + v_y r\cos\theta = r(-v_x \sin\theta + v_y \cos\theta).$$
Substituting the Cauchy-Riemann equations $u_x = v_y$, $u_y = -v_x$ into the above equations leads to
$$u_r = -v_x \sin\theta + v_y \cos\theta, \quad u_\theta = r(-v_y \sin\theta - v_x \cos\theta),$$
which implies that
$$v_\theta = ru_r, \quad u_\theta = -rv_r.$$
Conversely, if $v_\theta = ru_r$, $u_\theta = -rv_r$, it is straightforward to show that $u_x = v_y$, $u_y = -v_x$ must hold. Therefore, we have obtained an alternative form of the Cauchy-Riemann equations in polar coordinates. Thus we are now able to restate Theorem 2.1.2 using r and θ.

Theorem 2.1.3 Suppose that the function $f(z) = u(r,\theta) + iv(r,\theta)$ is defined throughout some neighborhood of a nonzero point $z_0 = r_0 \exp(i\theta_0)$, and suppose that u_r, u_θ, v_r, v_θ exist everywhere in that neighborhood. If u_r, u_θ, v_r, v_θ are continuous at (r_0, θ_0) and satisfy the polar form
$$v_\theta = ru_r, \quad u_\theta = -rv_r \tag{2.1.16}$$
of the Cauchy-Riemann equations at (r_0, θ_0), then $f'(z_0)$ exists, and
$$f'(z_0) = e^{-i\theta_0}(u_r(r_0, \theta_0) + iv_r(r_0, \theta_0)). \tag{2.1.17}$$

Example 2.1.9 Consider the function
$$f(z) = \frac{1}{z} = \frac{1}{re^{i\theta}} = \frac{1}{r}(\cos\theta - i\sin\theta), \quad (z \neq 0).$$

Since
$$u(r,\theta)=\frac{\cos\theta}{r} \quad \text{and} \quad v(r,\theta)=-\frac{\sin\theta}{r},$$
the Cauchy-Riemann equations
$$v_\theta=-\frac{\cos\theta}{r}=ru_r \quad \text{and} \quad u_\theta=-\frac{\sin\theta}{r}=-rv_r$$
are satisfied, and u_r, u_θ, v_r, v_θ are continuous at every $z\neq 0$, so the derivative of $f(z)$ exists for all nonzero points, and
$$f'(z)=e^{-i\theta}\left(-\frac{\cos\theta}{r^2}+i\frac{\sin\theta}{r^2}\right)=-e^{-i\theta}\frac{e^{-i\theta}}{r^2}=-\frac{1}{z^2}.$$

Example 2.1.10 Consider the function
$$f(z)=\sqrt[3]{r}\exp\left(\frac{i\theta}{3}\right), \quad (r>0, \alpha<\theta<\alpha+2\pi),$$
where α is a fixed real number. Here
$$u(r,\theta)=\sqrt[3]{r}\cos\left(\frac{\theta}{3}\right), \quad v(r,\theta)=\sqrt[3]{r}\sin\left(\frac{\theta}{3}\right).$$
Direct calculations show that
$$v_\theta=\frac{\sqrt[3]{r}}{3}\cos\frac{\theta}{3}=ru_r, \quad u_\theta=-\frac{\sqrt[3]{r}}{3}\sin\frac{\theta}{3}=-rv_r,$$
and u_r, u_θ, v_r, v_θ are continuous at each point in the specified domain, so $f(z)$ is differentiable at each point where $f(z)$ is defined. Theorem 2.1.3 further tells us that
$$f'(z)=e^{-i\theta}\left(\frac{\sqrt[3]{r}}{3r}\cos\frac{\theta}{3}+i\frac{\sqrt[3]{r}}{3r}\sin\frac{\theta}{3}\right)$$
$$=e^{-i\theta}\frac{e^{\frac{i\theta}{3}}}{3(\sqrt[3]{r})^2}=\frac{1}{3(\sqrt[3]{r}e^{\frac{i\theta}{3}})^2}=\frac{1}{3(f(z))^2}.$$

Note that when a specific point z is taken in the specified domain, $f(z)$ is one value of $z^{\frac{1}{3}}$. Hence the expression above can be put in the form
$$\frac{d}{dz}z^{\frac{1}{3}}=\frac{1}{3(z^{\frac{1}{3}})^2},$$
when that value is taken.

Exercises 2.1

1. Use the definition of derivatives to show that the function $f(z)=\dfrac{1}{z}$ has a derivative at any nonzero point, and $f'(z)=-\dfrac{1}{z^2}$ $(z\neq 0)$.

2. Find $f'(z)$, when

   (a) $f(z)=3z^3+4z$; (b) $f(z)=\dfrac{z+1}{2z^2}$ $(z\neq 0)$.

3. Show that for $f(z)=\text{Re}(z)$, $f'(z)$ does not exist at any point z, by using

(a) the definition of derivatives; (b) Cauchy-Riemann equations.

4. Determine where $f'(z)$ exists and find its value when it exists, when
(a) $f(z)=\bar{z}$;
(b) $f(z)=x^2+iy^2$;
(c) $f(z)=2x+ixy^2$;
(d) $f(z)=e^x e^{-iy}$;
(e) $f(z)=x^3+i(1-y)^3$;
(f) $f(z)=z\text{Im } z$.

5. Show that each of the functions is differentiable in the indicated domain of definition, and also find $f'(z)$:
(a) $f(z)=\sqrt{r}e^{\frac{i\theta}{2}}$ $(r>0, \alpha<\theta<\alpha+2\pi)$;
(b) $f(z)=e^{-\theta}\cos(\ln r)+ie^{-\theta}\sin(\ln r)$ $(r>0, 0<\theta<2\pi)$.

2.2 Analytic functions

We are now ready to introduce the concept of an analytic function, and some of its properties.

2.2.1 Analytic functions

Definition 2.2.1 A function f of the complex variable z is **analytic in an open set** if it has a derivative at each point in the set. A function f is **analytic at a point**, if it is analytic throughout some neighborhood of the point. That a function f is analytic in a set S which is not open should be understood that f is analytic in some open set containing S.

We have already shown before that the function $f(z)=\dfrac{1}{z}$ is differentiable at each nonzero point. This means that it is analytic at each nonzero point. But the function $f(z)=|z|^2$ is not analytic at any point, since its derivative exists only at $z=0$.

Definition 2.2.2 An **entire function** is a function that is analytic at each point in the entire complex plane.

Since the derivative of a polynomial exists everywhere, every polynomial is an entire function.

Definition 2.2.3 If a function f fails to be analytic at a point z_0, but is analytic at some point in every neighborhood of z_0, then z_0 is called a **singular point**, or **singularity** of the function.

Obviously, the function $f(z)=\dfrac{1}{z}$ has a singular point $z=0$. However, the function $f(z)=|z|^2$ has no singular points, since it is nowhere analytic.

A necessary condition for a function to be analytic in a domain D is the continuity of the function throughout the domain. Satisfying the Cauchy-Riemann equations is another necessary condition. But they are not sufficient. Sufficient conditions for analyticity are provided by Theorems 2.1.2 and 2.1.3 in the above Section 2.1.

From the differentiation formulas we know that if $f(z)$ and $g(z)$ are analytic in a domain D, then $f(z) \pm g(z)$ and $f(z)g(z)$ are all analytic in D. If furthermore $g(z) \neq 0$, $z \in D$, then $\dfrac{f(z)}{g(z)}$ is also analytic in D. As a special case, if $p(z)$ and $q(z)$ are polynomials, then $\dfrac{p(z)}{q(z)}$ is analytic in any domain throughout which $q(z) \neq 0$.

From the chain rule for the derivative of a composite function, we know that a composition of two analytic functions is also analytic. Specifically, suppose that $f(z)$ is analytic in a domain D, and that the image of D under the transformation $w = f(z)$ is contained in the domain of definition of a function $g(w)$. Then $g[f(z)]$ is analytic in D, and

$$\frac{\mathrm{d}}{\mathrm{d}z} g[f(z)] = g'[f(z)] f'(z). \tag{2.2.1}$$

The following property of analytic functions is similar as that of functions of a real variable.

Theorem 2.2.1 If $f'(z) = 0$ everywhere in a domain D, then $f(z)$ must be constant throughout D.

Proof Write $f(z) = u(x, y) + iv(x, y)$, then from

$$f'(z) = u_x + iv_x = 0,$$

and the Cauchy-Riemann equations

$$u_x = v_y, \quad u_y = -v_x,$$

we have for each point in D,

$$u_x = u_y = 0, \quad v_x = v_y = 0.$$

For any line segment lying entirely in D, from the knowledge in Calculus, we know that the directional derivative of u and v along the segment is zero, which means that both u and v are constants along the segment.

For any two points in D, there is always a finite number of such line segments, joined end to end, connecting these two points. So the values of u and v at these two points must be the same. We may then conclude that there are real constants a and b, such that

$$u(x,y) = a, \quad v(x,y) = b$$

throughout D, which implies that $f(z) = a + bi$ at each point z in D.

Example 2.2.1 The quotient

$$f(z) = \frac{z^3 + 4}{(z^2 - 3)(z^2 + 1)}$$

is analytic except for the four singular points $z = \pm\sqrt{3}, \pm i$ at which the denominator vanishes.

Example 2.2.2 Consider the function

$$f(z) = \cosh x \cos y + i \sinh x \sin y.$$

Note that $u_x = \sinh x \cos y = v_y$, $u_y = -\cosh x \sin y = -v_x$ everywhere and these derivatives are continuous. So f is an entire function.

Example 2.2.3 Consider the function

$$g(z) = \ln r + i\theta, \quad (r > 0, 0 < \theta < 2\pi),$$

where $u(r,\theta)=\ln r$, $v(r,\theta)=\theta$. Because $v_\theta=1=ru_r$, $u_\theta=0=-rv_r$ everywhere in the domain and these partial derivatives are continuous, so g is analytic in the domain, and
$$g'(z)=e^{-i\theta}\left(\frac{1}{r}+0i\right)=\frac{1}{r}e^{-i\theta}=\frac{1}{z}.$$

Example 2.2.4 Suppose that a function
$$f(z)=u(x,y)+iv(x,y)$$
and its conjugate
$$\overline{f(z)}=u(x,y)-iv(x,y)$$
are both analytic in a domain D. Then the Cauchy-Riemann equations give
$$u_x=v_y, \quad u_y=-v_x,$$
since $f(z)$ is supposed to be analytic, and
$$u_x=(-v)_y=-v_y, \quad u_y=-(-v)_x=v_x,$$
since $\overline{f(z)}$ is analytic. These implies that
$$u_x=u_y=v_x=v_y=0.$$
Then similar as in the proof of Theorem 2.2.1, we know that $f(z)$ must be constant throughout the domain D.

2.2.2 Harmonic functions

Now let us take a further look at analytic functions. Suppose that $f(z)=u(x,y)+iv(x,y)$ is analytic in a domain D, then its component functions $u(x,y)$ and $v(x,y)$ satisfy the Cauchy-Riemann equations
$$u_x=v_y, \quad u_y=-v_x.$$
Further assume that the second-order partial derivatives of the component functions $u(x,y)$ and $v(x,y)$ exist and are continuous throughout the domain D. Differentiating the equation above with respect to x and y leads to
$$u_{xx}=v_{yx}, \quad u_{yx}=-v_{xx}, \quad u_{xy}=v_{yy}, \quad u_{yy}=-v_{xy}.$$
We have learnt in calculus that
$$u_{xy}=u_{yx}, \quad v_{xy}=v_{yx},$$
so we have
$$u_{xx}+u_{yy}=0, \quad v_{xx}+v_{yy}=0.$$

A real-valued function H of two variables x and y is said to be **harmonic** in a given domain of the xy plane if, throughout the domain, it has continuous partial derivatives of the first order and second order, and satisfies the equation
$$H_{xx}(x,y)+H_{yy}(x,y)=0, \tag{2.2.2}$$
known as *Laplace's equation*. With this definition, the discussion above can be summarized into the following theorem.

Theorem 2.2.2 If a function $f(z)=u(x,y)+iv(x,y)$ is analytic in a domain D, then its component functions $u(x,y)$ and $v(x,y)$ are harmonic in D.

Example 2.2.5 The function $f(z)=\dfrac{i}{z^2}$ is obviously analytic at any nonzero point. Since

$$f(z)=\frac{\mathrm{i}}{z^2}=\frac{\mathrm{i}\,\bar{z}^2}{z^2\,\bar{z}^2}=\frac{\mathrm{i}\,\bar{z}^2}{|z|^4}=\frac{2xy+\mathrm{i}(x^2-y^2)}{(x^2+y^2)^2},$$

the two functions

$$u(x,y)=\frac{2xy}{(x^2+y^2)^2},\quad v(x,y)=\frac{x^2-y^2}{(x^2+y^2)^2}$$

are both harmonic throughout any domain in the xy plane that does not contain the origin.

If both $u(x,y)$ and $v(x,y)$ are harmonic in a domain D and their first-order partial derivatives satisfy the Cauchy-Riemann equations throughout D

$$u_x=v_y,\quad u_y=-v_x,$$

then v is said to be a **harmonic conjugate** of u. We then easily have the following theorem.

Theorem 2.2.3 A function $f(z)=u(x,y)+\mathrm{i}v(x,y)$ is analytic in a domain D if and only if v is a harmonic conjugate of u.

Note that if v is a harmonic conjugate of u in some domain, it is, generally, not true that u is a harmonic conjugate of v. Actually, let us assume that u is also a harmonic conjugate of v. Recall that the fact that v is a harmonic conjugate of u implies that

$$u_x=v_y,\quad u_y=-v_x,$$

while the assumption that u is a harmonic conjugate of v implies that

$$v_x=u_y,\quad v_y=-u_x.$$

These equations, together, show that

$$u_x=u_y=v_x=v_y=0,$$

which means that both u and v must be constant.

We now illustrate a method of obtaining a harmonic conjugate of a given harmonic function.

Example 2.2.6 Given the function

$$u(x,y)=y^3-3x^2y,$$

it can be easily verified that

$$u_{xx}+u_{yy}=-6y+6y=0,$$

which means that u is harmonic throughout the entire xy plane.

If v is a harmonic conjugate of u, the Cauchy-Riemann equations must be satisfied:

$$u_x=v_y,\quad u_y=-v_x,$$

which lead to

$$v_y=u_x=-6xy,\quad v_x=-u_y=-3y^2+3x^2.$$

The second of the equations tells us that

$$v(x,y)=x^3-3xy^2+\varphi(y),$$

where $\varphi(y)$ is an arbitrary function of y. Substituting it into the first of the equations gives

$$v_y=-6xy+\varphi'(y)=-6xy,$$

which means that

$$\varphi'(y)=0,$$

or equivalently $\varphi(y)=C$, where C is an arbitrary real number. So we have found a harmonic conjugate of u as

$$v(x,y)=x^3-3xy^2+C.$$
The corresponding analytic function is
$$f(z)=(y^3-3xy^2)+i(x^3-3xy^2+C)=i(z^3+C),$$
and then
$$\begin{aligned}f'(z)&=u_x+iv_x=(y^3-3xy^2)_x+i(x^3-3xy^2)_x\\&=-6xy+i(3x^2-3y^2)=3iz^2.\end{aligned}$$

Exercises 2.2

1. Verify that each of these functions is entire:
   (a) $f(z)=3x+y+i(3y-x)$; (b) $f(z)=(z^2-2)e^{-x}e^{-iy}$.

2. Show that each of these functions is nowhere analytic:
   (a) $f(z)=xy+iy$; (b) $f(z)=\text{Arg } z$; (c) $f(z)=e^y e^{ix}$.

3. In each case, determine the singular points of the function and state why the function is analytic everywhere except at those points:
   (a) $f(z)=\dfrac{z^3+i}{z^2-3z+2}$; (b) $f(z)=\dfrac{z^2+1}{(z+2)(z^2+2z+2)}$.

4. Show that $u(x,y)$ is harmonic in some domain and find a harmonic conjugate $v(x,y)$ when
   (a) $u(x,y)=2x(1-y)$; (b) $u(x,y)=\dfrac{y}{x^2+y^2}$.

5. Let the function $f(z)=u(r,\theta)+iv(r,\theta)$ be analytic in a domain D that does not include the origin. Using the Cauchy-Riemann equations in polar coordinates and assuming continuity of partial derivatives, show that throughout D the function $u(r,\theta)$ satisfies the partial differential equation
$$r^2 u_{rr}(r,\theta)+ru_r(r,\theta)+u_{\theta\theta}(r,\theta)=0,$$
which is the polar form of Laplace's equation. Show that it also holds for $v(r,\theta)$.

2.3 Elementary functions

In this section, we are to extend various elementary functions, such as the exponential function, logarithmic function, trigonometric functions and hyperbolic functions, studied in calculus to functions of a complex variable. These functions of a complex variable z will reduce to the elementary functions in calculus when $z=x+i0$. We start by defining the exponential function of a complex variable, and then use it to develop the others.

2.3.1 Exponential functions

The ***exponential function*** of a complex variable $z=x+iy$ is defined as

$$e^z = e^{x+iy} = e^x e^{iy} = e^x(\cos y + i\sin y). \tag{2.3.1}$$

When $z = x + i0$, the exponential function e^z reduces to the normal exponential function in calculus. Following the convention used in calculus, we often write exp z for e^z.

With the definition above, some properties of e^x in calculus can be extended to the exponential function of a complex variable. The following properties can be easily verified:

$$e^{z_1} e^{z_2} = e^{z_1 + z_2}, \quad \frac{e^{z_1}}{e^{z_2}} = e^{z_1 - z_2}. \tag{2.3.2}$$

Noting the fact $e^0 = 1$, it follows that

$$\frac{1}{e^z} = e^{-z}. \tag{2.3.3}$$

From Example 2.1.7 we know that e^z is an entire function, and

$$\frac{d}{dz} e^z = e^z \tag{2.3.4}$$

everywhere in the complex plane. It is also true that

$$e^z \neq 0 \tag{2.3.5}$$

for any complex number z, which can be easily deduced by noting

$$|e^z| = |e^x(\cos y + i\sin y)| = e^x \neq 0.$$

However, some properties of e^x are not expected. For example, since

$$e^{z+2k\pi i} = e^z e^{2k\pi i} = e^z, \tag{2.3.6}$$

for any integer k, we know that e^z is periodic, with a pure imaginary period of $2k\pi i$.

It is well known that e^x is positive for any real number x. However, this property does not hold for e^z, by noting

$$e^{(2k+1)\pi i} = \cos(2k+1)\pi + i\sin(2k+1)\pi = -1.$$

Indeed, e^z can be any nonzero complex number. This is shown in the following section, where the logarithmic function is developed.

2.3.2 Logarithmic functions

As the inverse function of the exponential function, the logarithmic function is defined by solving the equation

$$e^w = z$$

for w, where $z \neq 0$. Denote $z = re^{i\Theta}$ ($-\pi < \Theta \leq \pi$), and write $w = u + iv$, then the equation above becomes

$$e^u e^{iv} = re^{i\Theta}.$$

Equaling two complex numbers in exponential form shows that

$$e^u = r, \quad v = \Theta + 2n\pi,$$

where n is any integer. So for any nonzero complex number z, the equation $e^w = z$ is satisfied if and only if w admits one of the values

$$w = \ln r + i(\Theta + 2n\pi), \quad (n = 0, \pm 1, \pm 2, \cdots).$$

Thus we define the **logarithmic function** as

$$\log z = \ln r + i(\Theta + 2n\pi), \quad (n = 0, \pm 1, \pm 2, \cdots). \tag{2.3.7}$$

Example 2.3.1 To find $\log(-1-\sqrt{3}\mathrm{i})$, we first write
$$-1-\sqrt{3}\mathrm{i}=2\exp\left(-\frac{2\pi\mathrm{i}}{3}\right),$$
then we have
$$\log(-1-\sqrt{3}\mathrm{i})=\ln 2+\mathrm{i}\left(-\frac{2}{3}\pi+2n\pi\right),\quad (n=0,\ \pm 1,\ \pm 2,\ \cdots).$$

It should be emphasized that the logarithmic function is a multiple-valued function, while the exponential function is a single-valued function. From the definition of $\log z$, we can easily found that
$$\mathrm{e}^{\log z}=z. \tag{2.3.8}$$
However, the above equation with the order of the exponential and logarithmic functions reversed will not reduce to just z. More precisely, write $z=x+\mathrm{i}y$, then
$$\log \mathrm{e}^z=\log \mathrm{e}^x \mathrm{e}^{\mathrm{i}y}=\ln \mathrm{e}^x+\mathrm{i}(y+2n\pi)=x+\mathrm{i}(y+2n\pi)=z+2n\pi\mathrm{i}, \tag{2.3.9}$$
where n is any integer.

The definition of $\log z$ can be also written as
$$\log z=\ln|z|+\mathrm{i}\arg z. \tag{2.3.10}$$
The ***principal value*** of $\log z$ is the value when the principal value of $\arg z$ is employed in the above equation, and is denoted by $\mathrm{Log}\ z$. Thus
$$\mathrm{Log}\ z=\ln|z|+\mathrm{i}\mathrm{Arg}\ z. \tag{2.3.11}$$
Note that $\mathrm{Log}\ z$ is well defined and single-valued for any $z\neq 0$, and
$$\log z=\mathrm{Log}\ z+2n\pi\mathrm{i},\quad (n=0,\ \pm 1,\ \pm 2,\ \cdots). \tag{2.3.12}$$
It reduces to the usual logarithm in calculus when z is a positive real number.

Example 2.3.2 When $z=1$,
$$\log 1=\ln 1+\mathrm{i}(0+2n\pi)=2n\pi\mathrm{i},\quad (n=0,\ \pm 1,\ \pm 2,\ \cdots).$$
As anticipated, $\mathrm{Log}\ 1=0$.

Example 2.3.3 When $z=-1$,
$$\log(-1)=\ln 1+\mathrm{i}(\pi+2n\pi)=(2n+1)\pi\mathrm{i},\quad (n=0,\ \pm 1,\ \pm 2,\ \cdots),$$
and $\mathrm{Log}(-1)=\mathrm{i}\pi$.

Example 2.3.3 shows that the domain of the logarithm function involves negative real numbers.

For $z\neq 0$, write $z=r\mathrm{e}^{\mathrm{i}\theta}$, where the argument θ has any one of the values
$$\theta=\Theta+2n\pi,\quad (n=0,\ \pm 1,\ \pm 2,\ \cdots)$$
with $\Theta=\mathrm{Arg}\ z$. Hence the multiple-valued logarithmic function can be written as
$$\log z=\ln r+\mathrm{i}\theta.$$
If we restrict the value of θ so that $\alpha<\theta<\alpha+2\pi$ for a fixed real number α, the function
$$\log z=\ln r+\mathrm{i}\theta,\quad (r>0,\ \alpha<\theta<\alpha+2\pi) \tag{2.3.13}$$
with components
$$u(r,\theta)=\ln r,\quad v(r,\theta)=\theta$$
is single-valued and continuous in the specified domain. Note that if the function were to be defined on the ray $\theta=\alpha$, it would not be continuous along the ray. For if z is a point on the

ray, there are points arbitrarily close to z at which the values of v are near α, but also points at which the values of v are near $\alpha+2\pi$.

Throughout the domain $r>0$, $\alpha<\theta<\alpha+2\pi$, the first-order partial derivatives of u and v are continuous and satisfy the polar form

$$v_\theta = 1 = ru_r, \quad u_\theta = 0 = -rv_r$$

of the Cauchy-Riemann equations, so the logarithmic function defined in the domain $r>0$, $\alpha<\theta<\alpha+2\pi$ is analytic in that domain, and

$$\frac{d}{dz}\log z = e^{-i\theta}\left(\frac{1}{r}+0i\right) = \frac{1}{r}e^{-i\theta} = \frac{1}{z}.$$

That is,

$$\frac{d}{dz}\log z = \frac{1}{z}, \quad (z\neq 0, \alpha<\arg z<\alpha+2\pi). \tag{2.3.14}$$

In particular,

$$\frac{d}{dz}\text{Log } z = \frac{1}{z}, \quad (z\neq 0, -\pi<\text{Arg } z<\pi). \tag{2.3.15}$$

A **branch** of a multiple-valued function $f(z)$ is any single-valued function $F(z)$ that is analytic in some domain and the value $F(z)$ is one of the values $f(z)$. The requirement of analyticity prevents $F(z)$ from taking on a random selection of the values of $f(z)$. For any fixed real α, the single-valued function

$$\log z = \ln r + i\theta, \quad (r>0, \alpha<\theta<\alpha+2\pi)$$

is then a branch of the multiple-valued logarithmic function

$$\log z = \ln r + i\theta.$$

And the function

$$\text{Log } z = \ln r + i\Theta, \quad (r>0, -\pi<\Theta<\pi)$$

is called the principal branch.

A **branch cut** is a portion of a line or curve that is introduced to define a branch F of a multiple-valued function f. Points on the branch cut are singular points. Any point that is common to all branch cuts is called a **branch point**. For the multiple-valued logarithmic function, the origin and the ray $\theta=\alpha$ make up the branch cut for the branch

$$\log z = \ln r + i\theta, \quad (r>0, \alpha<\theta<\alpha+2\pi).$$

In particular, the origin and the ray $\Theta=\pi$ make up the branch cut for the principal branch. The origin is obviously a branch point.

Now we will introduce some identities involving logarithmic functions of a complex variable. The first one is in the same form of that for a real variable. From the definition of the logarithmic function of a complex variable, for any two nonzero complex numbers z_1 and z_2, we have

$$\log(z_1 z_2) = \log z_1 + \log z_2. \tag{2.3.16}$$

But it should be understood in a different manner, since each term in the above identity involves a multiple-valued function. It should be interpreted as, if values of two of the three logarithms are specified, there exists a value of the third one such that the above identity holds.

We also have
$$\log\left(\frac{z_1}{z_2}\right) = \log z_1 - \log z_2, \qquad (2.3.17)$$
which should be interpreted similarly as above.

It should be pointed out that in the above identities, when log are replaced by Log, they may not hold anymore. As an illustration, let $z_1 = z_2 = -1$, then
$$\text{Log}(z_1 z_2) = \text{Log } 1 = 0,$$
while
$$\text{Log } z_1 + \text{Log } z_2 = \pi i + \pi i = 2\pi i.$$

We include here two other properties of logarithmic functions. If z is a nonzero complex number, then for any integer n, it holds that
$$z^n = e^{n \log z}, \quad (n = 0, \pm 1, \pm 2, \cdots), \qquad (2.3.18)$$
for any value of $\log z$ that is taken. It is also true that
$$z^{\frac{1}{n}} = \exp\left(\frac{1}{n} \log z\right), \quad (n = \pm 1, \pm 2, \cdots). \qquad (2.3.19)$$
But keep in mind that in the former equation, although $\log z$ is multiple-valued, both z^n and $e^{n\log z}$ are single-valued, while in the latter equation, both terms are multiple-valued, and each term has exact n distinct values.

2.3.3 Complex exponents

When $z \neq 0$, for any given complex number c, the **power function** is defined by
$$z^c = e^{c \log z}, \qquad (2.3.20)$$
where $\log z$ denotes the multiple-valued logarithmic function. When $c = n$ ($n = 0, \pm 1, \pm 2, \cdots$) and $c = \frac{1}{n}$ ($n = \pm 1, \pm 2, \cdots$), it provides a consistent definition in the sense that it is already known to be valid in the above section. Keep in mind that the function might be a multiple-valued function.

Example 2.3.4 Let us compute
$$i^{-2i} = \exp(-2i \log i).$$
Since
$$\log i = \ln 1 + i\left(\frac{\pi}{2} + 2n\pi\right) = \left(2n + \frac{1}{2}\right)\pi i, \quad (n = 0, \pm 1, \pm 2, \cdots),$$
we have
$$i^{-2i} = \exp[(4n+1)\pi], \quad (n \neq 0, \pm 1, \pm 2, \cdots).$$
Since it holds that $e^{-z} = \frac{1}{e^z}$, we have
$$z^{-c} = e^{-c \log z} = \frac{1}{e^{c \log z}} = \frac{1}{z^c}. \qquad (2.3.21)$$
We have known that the branch
$$\log z = \ln r + i\theta, \quad (r > 0, \ \alpha < \theta < \alpha + 2\pi)$$

is single-valued and analytic in the specified domain. When this branch is used, the function $z^c = e^{c\log z}$ is also single-valued and analytic in the same domain. The derivative of such a branch of z^c is found by using the chain rule as

$$\frac{d}{dz}z^c = \frac{d}{dz}e^{c\log z} = \exp(c\log z)\frac{c}{z}. \tag{2.3.22}$$

Recalling that $\exp(\log z) = z$, we have

$$\frac{d}{dz}z^c = c\frac{\exp(c\log z)}{\exp(\log z)} = c\exp((c-1)\log z),$$

or

$$\frac{d}{dz}z^c = cz^{c-1}, \quad (z \neq 0, \ \alpha < \arg z < \alpha + 2\pi). \tag{2.3.23}$$

The **principle value** of z^c is defined as

$$\text{P. V. } z^c = \exp(c\text{Log } z), \tag{2.3.24}$$

where $\log z$ is replaced by Log z. It also serves to define the principle branch of the function z^c in the domain $z \neq 0$, $-\pi < \text{Arg } z < \pi$.

Example 2.3.5 The principle value of $(-i)^i$ is

$$\exp[i\text{Log}(-i)] = \exp\left[i(\ln 1 + i\frac{-\pi}{2})\right] = \exp\frac{\pi}{2}.$$

That is,

$$\text{P. V. }(-i)^i = \exp\frac{\pi}{2}.$$

Example 2.3.6 The principal branch of $z^{\frac{2}{3}}$ is

$$\exp\left(\frac{2}{3}\text{Log } z\right) = \exp\left(\frac{2}{3}\ln r + \frac{2}{3}i\Theta\right) = \sqrt[3]{r^2}\exp\left(\frac{2}{3}i\Theta\right).$$

Thus

$$\text{P. V. } z^{\frac{2}{3}} = \sqrt[3]{r^2}\cos\frac{2}{3}\Theta + i\sqrt[3]{r^2}\sin\frac{2}{3}\Theta.$$

This function is analytic in the domain $r > 0$, $-\pi < \Theta < \pi$.

The exponential function with base c, where c is any nonzero complex number, is defined as

$$c^z = e^{z\log c}. \tag{2.3.25}$$

According to this definition, e^z should be a multiple-valued function. But generally, we refer to the principal value Log $e(=1)$.

When a value of $\log c$ is specified, c^z is an entire function, and

$$\frac{d}{dz}c^z = \frac{d}{dz}e^{z\log c} = e^{z\log c}\log c = c^z \log c. \tag{2.3.26}$$

2.3.4 Trigonometric functions

We have learned that for any real number x,

$$e^{ix} = \cos x + i\sin x, \quad e^{-ix} = \cos x - i\sin x,$$

which imply that

$$\sin x = \frac{e^{ix} - e^{-ix}}{2i}, \quad \cos x = \frac{e^{ix} + e^{-ix}}{2}.$$

It is therefore natural to define the **sine and cosine functions** of a complex variable as

$$\sin z = \frac{e^{iz} - e^{-iz}}{2i}, \quad \cos z = \frac{e^{iz} + e^{-iz}}{2}. \tag{2.3.27}$$

It is easy to see that they are entire, and their derivatives are

$$\frac{d}{dz}\sin z = \frac{ie^{iz} + ie^{-iz}}{2i} = \cos z, \tag{2.3.28}$$

and

$$\frac{d}{dz}\cos z = \frac{ie^{iz} - ie^{-iz}}{2} = -\sin z. \tag{2.3.29}$$

From the definitions of the sine and cosine functions, we can easily see that

$$\sin(-z) = -\sin z, \quad \cos(-z) = \cos z. \tag{2.3.30}$$

We state some identities carrying over from trigonometry without proof. For instance,

$$\sin(z_1 + z_2) = \sin z_1 \cos z_2 + \cos z_1 \sin z_2,$$
$$\cos(z_1 + z_2) = \cos z_1 \cos z_2 - \sin z_1 \sin z_2. \tag{2.3.31}$$

It follows readily from these that

$$\sin^2 z + \cos^2 z = 1,$$
$$\sin 2z = 2\sin z \cos z, \quad \cos 2z = \cos^2 z - \sin^2 z, \tag{2.3.32}$$
$$\sin\left(z + \frac{\pi}{2}\right) = \cos z, \quad \sin\left(z - \frac{\pi}{2}\right) = -\cos z.$$

The periodic character of the sine and cosine functions also follows readily:

$$\sin(z + 2\pi) = \sin z, \quad \sin(z + \pi) = -\sin z,$$
$$\cos(z + 2\pi) = \cos z, \quad \cos(z + \pi) = -\cos z. \tag{2.3.33}$$

Recall for a real number y, the hyperbolic functions are

$$\sinh y = \frac{e^y - e^{-y}}{2}, \quad \cosh y = \frac{e^y + e^{-y}}{2}.$$

It then follows that

$$\sin(iy) = i\sinh y, \quad \cos(iy) = \cosh y.$$

And hence for $z = x + iy$,

$$\sin z = \sin x \cosh y + i\cos x \sinh y,$$
$$\cos z = \cos x \cosh y - i\sin x \sinh y. \tag{2.3.34}$$

Taking the modulus on both sides of the above equations gives

$$|\sin z|^2 = \sin^2 x + \sinh^2 y, \quad |\cos z|^2 = \cos^2 x + \sinh^2 y. \tag{2.3.35}$$

As y tends to infinity, $\sinh y$ tends to infinity. So from the above equations we know that both $\sin z$ and $\cos z$ are not bounded on the complex plane, whereas the absolute values of $\sin x$ and $\cos x$ are no more than unity for all real number x.

A **zero** of a given function $f(z)$ is a complex number z_0 such that $f(z_0) = 0$. To find zeros of the sine function, assume that $\sin z = 0$, or equivalently $|\sin z| = 0$. Noting

$$|\sin z|^2 = \sin^2 x + \sinh^2 y,$$

it then holds that

$$\sin x = 0, \quad \sinh y = 0,$$

which implies that $x = n\pi$ ($n = 0, \pm 1, \pm 2, \cdots$) and $y = 0$. This means that all zeros of the sine function are

$$z = n\pi, \quad (n = 0, \pm 1, \pm 2, \cdots). \tag{2.3.36}$$

Similarly, we can obtain all zeros of the cosine function as

$$z = \frac{\pi}{2} + n\pi, \quad (n = 0, \pm 1, \pm 2, \cdots). \tag{2.3.37}$$

As in calculus, the other four trigonometric functions are defined as

$$\tan z = \frac{\sin z}{\cos z}, \quad \cot z = \frac{\cos z}{\sin z},$$
$$\sec z = \frac{1}{\cos z}, \quad \csc z = \frac{1}{\sin z}. \tag{2.3.38}$$

From these definitions we can easily see that the functions $\tan z$ and $\sec z$ are analytic everywhere except at the singular points

$$z = \frac{\pi}{2} + n\pi, \quad (n = 0, \pm 1, \pm 2, \cdots),$$

which are the zeros of $\cos z$. Likewise, the functions $\cot z$ and $\csc z$ are analytic except at the singular points

$$z = n\pi, \quad (n = 0, \pm 1, \pm 2, \cdots),$$

the zeros of $\sin z$.

In the domain where these four trigonometric functions are analytic, their derivatives are

$$\frac{d}{dz} \tan z = \sec^2 z, \quad \frac{d}{dz} \cot z = -\csc^2 z,$$
$$\frac{d}{dz} \sec z = \sec z \tan z, \quad \frac{d}{dz} \csc z = -\csc z \cot z. \tag{2.3.39}$$

The periodicity of each of these four trigonometric functions follows readily from that of $\sin z$ and $\cos z$. Specifically,

$$\tan(z + \pi) = \tan z, \quad \cot(z + \pi) = \cot z,$$
$$\sec(z + 2\pi) = \sec z, \quad \csc(z + 2\pi) = \csc z. \tag{2.3.40}$$

2.3.5 Hyperbolic functions

The **hyperbolic sine and cosine functions** of a complex variable are defined in the same form as those of a real variable, that is,

$$\sinh z = \frac{e^z - e^{-z}}{2}, \quad \cosh z = \frac{e^z + e^{-z}}{2}. \tag{2.3.41}$$

Recalling the definitions of sine and cosine functions

$$\sin z = \frac{e^{iz} - e^{-iz}}{2i}, \quad \cos z = \frac{e^{iz} + e^{-iz}}{2},$$

the hyperbolic sine and cosine functions are closely related to these trigonometric functions as

$$-i\sinh(iz) = \sin z, \quad \cosh(iz) = \cos z,$$
$$-i\sin(iz) = \sinh z, \quad \cos(iz) = \cosh z. \tag{2.3.42}$$

The following properties of the hyperbolic sine and cosine functions then can be easily obtained by using their definitions or their relationship with the sine and cosine functions described above. Obviously, they are entire functions, and their derivatives are

$$\frac{d}{dz}\sinh z = \cosh z, \quad \frac{d}{dz}\cosh z = \sinh z. \tag{2.3.43}$$

It also holds that
$$\sinh(-z) = -\sinh z, \quad \cosh(-z) = \cosh z,$$
$$\sinh(z_1 + z_2) = \sinh z_1 \cosh z_2 + \cosh z_1 \sinh z_2,$$
$$\cosh(z_1 + z_2) = \cosh z_1 \cosh z_2 + \sinh z_1 \sinh z_2, \tag{2.3.44}$$
$$\cosh^2 z - \sinh^2 z = 1.$$

With $z = x + iy$, we have
$$\sinh z = \sinh x \cos y + i \cosh x \sin y,$$
$$\cosh z = \cosh x \cos y + i \sinh x \sin y. \tag{2.3.45}$$

These imply that
$$|\sinh z|^2 = \sinh^2 x + \sin^2 y, \quad |\cosh z|^2 = \sinh^2 x + \cos^2 y. \tag{2.3.46}$$

Hence all zeros of $\sinh z$ are
$$z = n\pi i, \quad (n = 0, \pm 1, \pm 2, \cdots), \tag{2.3.47}$$

while all zeros of $\cosh z$ are
$$z = \left(\frac{\pi}{2} + n\pi\right)i, \quad (n = 0, \pm 1, \pm 2, \cdots). \tag{2.3.48}$$

They are periodic with period $2\pi i$, that is,
$$\sinh(z + 2\pi i) = \sinh z, \quad \cosh(z + 2\pi i) = \cosh z. \tag{2.3.49}$$

The other hyperbolic functions are defined as
$$\tanh z = \frac{\sinh z}{\cosh z}, \quad \coth z = \frac{\cosh z}{\sinh z},$$
$$\operatorname{sech} z = \frac{1}{\cosh z}, \quad \operatorname{csch} z = \frac{1}{\sinh z}. \tag{2.3.50}$$

They are analytic in the domain where the denominator does not vanish. And
$$\frac{d}{dz}\tanh z = \operatorname{sech}^2 z, \quad \frac{d}{dz}\coth z = -\operatorname{csch}^2 z,$$
$$\frac{d}{dz}\operatorname{sech} z = -\operatorname{sech} z \tanh z, \quad \frac{d}{dz}\operatorname{csch} z = -\operatorname{csch} z \coth z. \tag{2.3.51}$$

2.3.6 Inverse trigonometric and hyperbolic functions

To define the inverse sine function, we write
$$z = \sin w = \frac{e^{iw} - e^{-iw}}{2i}.$$

Then we can solve for w by
$$e^{iw} = iz + (1 - z^2)^{\frac{1}{2}},$$

which leads to

$$w = \sin^{-1} z = -i\log[iz + (1-z^2)^{\frac{1}{2}}]. \tag{2.3.52}$$

This is generally a multiple-valued function. The **inverse sine function** $\sin^{-1} z$ can be also written as $\arcsin z$.

Example 2.3.7 Let us compute $\sin^{-1}(0) = -i\log[i(0) + (1)^{\frac{1}{2}}] = -i\log(\pm 1)$. Note
$$\log(1) = 2n\pi i, \quad (n=0, \pm 1, \pm 2, \cdots),$$
$$\log(-1) = (2n+1)\pi i, \quad (n=0, \pm 1, \pm 2, \cdots).$$
These lead to
$$\sin^{-1}(0) = i\log(\pm 1) = n\pi, \quad (n=0, \pm 1, \pm 2, \cdots),$$
which are all the zeros of $\sin z$ as shown earlier.

Similarly, we have
$$\cos^{-1} z = -i\log[z + i(1-z^2)^{\frac{1}{2}}], \tag{2.3.53}$$
and
$$\tan^{-1} z = \frac{i}{2}\log\frac{i+z}{i-z}. \tag{2.3.54}$$

They are also multiple-valued functions. When specific branches of the square root and logarithmic functions are used, these inverse functions are single-valued and analytic, and
$$\frac{d}{dz}\sin^{-1} z = \frac{1}{(1-z^2)^{\frac{1}{2}}},$$
$$\frac{d}{dz}\cos^{-1} z = \frac{-1}{(1-z^2)^{\frac{1}{2}}}, \tag{2.3.55}$$
$$\frac{d}{dz}\tan^{-1} z = \frac{1}{1+z^2}.$$

In a similar manner, we can obtain the inverse hyperbolic functions as
$$\sinh^{-1} z = \log[z + (z^2+1)^{\frac{1}{2}}],$$
$$\cosh^{-1} z = \log[z + (z^2-1)^{\frac{1}{2}}], \tag{2.3.56}$$
$$\tanh^{-1} z = \frac{1}{2}\log\frac{1+z}{1-z}.$$

Exercises 2.3

1. Show that $|\exp(z^2)| \leq \exp(|z|^2)$.
2. Prove that $|\exp(-2z)| < 1$ if and only if $\operatorname{Re} z > 0$.
3. Let the function $f(z) = u(x,y) + iv(x,y)$ be analytic in a domain D. State why the functions
$$U(x,y) = e^{u(x,y)}\cos v(x,y), \quad V(x,y) = e^{u(x,y)}\sin v(x,y)$$
are harmonic in D and why $V(x,y)$ is a harmonic conjugate of $U(x,y)$.
4. Find all values of z such that
   (a) $e^z = 1 + \sqrt{3}i$; (b) $e^{2z-1} = -1$.
5. Show that

(a) Log $(1+i)^2 = 2\text{Log}(1+i)$; (b) Log $(-1+i)^2 \neq 2\text{Log}(-1+i)$.

6. Show that $f(z) = \text{Log}(z-i)$ is analytic everywhere except on the portion $x \leq 0$ of the line $y=1$.

7. Show that if Re $z_1 > 0$ and Re $z_2 > 0$, then $\text{Log}(z_1 z_2) = \text{Log } z_1 + \text{Log } z_2$.

8. Show that $(1+i)^i = \exp\left(-\frac{\pi}{4} + 2n\pi\right) \exp\left(i\frac{\ln 2}{2}\right)$, $(n = 0, \pm 1, \pm 2, \cdots)$.

9. Find the principal value of $(1-i)^{4i}$.

10. Find all solutions of the equation $\sin z = \cosh 4$.

11. Show that

(a) $\overline{\cos(iz)} = \cos(i\bar{z})$ for all z;

(b) $\overline{\sin(iz)} = \sin(i\bar{z})$ if and only if $z = n\pi i$ $(n = 0, \pm 1, \pm 2, \cdots)$.

12. Show that $|\sinh x| \leq |\cosh z| \leq \cosh x$, where $z = x + iy$.

13. Find all solutions of the equation

(a) $\cos z = \sqrt{2}$; (b) $\sinh z = i$.

Chapter 3 Integral of Complex Function

We now develop a theory of integration for the functions of a complex variable. As the theory of differentiation, the complex function integrals play an important role in the study of functions of a complex variable. The theorems are generally concise and powerful for solving some problems in the nature.

3.1 Derivatives and definite integrals of functions $w(t)$

In order to introduce integrals of $f(z)$ in a fairly simple way, we need to consider derivatives and definite integrals of complex valued functions w of a real variable t.

3.1.1 Derivatives of functions $w(t)$

Definition 3.1.1 We write
$$w(t) = u(t) + iv(t), \qquad (3.1.1)$$
where the functions u and v are **real-valued** functions of t. The derivative $w'(t)$ or $\dfrac{d[w(t)]}{dt}$, of the function (3.1.1) at a point t is defined as
$$w'(t) = u'(t) + iv'(t), \qquad (3.1.2)$$
provided each of the derivatives u' and v' exists at t.

Example 3.1.1 Let $z_0 = x_0 + iy_0$. Then show that
$$\frac{d}{dt}[z_0 w(t)] = z_0 w'(t) \quad \text{and} \quad \frac{d}{dt}e^{z_0 t} = z_0 e^{z_0 t}.$$

Proof (a) Let $w(t) = u(t) + iv(t)$, then we have
$$\frac{d}{dt}[z_0 w(t)] = [(x_0 + iy_0)(u + iv)]'$$
$$= [(x_0 u - y_0 v) + i(y_0 u + x_0 v)]'$$
$$= (x_0 u - y_0 v)' + i(y_0 u + x_0 v)' = (x_0 u' - y_0 v') + i(y_0 u' + x_0 v').$$

Since
$$(x_0 u' - y_0 v') + i(y_0 u' + x_0 v') = (x_0 + iy_0)(u' + iv') = z_0 w'(t),$$
then

$$\frac{d}{dt}[z_0 w(t)] = z_0 w'(t).$$

(b) We write
$$e^{z_0 t} = e^{x_0 t} e^{iy_0 t} = e^{x_0 t}\cos y_0 t + i e^{x_0 t}\sin y_0 t.$$
Then (3.1.2), familiar rules from calculus and some simple algebra lead us to the expression
$$\frac{d}{dt} e^{z_0 t} = (e^{x_0 t}\cos y_0 t)' + i(e^{x_0 t}\sin y_0 t)'$$
$$= (x_0 + iy_0)(e^{x_0 t}\cos y_0 t + i e^{x_0 t}\sin y_0 t)$$
$$= (x_0 + iy_0) e^{x_0 t} e^{iy_0 t}$$
$$= z_0 e^{z_0 t}.$$

Various other rules learned in calculus, such as the ones for differentiating sums and products, apply just as they do for real-valued functions of t. As example 3.1.1, verifications may be based on corresponding rules in calculus. It should be pointed out, however, that not every rule for derivatives in calculus carries over to functions of type (3.1.1).

Suppose that $w(t)$ is continuous on an interval $[a,b]$; that is, its component functions $u(t)$ and $v(t)$ are continuous there. Even if $w'(t)$ exists when $a<t<b$, the mean value theorem for derivatives no longer applies. To be precise, it is not necessarily true that there is a number c in the interval (a,b) such that
$$w'(c) = \frac{w(b) - w(a)}{b - a}.$$

Example 3.1.2 Let $w(t) = e^{it}$, $t \in [0, 2\pi]$. Whether the mean value theorem dose still apply for $w(t) = e^{it}$?

Solution It is easy to see that $|w'(t)| = |ie^{it}| = 1$; and this means that the derivative $w'(t)$ is never zero, while $w(2\pi) - w(0) = 0$. So, the mean value theorem doesn't apply for the function e^{it} on the interval $[0, 2\pi]$.

3.1.2 Definite integrals of functions $w(t)$

Definition 3.1.2 When $w(t)$ is a ***complex-valued function*** of a real variable t and is written
$$w(t) = u(t) + iv(t),$$
where u and v are real-valued, the definite integral of $w(t)$ over an interval $[a, b]$ is defined as
$$\int_a^b w(t) dt = \int_a^b u(t) dt + i \int_a^b v(t) dt, \qquad (3.1.3)$$
when the individual integrals on the right exist. Thus
$$\operatorname{Re}\int_a^b w(t) dt = \int_a^b \operatorname{Re}[w(t)] dt \quad \text{and} \quad \operatorname{Im}\int_a^b w(t) dt = \int_a^b \operatorname{Im}[w(t)] dt. \qquad (3.1.4)$$

Example 3.1.3 From (3.1.3), it is easy to see that
$$\int_0^1 (1 + it)^2 dt = \int_0^1 (1 - t^2) dt + i \int_0^1 2t\, dt = \frac{2}{3} + i.$$

Existence

(a) The existence of the integrals of u and v in definition (3.1.3) is ensured if those functions are **piecewise continuous** on the interval $[a,b]$. Such a function is continuous everywhere in the stated interval except possibly for a finite number of points where, although discontinuous, it has one-sided limits. Of course, only the right-hand limit is required at a; and only the left-hand limit is required at b. When both u and v are piecewise continuous, the function w is said to have that property.

(b) Improper integrals of $w(t)$ over unbounded intervals are defined in a similar way.

Properties

(a) $\int_a^b [z_1 w_1(t) + z_2 w_2(t)]\,dt = z_1 \int_a^b w_1(t)\,dt + z_2 \int_a^b w_2(t)\,dt.$

(b) $\lim\limits_{n\to\infty} \int_a^b w_n(t)\,dt = \int_a^b \lim\limits_{n\to\infty} w_n(t)\,dt.$

(c) $\int_a^b w(t)\,dt = \int_a^c w(t)\,dt + \int_c^b w(t)\,dt.$

(d) **Fundamental theorem of calculus.**

Suppose that the functions
$$w(t) = u(t) + iv(t) \quad \text{and} \quad W(t) = U(t) + iV(t)$$
are continuous on the interval $[a,b]$. If $W'(t) = w(t)$ when $a \leqslant t \leqslant b$, then
$$\int_a^b w(t)\,dt = W(b) - W(a) = W(t)\Big|_a^b. \tag{3.1.5}$$

Proof Since $W'(t) = w(t)$ when $a \leqslant t \leqslant b$, then
$$U'(t) = u(t) \quad \text{and} \quad V'(t) = v(t).$$
Hence, in view of definition (3.1.3),
$$\int_a^b w(t)\,dt = U(t)\Big|_a^b + iV(t)\Big|_a^b = [U(b) + iV(b)] - [U(a) + iV(a)].$$
That is,
$$\int_a^b w(t)\,dt = W(b) - W(a) = W(t)\Big|_a^b.$$

(e) **Inequality of moduli of integrals.**
$$\left| \int_a^b w(t)\,dt \right| \leqslant \int_a^b |w(t)|\,dt, \quad (a \leqslant b). \tag{3.1.6}$$

Proof This inequality clearly holds when value of the integral on the left is zero, in particular when $a = b$. Thus, in the verification, we may assume that its value is a **nonzero** complex number. If r_0 is the modulus and θ_0 is an argument of the value of the integral, then
$$\int_a^b w(t)\,dt = r_0 e^{i\theta_0}.$$
Solving for r_0, we write
$$r_0 = \int_a^b e^{-i\theta_0} w(t)\,dt. \tag{3.1.7}$$

Now the left-hand side of this equation is a real number, and so the right-hand side is too. Thus, using the fact that the real part of a real number is the number itself and referring to the first of properties (3.1.4), we see that the right-hand side of equation (3.1.7) can be rewritten in the following way:

$$\int_a^b e^{-i\theta_0} w(t) dt = \operatorname{Re} \int_a^b e^{-i\theta_0} w(t) dt = \int_a^b \operatorname{Re}(e^{-i\theta_0} w(t)) dt.$$

Equation (3.1.7) then takes the form

$$r_0 = \int_a^b \operatorname{Re}(e^{-i\theta_0} w(t)) dt. \tag{3.1.8}$$

But

$$\operatorname{Re}(e^{-i\theta_0} w(t)) \leqslant |e^{-i\theta_0} w(t)| = |e^{-i\theta_0}| |w(t)| = |w(t)|;$$

and so, according to equation (3.1.8),

$$r_0 \leqslant \int_a^b |w(t)| dt.$$

Because r_0, in fact, the left-hand side of inequality (3.1.6) when the value of the integral there is nonzero, the verification is now complete.

With only minor modifications, the above discussion yields inequalities such as

$$\left| \int_a^\infty w(t) dt \right| \leqslant \int_a^\infty |w(t)| dt,$$

provided both improper integrals exist.

Example 3.1.4 Evaluate the integral

$$\int_0^{\frac{\pi}{6}} e^{i2t} dt.$$

Solution Since $(e^{i2t})' = 2i e^{i2t}$, then we have

$$\int_0^{\frac{\pi}{6}} e^{i2t} dt = \frac{e^{i2t}}{2i} \bigg|_0^{\frac{\pi}{6}} = \frac{e^{i\frac{\pi}{3}} - 1}{2i}$$

$$= \frac{\left(\cos \frac{\pi}{3} + i \sin \frac{\pi}{3} \right) - 1}{2i}$$

$$= \frac{\left(\frac{1}{2} + i \frac{\sqrt{3}}{2} \right) - 1}{2i} = \frac{\sqrt{3}}{4} + \frac{i}{4}.$$

Example 3.1.5 According to the definition (3.1.3) of integrals of complex-valued functions of a real variable, to evaluate the integral

$$\int_0^\pi e^x \cos x dx \quad \text{and} \quad \int_0^\pi e^x \sin x dx.$$

Solution It is easy to see that

$$\int_0^\pi e^{(1+i)x} dx = \int_0^\pi e^x \cos x dx + i \int_0^\pi e^x \sin x dx.$$

Since

$$\int_0^\pi e^{(1+i)x} dx = \frac{e^{(1+i)x}}{1+i} \bigg|_0^\pi = \frac{e^{(1+i)\pi} - 1}{1+i} = \frac{-e^\pi - 1}{1+i} = -\frac{e^\pi + 1}{2} + i \frac{e^\pi + 1}{2}.$$

Then from (3.1.3) and (3.1.4), we have

$$\int_0^\pi e^x \cos x dx = -\frac{e^\pi + 1}{2},$$

and
$$\int_0^\pi e^x \cos x \, dx = \frac{e^\pi + 1}{2}.$$

Exercises 3.1

1. Use the corresponding rules in calculus to establish the following rules when
$$w(t) = u(t) + iv(t)$$
is a complex-valued function of a real variable t and $w'(t)$ exists:

   (a) $\frac{d}{dt} w(-t) = -w'(-t)$, where $w'(-t)$ denotes the derivative of $w(t)$ with respect to t, evaluated at $-t$;

   (b) $\frac{d}{dt} [w(t)]^2 = 2w(t) w'(t)$.

2. Show that if $w(t) = u(t) + iv(t)$ is continuous on an interval $[a, b]$, then

   (a) $\int_{-b}^{-a} w(-t) \, dt = \int_a^b w(\tau) \, d\tau$;

   (b) $\int_a^b w(t) \, dt = \int_\alpha^\beta w[\varphi(\tau)] \varphi'(\tau) \, d\tau$, where $\varphi(\tau)$ is a real-valued function mapping an interval $[\alpha, \beta]$ onto the interval $[a, b]$.

   Suggestion The identities can be obtained by noting that they are valid for real-valued functions of t.

3. Evaluate the following integrals:

   (a) $\int_1^2 \left(\frac{1}{t} - i \right)^2 dt$; (b) $\int_0^{\frac{\pi}{4}} e^{it} \, dt$; (c) $\int_0^\infty e^{-zt} \, dt$ (Re $z > 0$).

3.2 Contour integral

Integrals of complex-valued functions of a **complex** variable are defined on curves in the complex plane, rather than on just intervals of the real line. Classes of curves that are adequate for the study of such integrals are introduced in the following sections.

3.2.1 Contour

A set of points $z = (x, y)$ in the complex plane is said to be an **arc** if
$$x = x(t), \quad y = y(t), \quad (a \leq t \leq b),$$
where $x(t)$ and $y(t)$ are continuous functions of the real parameter t. This definition establishes a continuous mapping of the interval $[a, b]$ into the xOy, or z plane; and the image points are ordered according to increasing values of t. It is convenient to describe the points of C by means of the equation

Chapter 3 Integral of Complex Function

$$z = z(t), \quad (a \leqslant t \leqslant b), \tag{3.2.1}$$

where

$$z(t) = x(t) + iy(t).$$

The arc C is a ***simple arc***, or a Jordan arc, if it does not cross itself; that is, C is simple if $z(t_1) \neq z(t_2)$, when $t_1 \neq t_2$. When the arc C is simple except for the fact that $z(b) = z(a)$, we say that C is a ***simple closed curve***, or a Jordan curve.

The geometric nature of a particular arc often suggests different notation for the parameter t in equation (3.2.1). This is, in fact, the case in the examples below.

Example 3.2.1 The polygonal line defined by means of the equations

$$z = \begin{cases} x + ix & \text{when } 0 \leqslant x \leqslant 1, \\ x + i & \text{when } 1 \leqslant x \leqslant 2, \end{cases}$$

and consisting of a line segment from 0 to $1+i$ followed by one from $1+i$ to $2+i$ is a simple arc (see Fig. 3.2.1).

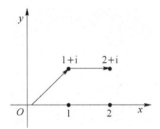

Fig. 3.2.1

Example 3.2.2 The unit circle

$$z = e^{i\theta}, \quad (0 \leqslant \theta \leqslant 2\pi) \tag{3.2.2}$$

about the origin is a simple closed curve, oriented in the counterclockwise direction.

Similarly, the circle

$$z = z_0 + Re^{i\theta}, \quad (0 \leqslant \theta \leqslant 2\pi),$$

is centered at the point z_0 and with radius R.

The same set of points can make up different arcs.

Example 3.2.3 The arc

$$z = e^{-i\theta}, \quad (0 \leqslant \theta \leqslant 2\pi) \tag{3.2.3}$$

is not the same as the arc described by equation (3.2.2). The set of points is the same but now the circle is traversed in the ***clockwise*** direction.

Example 3.2.4 The points on the arc

$$z = e^{i2\theta}, \quad (0 \leqslant \theta \leqslant 2\pi)$$

are the same as those making up the arcs (3.2.2) and (3.2.3). The arc here differs, however, from each of those arcs since the circle is traversed ***twice*** in the counterclockwise direction.

Let C by means of the equation

$$z = z(t), \quad (a \leqslant t \leqslant b).$$

Suppose now that the components $x'(t)$ and $y'(t)$ of the derivative

$$z'(t) = x'(t) + iy'(t)$$

are continuous on the entire interval $[a,b]$. The arc is then called a ***differentiable arc***, and the real-valued function

$$|z'(t)| = \sqrt{[x'(t)]^2 + [y'(t)]^2}$$

is integrable over the interval $[a,b]$. Then the **arc length** of C is the number

$$L = \int_a^b |z'(t)|\,dt = \int_a^b \sqrt{[x'(t)]^2 + [y'(t)]^2}\,dt. \qquad (3.2.4)$$

3.2.2 Definition of contour integral

We turn now to integrals of complex-valued functions f of the complex variable z. Such an integral is defined in terms of the values $f(z)$ along a given contour C, extending from a point $z = z_1$ to a point $z = z_2$ in the complex plane.

Definition 3.2.1 Suppose that C is an oriented smooth curve from A to B, and $\omega = f(z)$ is defined along the given contour C. Insert arbitrarily $n-1$ points $z_1, z_2, \cdots, z_{n-1}$ on C successively from A (denote by z_0) to B (denote by z_n), such that C is divided into n small oriented subarcs $\widehat{z_{k-1}z_k}$. Let ξ_k be any point on the subarc $\widehat{z_{k-1}z_k}$.

If the limit of the sum

$$S_n = \sum_{k=1}^n f(\xi_k)(z_k - z_{k-1}) = \sum_{k=1}^n f(\xi_k)\Delta z_k,$$

exists as $d = \max\limits_{1 \leqslant k \leqslant n}|\Delta z_k| \to 0$, no matter how we partition C and how we select the point ξ_k on the subarc $\widehat{z_{k-1}z_k}$. Then this limit is called the integral of $\omega = f(z)$ along the given contour C, which is denoted by

$$\int_C f(z)\,dz = \lim_{n \to \infty} \sum_{k=1}^n f(\xi_k)\Delta z_k.$$

The line integral depends, in general, on the contour C as well as on the function f. It is written

$$\int_C f(z)\,dz \quad \text{or} \quad \int_{z_1}^{z_2} f(z)\,dz,$$

the latter notation often being used when the value of the integral is independent of the choice of the contour taken between two fixed end points.

Existence When C is a contour and $f'(z(t))z'(t)$ is piecewise continuous on the interval $[a,b]$, then the existence of contour integral $\int_C f(z)\,dz$ is ensured.

Computation Suppose that the oriented contour C is given by

$$z = z(t), \quad (a \leqslant t \leqslant b),$$

and extends from a point $z_1 = z(a)$ to a point $z_2 = z(b)$. Let the function $f(z)$ be piecewise continuous on C; that is, $f[z(t)]$ is piecewise continuous on the interval $[a,b]$. Then

$$\int_C f(z)\,dz = \int_a^b f[z(t)]z'(t)\,dt. \qquad (3.2.5)$$

Proof Firstly, since $z = x+iy$, $dz = dx+idy$, $f(z) = u(x,y)+iv(x,y)$, then

$$\int_C f(z)\,\mathrm{d}z = \int_C [u(x,y)+\mathrm{i}v(x,y)][\mathrm{d}x+\mathrm{i}\mathrm{d}y]$$
$$= \int_C u(x,y)\,\mathrm{d}x - v(x,y)\,\mathrm{d}y + \mathrm{i}\int_C v(x,y)\,\mathrm{d}x + u(x,y)\,\mathrm{d}y.$$

Moreover, the oriented contour C is given by
$$z=z(t)=x(t)+\mathrm{i}y(t), \quad (a\leqslant t\leqslant b), \quad z_1=z(a) \text{ and } z_2=z(b).$$
And, denote $u(x(t),y(t))=u(t)$, $v(x(t),y(t))=v(t)$, then we have
$$\mathrm{d}x=x'(t)\mathrm{d}t, \quad \mathrm{d}y=y'(t)\mathrm{d}t, \quad \mathrm{d}z=z'(t)\mathrm{d}t, \quad z'(t)=x'(t)+y'(t).$$
Therefore,
$$\int_C f(z)\,\mathrm{d}z = \int_a^b [u(t)x'(t) - v(t)y'(t)]\,\mathrm{d}t +$$
$$\mathrm{i}\int_a^b [v(t)x'(t) + u(t)y'(t)]\,\mathrm{d}t$$
$$= \int_a^b [u(t)+\mathrm{i}v(t)][x'(t)+\mathrm{i}y'(t)]\,\mathrm{d}t$$
$$= \int_a^b f[z(t)]z'(t)\,\mathrm{d}t.$$

Properties

(a) Linearity
$$\int_C [z_1 f_1(z) + z_2 f_2(z)]\,\mathrm{d}z = z_1 \int_C f_1(z)\,\mathrm{d}z + z_2 \int_C f_2(z)\,\mathrm{d}z.$$

(b) Orientation If $-C$ denotes the curve consisting of the same points as C but with the opposite orientation, then
$$\int_{-C} f(z)\,\mathrm{d}z = -\int_C f(z)\,\mathrm{d}z.$$

Proof Suppose that the oriented contour C is given by
$$z=z(t), \quad (a\leqslant t\leqslant b),$$
and extends from a point $z_1=z(a)$ to a point $z_2=z(b)$. Then the contour $-C$ has parametric representation
$$z=z(\tau), \quad (a\leqslant \tau\leqslant b),$$
and extends from the point $z_2(b)$ to the point $z_1(a)$ (see Fig. 3.2.2).

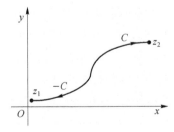

Fig. 3.2.2

Thus, from (3.2.5), it follows that
$$\int_{-C} f(z)\,\mathrm{d}z = \int_b^a f[z(\tau)]z'(\tau)\,\mathrm{d}\tau = -\int_a^b f[z(\tau)]z'(\tau)\,\mathrm{d}\tau,$$
which is the same as

$$\int_{-C} f(z)\,dz = -\int_{C} f(z)\,dz.$$

(c) Additivity Suppose that the oriented curve C consists of two oriented curves C_1 and C_2. Then
$$\int_C f(z)\,dz = \int_{C_1} f(z)\,dz + \int_{C_2} f(z)\,dz.$$

Proof Without loss of generality, we can assume that the contour C consists of a contour C_1 from z_1 to z_2 followed by a contour C_2 from z_2 to z_3, and the initial point of C_2 being the final point of C_1 (Fig. 3.2.3).

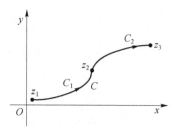

Fig. 3.2.3

Let contour C be given by
$$z = z(t), \quad (a \leqslant t \leqslant b)$$
and extends from a point $z_1 = z(a)$ to a point $z_3 = z(b)$. And there is a value c of t, where $a < c < b$, such that $z(c) = z_2$. Consequently, C_1 is represented by
$$z = z(t), \quad (a \leqslant t \leqslant c)$$
and C_2 is represented by
$$z = z(t), \quad (c \leqslant t \leqslant b).$$
Also, by a rule for integrals of functions $w(t)$ that was noted in Section 3.1.2,
$$\int_a^b f[z(t)]z'(t)\,dz = \int_a^c f[z(t)]z'(t)\,dt + \int_c^b f[z(t)]z'(t)\,dt.$$
Evidently, then
$$\int_C f(z)\,dz = \int_{C_1} f(z)\,dz + \int_{C_2} f(z)\,dz.$$
Sometimes the contour C is called the sum of its legs C_1 and C_2 and is denoted by $C_1 + C_2$. The sum of two contours C_1 and $-C_2$ is well defined when C_1 and C_2 have the same final points, and it is written $C_1 - C_2$.

Remark 3.2.1 Definite integrals in calculus can be interpreted as areas, and they have other interpretations as well. Except in special cases, no corresponding helpful interpretation, geometric or physical, is available for integrals in the complex plane.

(d) Upper bounds for moduli of contour integrals

Suppose that C is an oriented smooth curve and $w = f(z)$ is defined along the given contour C.

If for any nonnegative constant M such that the values of f on C satisfy the inequality $|f(z)| \leqslant M$, then we have

$$\left|\int_C f(z)\,\mathrm{d}z\right| \leqslant ML, \tag{3.2.6}$$

where L represents the length of the contour and $a<b$. Note that (3.2.6) means that the modulus of the integral of f along C does not exceed ML.

Proof Suppose that the contour C is given by
$$z=z(t), \quad (a \leqslant t \leqslant b).$$
Then from (3.2.5) and inequality (3.2.6), it follows that
$$\left|\int_C f(z)\,\mathrm{d}z\right| = \left|\int_a^b f[z(t)]z'(t)\,\mathrm{d}t\right| \leqslant \int_a^b |f[z(t)]|\,|z'(t)|\,\mathrm{d}t.$$
So, for any nonnegative constant M such that the values of f on C satisfy the inequality $|f(z)| \leqslant M$,
$$\left|\int_C f(z)\,\mathrm{d}z\right| \leqslant M\int_a^b |z'(t)|\,\mathrm{d}t.$$
Since the integral on the right here represents the length L of the contour, it follows that the modulus of the integral of f along C does not exceed ML, that is,
$$\left|\int_C f(z)\,\mathrm{d}z\right| \leqslant ML.$$
This is, of course, a strict inequality when the values of f on C are such that $|f(z)| < M$.

Remark 3.2.2 Note that since all of the paths of integration to be considered here are contours and the integrands are piecewise continuous functions defined on those contours, a number M such as the one appearing in inequality (3.2.6) will always exist. This is because the real-valued function $|f[z(t)]|$ is continuous on the closed bounded interval $[a,b]$ when f is continuous on C; and such a function always reaches a maximum value M on that interval. Hence $|f(z)|$ has a maximum value on C when f is continuous on it. It now follows immediately that the same is true when f is piecewise continuous on C.

Example 3.2.5 Let us find the value of the integral
$$I = \int_C \bar{z}\,\mathrm{d}z,$$
where the contour of the integration is:
(a) C_1 is the upper circle $|z|=1$ from -1 to 1;
(b) C_2 is the lower circle $|z|=1$ from -1 to 1;
(c) C is the positively oriented circle $|z|=1$.

Solution (a) When C_1 is the upper circle $|z|=1$ from -1 to 1, then C_1 may be represented parametrically as
$$z = e^{i\theta}, \quad (\theta: \pi \to 0).$$
According to (3.2.5), we have
$$I = \int_\pi^0 \overline{e^{i\theta}}(e^{i\theta})'\,\mathrm{d}\theta.$$
And, since
$$\overline{e^{i\theta}} = e^{-i\theta} \quad \text{and} \quad (e^{i\theta})' = ie^{i\theta}.$$
this means that

$$I = \int_\pi^0 e^{-i\theta} i e^{i\theta} d\theta = i\int_\pi^0 d\theta = -\pi i.$$

(b) When C is the lower circle $|z|=1$ from -1 to 1, then the contour may be represented parametrically as
$$z = e^{i\theta}, \quad (\theta: \pi \to 2\pi).$$

According to (3.2.5), we have
$$I = \int_\pi^{2\pi} \overline{e^{i\theta}} (e^{i\theta})' d\theta = \int_\pi^{2\pi} e^{-i\theta} i e^{i\theta} d\theta = i\int_\pi^{2\pi} d\theta = \pi i.$$

(c) When C_2 is the positively oriented circle $|z|=1$, then the contour may be represented parametrically as
$$z = e^{i\theta}, \quad (\theta: 0 \to 2\pi).$$

According to (3.2.5), we have
$$I = \int_0^{2\pi} \overline{e^{i\theta}} (e^{i\theta})' d\theta = \int_0^{2\pi} e^{-i\theta} i e^{i\theta} d\theta = i\int_0^{2\pi} d\theta = 2\pi i.$$

Example 3.2.6 Let us find the value of the integral
$$\int_C f(z) dz,$$
where $f(z) = y - x - i3x^2$, ($z = x + iy$) and the contour of the integration is:

(a) C_1 denotes the contour $\overline{OAB}$ shown in Fig. 3.2.4;

(b) C_2 denotes the segment $\overline{OB}$ shown in Fig. 3.2.4.

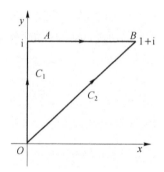

Fig. 3.2.4

Solution

(a) By the additivity of contour integral, we have
$$\int_{C_1} f(z) dz = \int_{\overline{OA}} f(z) dz + \int_{\overline{AB}} f(z) dz.$$

The segment $\overline{OA}$ may be represented parametrically as $z = 0 + iy$, $(y: 0 \to 1)$; and since $x = 0$ at points on that segment, the values of f there vary with the parameter y according to the equation $f(z) = y$, $(0 \leqslant y \leqslant 1)$. Consequently,
$$\int_{\overline{OA}} f(z) dz = \int_0^1 y i dy = i\int_0^1 y dy = \frac{i}{2}.$$

On the segment $\overline{AB}$, $z = x + i$, $(x: 0 \to 1)$, and so $f(z) = (1 - x - i3x^2)$, $(0 \leqslant x \leqslant 1)$. Thus,

$$\int_{AB} f(z)\,dz = \int_0^1 (1-x-i3x^2)\cdot 1\,dx$$
$$= \int_0^1 (1-x)\,dx - 3i\int_0^1 x^2\,dx = \frac{1}{2} - i.$$

Therefore,
$$\int_{C_1} f(z)\,dz = \int_{OA} f(z)\,dz + \int_{AB} f(z)\,dz = \frac{i}{2} + \left(\frac{1}{2}-i\right) = \frac{1-i}{2}.$$

(b) The segment OB may be represented parametrically as $z=x+ix$ $(x:0\to 1)$.

Then the values of f there vary with the parameter x according to the equation $f(z)=-i3x^2$, $(0\leqslant x\leqslant 1)$. So,
$$\int_{C_2} f(z)\,dz = \int_0^1 -i3x^2(1+i)\,dx = 3(1-i)\int_0^1 x^2\,dx = 1-i.$$

Evidently, then, the integrals of $f(z)$ along the two paths C_1 and C_2 have different values even though those paths have the same initial and the same final points.

Observe how it follows that the integral of $f(z)$ over the simple closed contour $OABO$, or C_1-C_2, has the nonzero value
$$\int_{C_1} f(z)\,dz - \int_{C_2} f(z)\,dz = \frac{-1+i}{2}.$$

Example 3.2.7 Let C be the positively oriented circle $|z-z_0|=R$. Evaluate the integral
$$\int_C \frac{1}{(z-z_0)^{n+1}}\,dz.$$

Solution C may be represented parametrically as
$$z=z_0+Re^{i\theta},\quad (\theta:0\to 2\pi).$$

Then,
$$\int_C \frac{1}{(z-z_0)^{n+1}}\,dz = \int_0^{2\pi} \frac{iRe^{i\theta}}{R^{n+1}e^{i(n+1)\theta}}\,d\theta$$
$$= \frac{i}{R^n}\int_0^{2\pi} e^{-in\theta}\,d\theta$$
$$= \begin{cases} i\int_0^{2\pi} d\theta = 2\pi i, & n=0, \\ \dfrac{i}{R^n}\int_0^{2\pi}(\cos n\theta - i\sin n\theta)\,d\theta = 0, & n\neq 0. \end{cases}$$

Example 3.2.8 Let C denote an arbitrary smooth arc
$$z=z(t),\quad (a\leqslant t\leqslant b)$$
from a fixed point $z_1=z(a)$ to a fixed point $z_2=z(b)$ (Fig. 3.2.5). Evaluate the integral
$$I = \int_C z\,dz.$$

Solution C may be represented parametrically as
$$z=z(t),\quad (t:a\to b).$$

Then
$$I = \int_C z\,dz = \int_a^b z(t)z'(t)\,dt.$$

We note that,

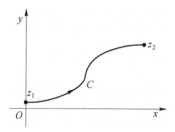

Fig. 3.2.5

$$\frac{d}{dt}\frac{[z(t)]^2}{2} = z(t)z'(t).$$

Thus

$$I = \frac{[z(t)]^2}{2}\bigg|_a^b = \frac{[z(b)]^2 - [z(a)]^2}{2}.$$

But $z(b) = z_2$ and $z(a) = z_1$; and so $I = \frac{z_2^2 - z_1^2}{2}$. Note that the value of I depends only on the end points of C, and is otherwise independent of arc that is taken, we may write

$$\int_{z_1}^{z_2} z\, dz = \frac{z_2^2 - z_1^2}{2}. \tag{3.2.7}$$

(Compare Example 3.2.5 and 3.2.6, where the value of an integral from one fixed point to another depended on the path that was taken.)

Expression (3.2.7) is also valid when C is a contour that is not necessarily smooth since a contour consists of a finite number of smooth arcs C_k ($k=1, 2, \cdots, n$), joined end to end. More precisely, suppose that each C_k extend from z_k to z_{k+1}. Then

$$\int_C z\, dz = \sum_{k=1}^n \int_{C_k} z\, dz = \sum_{k=1}^n \frac{z_{k+1}^2 - z_k^2}{2} = \frac{z_{n+1}^2 - z_1^2}{2}, \tag{3.2.8}$$

z_1 being the initial point of C and z_{n+1} its final point.

It follows from expression (3.2.8) that the integral of the function $f(z) = z$ around each closed contour in the plane has value zero. (Once again, compare to Example 3.2.6, where the value of the integral on $OABO$ of a given function around a certain closed path was not zero.) The question of predicting when an integral around a closed contour has value zero will be discussed in Sections 3.3 and 3.4.

Example 3.2.9 Let C denote the semicircular path $z = 3e^{i\theta}$ ($0 \leqslant \theta \leqslant \pi$) from the point $z = 3$ to the point $z = -3$ (Fig. 3.2.6). Then evaluate the integral

$$I = \int_C f(z)\, dz,$$

where $f(z) = z^{\frac{1}{2}} = \sqrt{r}e^{\frac{i\theta}{2}}$, $(r>0, 0<\theta<2\pi)$.

Solution Although the branch

$$f(z) = \sqrt{r}e^{\frac{i\theta}{2}}, \quad (r>0,\ 0<\theta<2\pi)$$

of the multiple-valued function $z^{\frac{1}{2}}$ is not defined at the initial point $z=3$ of the contour C, the integral $I = \int_C z^{\frac{1}{2}}\, dz$ of that branch nevertheless exists. For the integrand is piecewise

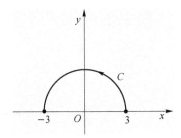

Fig. 3.2.6

continuous on C. To see that this is so, we observe that when $z(\theta)=3e^{i\theta}$, the right-hand limits of the real and imaginary components of the function

$$f[z(\theta)] = \sqrt{3}e^{\frac{i\theta}{2}} = \sqrt{3}\cos\frac{\theta}{2} + i\sqrt{3}\sin\frac{\theta}{2}, \quad 0 < \theta \leqslant \pi,$$

at $\theta=0$ are $\sqrt{3}$ and 0, respectively. Hence $f[z(\theta)]$ is continuous on the closed interval $0 \leqslant \theta \leqslant \pi$ when its value at $\theta=0$ is defined as $\sqrt{3}$. Consequently,

$$I = \int_0^\pi \sqrt{3}e^{\frac{i\theta}{2}} 3ie^{i\theta} d\theta = 3\sqrt{3}i \int_0^\pi e^{\frac{i3\theta}{2}} d\theta$$

$$= 3\sqrt{3}i \left[\frac{2}{3i} e^{\frac{i3\theta}{2}} \Big|_0^\pi \right] = -2\sqrt{3}(1+i).$$

Example 3.2.10 Let C be the arc of the circle $|z|=2$ from $z=2$ to $z=2i$ that lies in the first quadrant (Fig. 3.2.7). Please show that

$$\left| \int_C \frac{z+4}{z^3-1} dz \right| \leqslant \frac{6\pi}{7}.$$

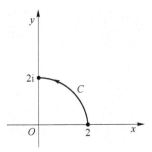

Fig. 3.2.7

Proof This is done by noting first that if z is a point on C, so that $|z|=2$, then
$$|z+4| \leqslant |z|+4=6,$$
and
$$|z^3-1| \geqslant ||z|^3-1|=7.$$
Thus, when z lies on C, we have
$$\left| \frac{z+4}{z^3-1} \right| = \frac{|z+4|}{|z^3-1|} \leqslant \frac{6}{7}.$$

Writing $M=\frac{6}{7}$ and observing that $L=\pi$ is the length of C, we may now use inequality (3.2.6) to obtain inequality

$$\left|\int_C \frac{z+4}{z^3-1}dz\right| \leqslant \frac{6\pi}{7}.$$

Example 3.2.11 Let C_R be the semicircular path
$$z=Re^{i\theta}, \quad (0\leqslant\theta\leqslant\pi),$$
and $z^{\frac{1}{2}}$ denotes the branch
$$z^{\frac{1}{2}}=\sqrt{r}e^{\frac{i\theta}{2}}, \quad (r>0, \ -\frac{\pi}{2}<\theta<\frac{3\pi}{2})$$
of the square root function (See Fig. 3.2.8). Please show that
$$\lim_{R\to\infty}\int_{C_R}\frac{z^{\frac{1}{2}}}{z^2+3z+2}dz=0.$$

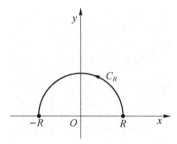

Fig. 3.2.8

Proof When $|z|=R>2$,
$$|z^{\frac{1}{2}}|=\left|\sqrt{R}e^{\frac{i\theta}{2}}\right|=\sqrt{R},$$
and
$$|z^2+3z+2|=|z+2||z+1|\geqslant||z|-2|||z|-1|=(R-2)(R-1).$$
Consequently, at points on C_R,
$$\left|\frac{z^{\frac{1}{2}}}{z^2+3z+2}\right|\leqslant M_R,$$
where $M_R=\dfrac{\sqrt{R}}{(R-2)(R-1)}$.

Since the length of C_R is the number $L=\pi R$, it follows from inequality (3.2.6) that
$$\left|\int_{C_R}\frac{z^{\frac{1}{2}}}{z^2+3z+2}dz\right|\leqslant M_R L.$$
But
$$M_R L=\frac{\pi R\sqrt{R}}{(R-2)(R-1)},$$
and it is clear that the term on the right here tends to zero as R tends to infinity. Therefore
$$\lim_{R\to\infty}\int_{C_R}\frac{z^{\frac{1}{2}}}{z^2+3z+2}dz=0.$$

3.2.3 Antiderivatives

Although the value of a contour integral of a function $f(z)$ from a fixed point z_1 to a fixed point z_2 depends, in general, on the path that is taken, there are certain functions whose integrals from z_1 to z_2 have values that are independent of path (Compare Examples 3.2.6 and 3.2.8 in Section 3.2.2). The examples just cited also illustrate the fact that the values of integrals around closed paths are sometimes, but not always, zero. The theorem below is useful in determining when integration is independent of path and, moreover, when an integral around a closed path has value zero.

In proving the theorem, we shall discover an extension of the fundamental theorem of calculus that simplifies the evaluation of many contour integrals. That extension involves the concept of an antiderivative of a continuous function f in a domain D, or a function F such that $F'(z) = f(z)$ for all z in D. Note, too, that an antiderivative of a given function f is unique except for an additive complex constant. This is because the derivative of the difference $F(z) - G(z)$ of any two such antiderivatives $F(z)$ and $G(z)$ is zero; and, according to the theorem 2.2.1 in Section 2.2, an analytic function is constant in a domain D, when its derivative is zero throughout D.

Theorem 3.2.1 Suppose that a function $f(z)$ is continuous on a domain D. If any one of the following statements is true, then so are the others:

(i) $f(z)$ has an antiderivative $F(z)$ in D;

(ii) the integrals of $f(z)$ along contours lying entirely in D and extending from any fixed point z_1 to any fixed point z_2 all have the same value;

(iii) the integrals of $f(z)$ around closed contours lying entirely in D all value zero.

It should be emphasized that the theorem does not claim that any of these statements is true for a given function f and a given domain D. It says only that all of them are true or that none of them is true. To prove the theorem, it is sufficient to show that statement (i) implies statement (ii), that statement (ii) implies statement (iii), and finally that statement (iii) implies statement (i).

Proof Let us assume that statement (i) is true. If a contour C from z_1 to z_2, lying in D, is just a smooth arc, with parametric representation $z = z(t)$ ($a \leqslant t \leqslant b$), then

$$\frac{d}{dt}F[z(t)] = F'[z(t)]z'(t) = f[z(t)]z'(t), \quad (a \leqslant t \leqslant b).$$

Because the fundamental theorem of calculus can be extended so as to apply to complex-valued functions of a real variable, it follows that

$$\int_C f(z)dz = \int_a^b f[z(t)]z'(t)dt = F[z(t)]\big|_a^b = F[z(b)] - F[z(a)].$$

Since $z(b) = z_2$ and $z(a) = z_1$, the value of this contour integral is, then, $F(z_2) - F(z_1)$.

And that value is evidently independent of the contour C as long as C extends from z_1 to z_2 and lies entirely in D. That is,

$$\int_{z_1}^{z_2} f(z)dz = F(z_2) - F(z_1) = F(z)\bigg|_{z_1}^{z_2}, \qquad (3.2.9)$$

when C is smooth. Expression (3.2.9) is also valid when C is any contour, not necessarily a smooth one, that lies in D. For, if C consists of a finite number of smooth arcs C_k ($k = 1$,

$2, \cdots, n$), each C_k extending from a point z_k to a point z_{k+1}, then
$$\int_C f(z)\mathrm{d}z = \sum_{k=1}^{n} \int_{C_k} f(z)\mathrm{d}z = \sum_{k=1}^{n} [F(z_{k+1}) - F(z_k)] = F(z_{n+1}) - F(z_1).$$
The fact that statement (ii) follows from statement (i) is now established.

To see that statement (ii) implies statement (iii), we let z_1 and z_2 denote any two points on closed contour C lying in D and form two paths, each with initial point z_1 and final point z_2, such that $C = C_1 - C_2$ (Fig. 3.2.9).

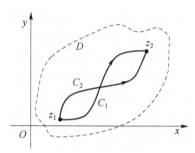

Fig. 3.2.9

Assuming that statement (ii) is true, one can write
$$\int_{C_1} f(z)\mathrm{d}z = \int_{C_2} f(z)\mathrm{d}z, \tag{3.2.10}$$
or
$$\int_{C_1} f(z)\mathrm{d}z + \int_{-C_2} f(z)\mathrm{d}z = 0. \tag{3.2.11}$$
That is, the integral of $f(z)$ around the closed contour $C = C_1 - C_2$ has value zero.

It remains to show that statement (iii) implies statement (i). We do this by assuming that statement (iii) is true, establishing the validity of statement (ii), and then arriving at statement (i). To see that statement (ii) is true, we let C_1 and C_2 denote any two contours, lying in D, from a point z_1 to a point z_2 and observe that, in view of statement (iii), equation (3.2.11) holds (see Fig. 3.2.9). Thus equation (3.2.10) holds. Integration is, therefore, independent of path in D and we can define the function
$$F(z) = \int_{z_0}^{z} f(s)\mathrm{d}s$$
in D. The proof of the theorem is complete once we show that $F'(z) = f(z)$ everywhere in D. We do this by letting $z + \Delta z$ be any point, distinct from z, lying in some neighborhood of z that is small enough to be contained in D. Then
$$F(z + \Delta z) - F(z) = \int_{z_0}^{z+\Delta z} f(s)\mathrm{d}s - \int_{z_0}^{z} f(s)\mathrm{d}s = \int_{z}^{z+\Delta z} f(s)\mathrm{d}s,$$
where the path of integration from z to $z + \Delta z$ may be selected as line segment (Fig. 3.2.10).
Since
$$\int_{z}^{z+\Delta z} \mathrm{d}s = \Delta z,$$
we can write

Chapter 3 Integral of Complex Function

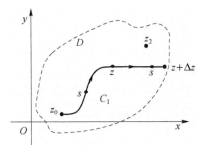

Fig. 3.2.10

$$f(z) = \frac{1}{\Delta z}\int_{z}^{z+\Delta z} f(z)\,\mathrm{d}s;$$

and it follows that

$$\frac{F(z+\Delta z)-F(z)}{\Delta z} - f(z) = \frac{1}{\Delta z}\int_{z}^{z+\Delta z}[f(s)-f(z)]\,\mathrm{d}s.$$

Since f is continuous at the point z. Hence, for each positive number ε, there exists a positive number δ such that

$$|f(s)-f(z)|<\varepsilon, \quad \text{whenever} \quad |s-z|<\delta.$$

Consequently, if the point $z+\Delta z$ is close enough to z so that $|\Delta z|<\delta$, then

$$\left|\frac{F(z+\Delta z)-F(z)}{\Delta z} - f(z)\right| < \frac{1}{|\Delta z|}\varepsilon|\Delta z| = \varepsilon.$$

That is,

$$\lim_{\Delta z \to 0}\frac{F(z+\Delta z)-F(z)}{\Delta z} = f(z),$$

or $F'(z)=f(z)$.

Example 3.2.12 Let C be the every contour from $z=0$ to $z=1$. Then evaluate the integral

$$\int_C z\cos z + \sin z\,\mathrm{d}z.$$

Solution The continuous function $f(z)=z\cos z+\sin z$ has an antiderivative $F(z)=z\sin z$ throughout the plane. Hence

$$\int_C z\cos z + \sin z\,\mathrm{d}z = \int_0^1 z\cos z + \sin z\,\mathrm{d}z = z\sin z\Big|_0^1 = \sin 1.$$

Example 3.2.13 Let C be the every contour from $z=0$ to $z=1+\mathrm{i}$. Then evaluate the integral

$$\int_C z^2\,\mathrm{d}z.$$

Solution The continuous function $f(z)=z^2$ has an antiderivative $F(z)=\dfrac{z^3}{3}$ throughout the plane. Hence

$$\int_C z^2\,\mathrm{d}z = \int_0^{1+\mathrm{i}} z^2\,\mathrm{d}z = \frac{z^3}{3}\Big|_0^{1+\mathrm{i}} = \frac{1}{3}(1+\mathrm{i})^3 = \frac{2}{3}(-1+\mathrm{i}).$$

Example 3.2.14 Let C be the positively oriented circle (Fig. 3.2.11) $z=2\mathrm{e}^{\mathrm{i}\theta}$, ($-\pi\leqslant\theta\leqslant\pi$) about the origin. Then evaluate the integral $\displaystyle\int_C \frac{\mathrm{d}z}{z^2}$.

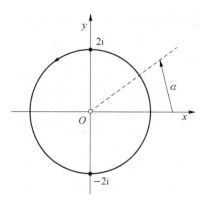

Fig. 3.2.11

Solution The function $f(z)=\dfrac{1}{z^2}$, which is continuous everywhere except at the origin, has an antiderivative $F(z)=-\dfrac{1}{z}$ in the domain $|z|>0$, consisting of the entire plane with the origin deleted. Consequently,

$$\int_C \frac{dz}{z^2} = 0,$$

when C is the positively oriented circle (Fig. 3.2.11) $z=2e^{i\theta}$, $(-\pi\leqslant\theta\leqslant\pi)$ about the origin.

Note that the integral of the function $f(z)=\dfrac{1}{z}$ around the same circle cannot be evaluated in a similar way. For, although the derivative of any branch $F(z)$ of $\log z$ is $\dfrac{1}{z}$. $F(z)$ is not differentiable, or even defined, along its branch cut. In particular, if a ray $\theta=\alpha$ from the origin is used to form the branch cut, $F'(z)$ fails to exist at the point where that ray intersects the circle C (see Fig. 3.2.12). So C does not lie in a domain throughout which $F'(z)=\dfrac{1}{z}$, and we cannot make direct use of an antiderivative. Example 3.2.15, just below, illustrates how a combination of two different antiderivatives can be used to evaluate $f(z)=\dfrac{1}{z}$ around C.

Example 3.2.15 Let C be the positively oriented circle (Fig. 3.2.12) $z=2e^{i\theta}$, $(-\pi\leqslant\theta\leqslant\pi)$ about the origin. Then evaluate the integral

$$\int_C \frac{dz}{z}.$$

Solution Firstly, let C_1 denote the right half

$$z=2e^{i\theta}, \quad \left(-\frac{\pi}{2}\leqslant\theta\leqslant\frac{\pi}{2}\right), \tag{3.2.12}$$

of the circle C in Example 3.2.14 from $-2i$ to $2i$.

Then the principal branch

$$\text{Log } z = \ln r + i\Theta, \quad (r>0, \ -\pi<\Theta<\pi)$$

of the logarithmic function serves as an antiderivative of the function $\dfrac{1}{z}$ in the evaluation of

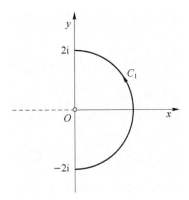

Fig. 3.2.12

the integral of $\dfrac{1}{z}$ along C_1 (Fig. 3.2.12):

$$\int_{C_1} \frac{dz}{z} = \int_{-2i}^{2i} \frac{dz}{z} = \text{Log } z \big|_{-2i}^{2i} = \text{Log}(2i) - \text{Log}(-2i)$$

$$= \left(\ln 2 + i\frac{\pi}{2}\right) - \left(\ln 2 - i\frac{\pi}{2}\right) = \pi i.$$

Next, let C_2 denote the left half

$$z = 2e^{i\theta}, \quad \left(\frac{\pi}{2} \leqslant \theta \leqslant \frac{3\pi}{2}\right), \tag{3.2.13}$$

from $2i$ to $-2i$.

Then the branch

$$\log z = \ln r + i\theta, \quad (r > 0, \ 0 < \theta < 2\pi)$$

of the logarithmic function (Fig. 3.2.13) serves as an antiderivative of the function $\dfrac{1}{z}$ in the evaluation of the integral of $\dfrac{1}{z}$ along C_2 (Fig. 3.2.13).

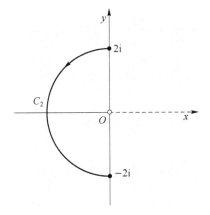

Fig. 3.2.13

One can write

$$\int_{C_2} \frac{dz}{z} = \int_{2i}^{-2i} \frac{dz}{z} = \log z \Big|_{2i}^{-2i} = \log(-2i) - \log(2i)$$

$$= \left(\ln 2 + i\frac{3\pi}{2}\right) - \left(\ln 2 + i\frac{\pi}{2}\right) = \pi i.$$

The value of the integral of $\dfrac{1}{z}$ around the entire circle $C = C_1 + C_2$ is thus obtained:

$$\int_C \frac{dz}{z} = \int_{C_1} \frac{dz}{z} + \int_{C_2} \frac{dz}{z} = \pi i + \pi i = 2\pi i.$$

Example 3.2.16 Let us use an antiderivative to evaluate the integral

$$\int_{C_1} z^{\frac{1}{2}} dz, \tag{3.2.14}$$

where the integral is the branch

$$z^{\frac{1}{2}} = \sqrt{r} e^{\frac{i\theta}{2}}, \quad (r > 0,\ 0 < \theta < 2\pi) \tag{3.2.15}$$

of the square root function and where C_1 is any contour from $z = -3$ to $z = 3$ that, except for its end points, lies above the x axis (Fig. 3.2.14).

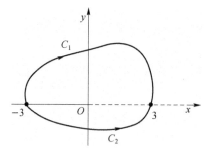

Fig. 3.2.14

Solution Although the integrand is piecewise continuous on C_1, and the integral therefore exists, the branch (3.2.15) of $z^{\frac{1}{2}}$ is not defined on the ray $\theta = 0$, in particular at the point $z = 3$. But another branch,

$$f_1(z) = \sqrt{r} e^{\frac{i\theta}{2}}, \quad \left(r > 0,\ -\frac{\pi}{2} < \theta < \frac{3\pi}{2}\right)$$

is defined and continuous everywhere on C_1. The values of $f_1(z)$ at all points on C_1 except $z = -3$ coincide with those of our integrand (3.2.14), so the integrand can be replaced by $f_1(z)$. Since an antiderivative of $f_1(z)$ is the function

$$F_1(z) = \frac{2}{3} z^{\frac{3}{2}} = \frac{2}{3} r \sqrt{r} e^{\frac{i3\theta}{2}}, \quad \left(r > 0,\ -\frac{\pi}{2} < \theta < \frac{3\pi}{2}\right),$$

we can now write

$$\int_{C_1} z^{\frac{1}{2}} dz = \int_{-3}^{3} f_1(z) dz = F_1(z) \Big|_{-3}^{3} = 2\sqrt{3}\left(e^{i0} - e^{\frac{i3\pi}{2}}\right) = 2\sqrt{3}(1 + i).$$

The integral

$$\int_{C_2} z^{\frac{1}{2}} dz \tag{3.2.16}$$

of the function (3.2.15) over any contour C_2 that extends from $z = -3$ to $z = 3$ below the real axis can be evaluated in a similar way. In this case, we can replace the integrand by the branch

$$f_2(z) = \sqrt{r} e^{\frac{i\theta}{2}}, \quad \left(r > 0, \frac{\pi}{2} < \theta < \frac{5\pi}{2}\right),$$

whose values coincide with those of the integrand at $z = -3$ and at all points on C_2 below the real axis. This enables us to use an antiderivative of $f_2(z)$ to evaluate integral (3.2.16). Details are left to the exercises.

Exercises 3.2

1. Use parametric representations for C to evaluate

$$\int_C f(z) \, dz,$$

where $f(z) = \dfrac{(z+2)}{z}$ and C is

   (a) the semicircle $z = 2e^{i\theta} (\theta : 0 \to \pi)$;

   (b) the semicircle $z = 2e^{i\theta} (\theta : \pi \to 2\pi)$;

   (c) the circle $z = 2e^{i\theta} (\theta : 0 \to 2\pi)$.

2. Evaluate

$$\int_C f(z) \, dz,$$

where $f(z) = z - 1$ and C is the arc from the point $z_1 = 0$ to the point $z_2 = 2$ consisting of

   (a) the semicircle $z = 1 + e^{i\theta} (\theta : \pi \to 2\pi)$;

   (b) the segment $z = x (x : 0 \to 2)$ of the real axis.

3. Evaluate

$$\int_C (x^2 + iy) \, dz,$$

where C denotes the segment of the curve $y = x^2$ from the point $z_1 = 0$ to the point $z_2 = 1 + i$.

4. Evaluate

$$\int_C (x - y + ix^2) \, dz,$$

where C denotes the segment of the straight line $y = x$ from the point $z_1 = 0$ to the point $z_2 = 1 + i$.

5. Evaluate

$$\int_C \pi \exp(\pi \bar{z}) \, dz,$$

where C is the boundary of the square with vertices at the points 0, 1, $1 + i$ and i, the orientation of C being in the counterclockwise direction.

6. Evaluate

$$\int_C dz,$$

where C is an arbitrary contour from any fixed point z_1 to any fixed point z_2 in the complex plane.

7. Suppose that the function is defined by $f(z)=\begin{cases}1, & \text{when } y<0,\\ 4y, & \text{when } y>0,\end{cases}$ and C is the arc from the point $z_1=-1-i$ to the point $z_2=1+i$ along the curve $y=x^3$. Then evaluate the integral
$$\int_C f(z)dz.$$

8. Without evaluating the integral, show that

(a) $\left|\int_{|z-1|=2}\dfrac{z+1}{z-1}dz\right|\leqslant 8\pi$;

(b) $\left|\int_C \dfrac{dz}{z^4}\right|\leqslant 4\sqrt{2}$, where C is the line segment from $z=i$ to $z=1$. By observing that, of all the points on that line segment, the midpoint is the closed to the origin;

(c) $\left|\int_{|z|=R}\dfrac{dz}{z^4-a^4}\right|\leqslant \dfrac{2\pi R}{R^4-a^4}$, $(R>a>0)$;

(d) $\left|\int_{C_R}\dfrac{2z^2-1}{z^4+5z^2+4}dz\right|\leqslant \dfrac{\pi R(2R^2+1)}{(R^2-1)(R^2-4)}$,

where C_R denotes the upper half of the circle $|z|=R$ $(R>2)$, taken in the counterclockwise direction. Then, by dividing the numerator and denominator on the right here by R^4, show that the value of the integral tends to zero as R tends to infinity.

9. Use an antiderivative to show that, for every contour C extending from a point z_1 to a point z_2,
$$\int_C z^n dz = \dfrac{1}{n+1}(z_2^{n+1}-z_1^{n+1}), \quad (n=0,1,2,\cdots).$$

10. By finding an antiderivative, evaluate each of these integrals, where the path is any contour between the indicated limits of integration:

(a) $\int_i^{\frac{i}{2}} e^{\pi z}dz$; (b) $\int_0^{\pi+2i}\cos\left(\dfrac{z}{2}\right)dz$; (c) $\int_1^3 (z-2)^3 dz$; (d) $\int_0^i z\cos z dz$.

11. Evaluate $\int_C (3z^2+2z+1)dz$, where C is the right half of the circle $|z|=1$ from the point $z_1=-i$ to the point $z_2=i$.

12. Show that
$$\int_{-1}^1 z^i dz = \dfrac{1+e^{-\pi}}{2}(1-i),$$
where z^i denotes the principal branch
$$z^i = \exp(i\text{Log } z), \quad (|z|>0, -\pi<\text{Arg } z<\pi),$$
and where the path of integration is any contour from $z=-1$ to $z=1$ that, except for its end points, lies above the real axis.

Suggestion Use an antiderivative of the branch
$$z^i = \exp(i\log z), \quad (|z|>0, -\dfrac{\pi}{2}<\arg z<\dfrac{3\pi}{2}).$$

3.3 Cauchy integral theorem

In this section, we present a theorem giving some conditions on a function f, which ensures that the value of the integral of $f(z)$ around a simple closed contour is zero.

3.3.1 Cauchy-Goursat theorem

Theorem 3.3.1 Let D be a domain and its boundary be a simple closed contour C. If a function f is analytic at all points interior to and on C and f' is continuous over D, then

$$\int_C f(z)\,dz = 0. \tag{3.3.1}$$

Proof Without loss of generality we let C denote a simple closed contour $z = z(t)$ ($t: a \to b$), described in the positive sense (counterclockwise), and we assume that f is analytic at each point interior to and on C. Then

$$\int_C f(z)\,dz = \int_a^b f(z(t))z'(t)\,dt. \tag{3.3.2}$$

And if

$$f(z) = u(x,y) + iv(x,y) \quad \text{and} \quad z(t) = x(t) + iy(t),$$

the integrand $f[z(t)]z'(t)$ in expression (3.3.2) is the product of the functions

$$u[x(t),y(t)] + iv[x(t),y(t)], \quad x'(t) + iy'(t)$$

of the real variable t. Thus

$$\int_C f(z)\,dz = \int_a^b (ux' - vy')\,dt + i\int_a^b (vx' + uy')\,dt.$$

In terms of line integrals of real-valued functions of two real variables, then,

$$\int_C f(z)\,dz = \int_C u\,dx - v\,dy + i\int_C v\,dx + u\,dy. \tag{3.3.3}$$

Expression (3.3.3) is, of course, also valid when C is any contour, not necessarily a simple closed one, and $f[z(t)]$ is only piecewise continuous on it.

We next recall a result from calculus that enabled us to express the line integrals on the right in equation (3.3.3) as double integrals. Suppose that two real-valued functions $P(x,y)$ and $Q(x,y)$, together with their first-order partial derivatives, are continuous throughout the closed region D consisting of all points interior to and on the simple closed contour C, which is described in the positive sense. According to Green's theorem,

$$\int_C P(x,y)\,dx + Q(x,y)\,dy = \iint_D (Q_x(x,y) - P_y(x,y))\,dx\,dy.$$

Now f is continuous in D, since it is analytic there. Hence the functions u and v are also continuous in D. Likewise, if the derivative f' of f is continuous in D, so are the first-order partial derivative of u and v. Green's theorem then enables us to rewrite equation (3.3.3) as

$$\int_C f(z)\,dz = \iint_D (-v_x - u_y)\,dx\,dy + \iint_D (u_x - v_y)\,dx\,dy.$$

But, in view of the Cauchy-Riemann equations
$$u_x = v_y, \quad u_y = -v_x,$$
then the integrands of these two double integrals are zero throughout D.

When C is taken in the clockwise direction, (3.3.1) is also true. Since
$$\int_C f(z)\,\mathrm{d}z = -\int_{-C} f(z)\,\mathrm{d}z.$$

This result was obtained by **Cauchy** in the early part of the nineteenth century.

Example 3.3.1 If C is any simple closed contour, in either direction, then show that
$$\int_C \exp(z^3)\,\mathrm{d}z = 0.$$

Solution Since the function $f(z) = \exp(z^3)$ is analytic everywhere and its derivative $f'(z) = 3z^2 \exp(z^3)$ is continuous everywhere, then from Theorem 3.3.1, we know that
$$\int_C \exp(z^3)\,\mathrm{d}z = 0.$$

Goursat was the first to prove that the condition of continuity on f' can be omitted. Its removal is important and will allow us to show, for example, that the derivative f' of an analytic function f is analytic without having to assume the continuity of f', which follows as consequence. We now state the revised form of Cauchy's result, known as the Cauchy-Goursat theorem.

Theorem 3.3.2 (Cauchy-Goursat theorem) If a function f is analytic at all points interior to and on a simple closed contour C, then
$$\int_C f(z)\,\mathrm{d}z = 0.$$

The proof is very complex, so we omit it here.

Example 3.3.2 Let C be the positively oriented circle $|z| = 1$. Evaluate the integral
$$\int_C \frac{\mathrm{d}z}{z^2 + 2z + 4}.$$

3.3.2 Simply and multiply connected domains

A simply connected domain D is a domain such that every simple closed contour within it encloses only points of D. The set of points interior to a simple closed contour is an example. The annular domain between two concentric circle is, however, not simply connected. A domain that is not simply connected is said to be multiply connected.

The Cauchy-Goursat theorem can be extended in the following way, involving a simply connected domain.

Theorem 3.3.3 If a function f is analytic throughout a simply connected domain D, then
$$\int_C f(z)\,\mathrm{d}z = 0, \tag{3.3.4}$$
for every closed contour C lying in D.

Proof The proof is easy if C is a simple closed contour or if it is a closed contour that

Chapter 3　Integral of Complex Function

intersects itself a finite number of times.

　　In fact, if C is simple and lies in D, the function f is analytic at each point interior to and on C; and the Cauchy-Goursat theorem ensures that equation (3.3.4) holds. Furthermore, if C is closed but intersects itself a finite number of times, it consists of a finite number of simple closed contours.

　　This is illustrated in Fig. 3.3.1, where the simple closed contours C_k is zero, according to the Cauchy-Goursat theorem, it follows that

$$\int_C f(z)\mathrm{d}z = \sum_{k=1}^{4} \int_{C_k} f(z)\mathrm{d}z = 0.$$

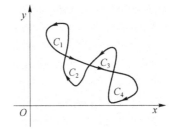

Fig 3.3.1

Subtleties arise if the closed contour has an infinite number of self-intersection points.

Corollary 3.3.1 (Path independence of contour integral)　A function f that is analytic throughout a simply connected domain D, then the value of the integral $\int_C f(z)\mathrm{d}z$ is independent of the path of integration in D.

Proof　Suppose that A and B are any two points in D. Now we travel from the point A to B by any two paths in D, say $\widehat{APB}$ and $\widehat{AQB}$ (see Fig. 3.3.2(a) and 3.3.2(b)).

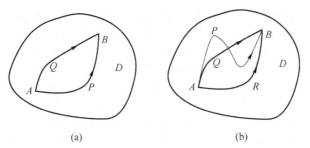

Fig. 3.3.2

If these two curves do not intersect except for A and B (see Fig. 3.3.2(a)), then from

$$\int_{\widehat{APB}} f(z)\mathrm{d}z + \int_{\widehat{BQA}} f(z)\mathrm{d}z = \int_{\widehat{APBQA}} f(z)\mathrm{d}z = 0,$$

we have

$$\int_{\widehat{APB}} f(z)\mathrm{d}z = -\int_{\widehat{BQA}} f(z)\mathrm{d}z = \int_{\widehat{ABQ}} f(z)\mathrm{d}z.$$

　　If $\widehat{APB}$ and $\widehat{AQB}$ have any intersections other than A and B (see Fig. 3.3.2(b)), draw another curve which doesn't intersect the curves $\widehat{APB}$ and $\widehat{AQB}$ except for A and B, then we have

$$\int_{\widehat{APB}} f(z)\mathrm{d}z = \int_{\widehat{ARB}} f(z)\mathrm{d}z = \int_{\widehat{AQB}} f(z)\mathrm{d}z.$$

Therefore, the conclusion is true.

Corollary 3.3.2 A function f that is analytic throughout a simply connected domain D must have an antiderivative everywhere in D.

This corollary follows immediately from Theorem 3.3.1 because of Theorem 3.2.1 in Section 3.2.3, which tells us that a continuous function f always has an antiderivative in a given domain when equation (3.3.4) holds for each closed contour C in that domain. Note that, since the finite plane is simply connected, Corollary 3.3.2 tells us that entire functions always possess antiderivatives.

The Cauchy-Goursat theorem can also be extended in a way that involves integrals along the boundary of a multiply connected domain. The following theorem is such an extension.

Theorem 3.3.4 Suppose that

(i) C is a simple closed contour, described in the counterclockwise direction;

(ii) $C_k(k=1, 2, \cdots, n)$ are simple closed contours interior to C, all described in the clockwise direction, that are disjoint and whose interiors have no points in common (Fig. 3.3.3).

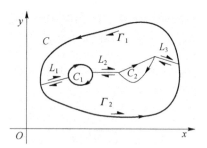

Fig. 3.3.3

If a function f is analytic on all of these contours and throughout the multiply connected domain consisting of all points inside C and exterior to each C_k, then

$$\int_C f(z)\mathrm{d}z + \sum_{k=1}^{n} \int_{C_k} f(z)\mathrm{d}z = 0. \qquad (3.3.5)$$

Note that, in equation (3.3.5), the direction of each path of integration is such that the multiply connected domain lies to the left of that path.

To prove the theorem, we introduce a polygonal path L_1, consisting of a finite number of line segments joined end to end, to connect the outer contour C to the inner contour C_1. We introduce another polygonal path L_2 which connects C_1 to C_2; and we continue in this manner, with L_{n+1} connecting C_n to C. As indicated by the single-barbed arrows in Fig. 3.3.3, two simple closed contours Γ_1 and Γ_2, and the sum of the values of the integrals over those contours is found to be zero. Since the integrals in opposite directions along each path L_k cancel, only the integrals along C and C_k remain; and we arrive at statement (3.3.5).

Corollary 3.3.3 (Principle of deformation of paths) Let C_1 and C_2 denote positively oriented simple closed contours, where C_2 is interior to C_1 (Fig. 3.3.4).

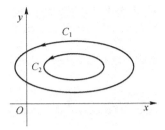

Fig. 3.3.4

If a function f is analytic in the closed region consisting of those contours and all points between them, then

$$\int_{C_1} f(z)\,dz = \int_{C_2} f(z)\,dz. \tag{3.3.6}$$

Proof From Theorem 3.3.4, we have

$$\int_{C_1} f(z)\,dz + \int_{-C_2} f(z)\,dz = 0.$$

And we note that this is just a different form of equation (3.3.6).

Corollary 3.3.3 is known as the principle of deformation of paths since it tells us that if C_1 is continuously deformed into C_2, always passing through points at which f is analytic, then the value of the integral of f over C_1 never changes.

Example 3.3.3 Let C be any positively oriented simple closed contour surrounding the origin, then show that

$$\int_C \frac{dz}{z} = 2\pi i.$$

Proof We need only construct a positively oriented circle C_0 with center at the origin and radius so small that C_0 lies entirely inside C (Fig. 3.3.5).

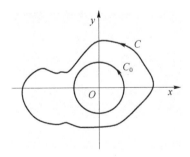

Fig. 3.3.5

Since $\dfrac{1}{z}$ is analytic everywhere except at $z=0$, then, from Exercise 3.2.15 in Section 3.2.3, the desired result follows.

$$\int_C \frac{dz}{z} = \int_{C_0} \frac{dz}{z} = 2\pi i.$$

Similarly, we can show that

$$\int_C \frac{dz}{z - z_0} = 2\pi i,$$

where C is any positively oriented simple closed contour surrounding the point z_0.

Example 3.3.4 Let C be the positively oriented circle $|z|=1$. Evaluate the integral
$$\int_C \frac{dz}{z(z+2)}.$$

Solution
$$\int_C \frac{dz}{z(z+2)} = \frac{1}{2}\int_C \left(\frac{1}{z} - \frac{1}{z+2}\right) dz$$
$$= \frac{1}{2}\int_C \frac{1}{z} dz - \frac{1}{2}\int_C \frac{1}{z+2} dz$$
$$= \pi i + 0 = \pi i.$$

Example 3.3.5 Let C be the positively oriented circle $|z|=3$. Then evaluate the integral
$$\int_C \frac{dz}{(z+1)(z+2)}.$$

Solution Let $f(z) = \frac{1}{(z+1)(z+2)}$. Then there are two singular points $z=-1$ and $z=-2$ in C.

Construct two small circles C_1 and C_2 enclosed in C and disjoint and have no common points, such that $z=-1$ is only interior to C_1 and $z=-2$ is only interior to C_2. Then according to Theorem 3.3.4, we obtain

$$\int_C \frac{dz}{(z+1)(z+2)}$$
$$= \int_{C_1} \frac{dz}{(z+1)(z+2)} + \int_{C_2} \frac{dz}{(z+1)(z+2)}$$
$$= \int_{C_1} \frac{dz}{z+1} - \int_{C_1} \frac{dz}{z+2} + \int_{C_2} \frac{dz}{z+1} - \int_{C_2} \frac{dz}{z+2}$$
$$= 2\pi i - 0 + 0 - 2\pi i = 0.$$

Exercises 3.3

1. Apply the Cauchy-Goursat theorem to show that
$$\int_C f(z) dz = 0$$
when the contour C is the circle $|z|=1$, in either direction, and when

(a) $f(z) = \frac{z^2}{z-3}$; (b) $f(z) = ze^{-z}$; (c) $f(z) = \frac{1}{z^2+2z+2}$;

(d) $f(z) = \operatorname{sech} z$; (e) $f(z) = \tan z$; (f) $f(z) = \operatorname{Log}(z+2)$;

(g) $f(z) = (z^2+1)\cos(z^5+1)$.

2. Let C_1 denote the positively oriented circle $|z|=4$ and let C_2 be the positively oriented boundary of the square whose sides lie along the lines $x=\pm 1$, $y=\pm 1$ (Fig. 3.3.6). Please point out why

$$\int_{C_1} f(z) dz = \int_{C_2} f(z) dz,$$

when

(a) $f(z)=\dfrac{1}{3z^2+1}$;   (b) $f(z)=\dfrac{z+2}{\sin\left(\dfrac{z}{2}\right)}$;   (c) $f(z)=\dfrac{z}{1-e^z}$.

3. Evaluate the following integrals, where the contour described in the positive sense.

(a) $\displaystyle\int_{|z|=3}\dfrac{2z+1+2i}{(z+1)(z+2i)}dz$;

(b) $\displaystyle\int_{|z-1|=3}\dfrac{2i}{1+z^2}dz$;

(c) $\displaystyle\int_{|z|=3}\dfrac{2z-1}{z^2-z}dz$;

(d) $\displaystyle\int_{|z|=1}\dfrac{2i}{z^2-2iz}dz$.

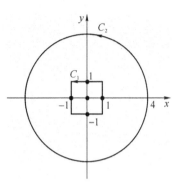

Fig. 3.3.6

4. If C_0 denotes a positively oriented circle $|z-z_0|=R$, then show that

$$\int_{C_0}(z-z_0)^{n-1}dz=\begin{cases}0, & \text{when } n=\pm1,\pm2,\cdots,\\ 2\pi i, & \text{when } n=0.\end{cases}$$

Use that result to show that if C is the boundary of the rectangle $0\leqslant x\leqslant 3$, $0\leqslant y\leqslant 2$, described in the positive sense, then

$$\int_C(z-2-i)^{n-1}dz=\begin{cases}0, & \text{when } n=\pm1,\pm2,\cdots,\\ 2\pi i, & \text{when } n=0.\end{cases}$$

5. Apply the Cauchy theorem to show that

$$\int_{|z|=1}\dfrac{1}{z+2}dz=0.$$

And use that result to show that

$$\int_0^{2\pi}\dfrac{1+2\cos\theta}{5+4\cos\theta}d\theta=0.$$

3.4 Cauchy integral formula and derivatives of analytic functions

3.4.1 Cauchy integral formula

Another fundamental result will now be established.

Theorem 3.4.1 Let f be analytic everywhere inside and on a simple closed contour C, taken in the positive sense. If z_0 is any point interior to C, then

$$f(z_0)=\dfrac{1}{2\pi i}\int_C\dfrac{f(z)dz}{z-z_0}, \quad (3.4.1)$$

or

$$\int_C\dfrac{f(z)dz}{z-z_0}=2\pi i f(z_0). \quad (3.4.2)$$

Formula (3.4.1) or (3.4.2) is called the Cauchy integral formula. It tells us that if a function f is analytic within and on a simple closed contour C, then the values of f interior to C are completely determined by the values of f on C.

Proof Let C_ρ denote a positively oriented circle $|z-z_0|=\rho$, where ρ is small enough that C_ρ is interior to C (see Fig. 3.4.1).

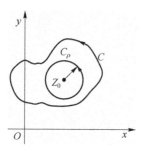

Fig. 3.4.1

Since the function $\dfrac{f(z)}{z-z_0}$ is analytic between and on the contours C and C_ρ, it follows from the principle of deformation of paths (Corollary 3.3.3 in Section 3.3.2) that

$$\int_C \frac{f(z)dz}{z-z_0} = \int_{C_\rho} \frac{f(z)dz}{z-z_0}.$$

And,

$$\int_C \frac{f(z)dz}{z-z_0} - 2\pi i f(z_0)$$

$$= \int_C \frac{f(z)dz}{z-z_0} - f(z_0)\int_{C_\rho} \frac{dz}{z-z_0} \qquad (3.4.3)$$

$$= \int_{C_\rho} \frac{f(z)-f(z_0)}{z-z_0}dz.$$

Now the fact that f is analytic, and therefore continuous at z_0 ensures that, corresponding to each positive number ε, however small, there is a positive number δ such that

$$|f(z)-f(z_0)|<\varepsilon \quad \text{whenever} \quad |z-z_0|<\delta. \qquad (3.4.4)$$

Let the radius ρ of the circle C_ρ be smaller than the number δ in the second of these inequalities. Since $|z-z_0|=\rho$ when z is on C_ρ, it follows that the first of inequalities (3.4.4) holds when z is such a point; and inequality (3.2.6) in Section 3.2.2, giving upper bounds for the moduli of contour integrals, tells us that

$$\left|\int_{C_\rho} \frac{f(z)-f(z_0)}{z-z_0}dz\right| < \frac{\varepsilon}{\rho}2\pi\rho = 2\pi\varepsilon.$$

In view of equation (3.4.3), then,

$$\left|\int_C \frac{f(z)dz}{z-z_0} - 2\pi i f(z_0)\right| < 2\pi\varepsilon.$$

Since the left-hand side of this inequality is a nonnegative constant that is less than an arbitrarily small positive number, it must equal to zero. Hence equation (3.4.2) is valid, and the theorem is proved.

Example 3.4.1 Let C be the positively oriented circle $|z|=2$. Evaluate the integral
$$\int_C \frac{z\,dz}{(9-z^2)(z+i)}.$$

Solution Since
$$\int_C \frac{z\,dz}{(9-z^2)(z+i)} = \int_C \frac{\frac{z}{9-z^2}}{z-(-i)}dz,$$

then we assume that
$$f(z) = \frac{z}{9-z^2}.$$

It is obvious that $f(z)$ is analytic within and on C and since the point $z_0 = -i$ is interior to C, formula (3.4.2) tells us that
$$\int_C \frac{z\,dz}{(9-z^2)(z+i)} = \int_C \frac{\frac{z}{9-z^2}}{z-(-i)}dz = 2\pi i\left(\frac{z}{9-z^2}\bigg|_{z=-i}\right) = 2\pi i\left(\frac{-i}{10}\right) = \frac{\pi}{5}.$$

Example 3.4.2 Let C be the positively oriented circle $|z|=3$. Evaluate the integral
$$\int_C \frac{e^z\,dz}{z^2+1}.$$

Solution It is obvious that $z=\pm i$ are two singular points of $g(z)=\frac{e^z}{z^2+1}$.

Construct two small circles C_1 and C_2 enclosed in C (see Fig. 3.4.2) and disjoint and have no common points. According to the Cauchy integral formula

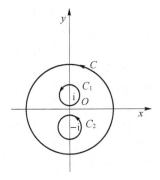

Fig. 3.4.2

$$\int_C \frac{e^z\,dz}{z^2+1}$$
$$= \int_{C_1} \frac{e^z\,dz}{z^2+1} + \int_{C_2} \frac{e^z\,dz}{z^2+1}$$
$$= \int_{C_1} \frac{\frac{e^z}{z+i}}{z-i}dz + \int_{C_2} \frac{\frac{e^z}{z-i}}{z+i}dz$$
$$= 2\pi i\frac{e^i}{i+i} + 2\pi i\frac{e^{-i}}{-i-i}$$
$$= \pi(e^i - e^{-i}) = 2\pi i\sin 1.$$

3.4.2 Higher-order derivatives formula of analytic functions

It follows from the Cauchy integral formula (Section 3.4.1) that if a function is analytic at a point, then its derivatives of all orders exist at that point and are themselves analytic there.

Lemma 3.4.1 Suppose that a function f is analytic everywhere inside and on a simple closed contour C, taken in the positive sense. If z is any point interior to C, then

$$f'(z) = \frac{1}{2\pi i} \int_C \frac{f(s)\,ds}{(s-z)^2}$$

and

$$f''(z) = \frac{1}{\pi i} \int_C \frac{f(s)\,ds}{(s-z)^3}. \qquad (3.4.5)$$

Note that expressions (3.4.5) can be obtained formally, or without rigorous verification, by differentiating with respect to z under the integral sign in the Cauchy integral formula

$$f(z) = \frac{1}{2\pi i} \int_C \frac{f(s)\,ds}{s-z}, \qquad (3.4.6)$$

where z is interior to C and s denotes points on C.

Proof To verify the first of expressions (3.4.5), we let d denote the smallest distance from z to points on C and use formula (3.4.6) to write

$$\frac{f(z+\Delta z)-f(z)}{\Delta z} = \frac{1}{2\pi i} \int_C \left(\frac{1}{s-z-\Delta z} - \frac{1}{s-z}\right) \frac{f(s)}{\Delta z}\,ds$$

$$= \frac{1}{2\pi i} \int_C \frac{f(s)\,ds}{(s-z-\Delta z)(s-z)},$$

where $0<|\Delta z|<d$ (see Fig. 3.4.3).

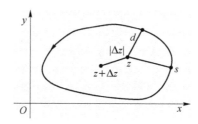

Fig. 3.4.3

Evidently, then

$$\frac{f(z+\Delta z)-f(z)}{\Delta z} - \frac{1}{2\pi i}\int_C \frac{f(s)\,ds}{(s-z)^2} = \frac{1}{2\pi i}\int_C \frac{\Delta z f(s)\,ds}{(s-z-\Delta z)(s-z)^2}. \qquad (3.4.7)$$

Next, we let M denote the maximum value of $|f(s)|$ on C and observe that, since $|s-z| \geq d$ and $|\Delta z| < d$,

$$|s-z-\Delta z| = |(s-z)-\Delta z| \geq ||s-z|-|\Delta z|| \geq d-|\Delta z| > 0.$$

Thus

$$\left|\int_C \frac{\Delta z f(s)\,ds}{(s-z-\Delta z)(s-z)^2}\right| \leq \frac{|\Delta z|M}{(d-|\Delta z|)d^2}L,$$

where L is the length of C. Upon letting Δz tend to zero, we find from this inequality that the right-hand side of equation (3.4.7) also tends to zero. Consequently,

$$\lim_{\Delta z \to 0} \frac{f(z+\Delta z)-f(z)}{\Delta z} - \frac{1}{2\pi i}\int_C \frac{f(s)}{(s-c)^2} = 0.$$

And the desired expression for $f'(z)$ is established.

The same technique can be used to verify the expression for $f''(z)$ in the statement of the lemma. The details are left to the reader.

Remark 3.4.1 (a) From Lemma 3.4.1, one can use mathematical induction to generalize formulas (3.4.5) to

$$f^{(n)}(z) = \frac{n!}{2\pi i}\int_C \frac{f(s)ds}{(s-z)^{n+1}}, \quad (n=1,2,\cdots). \tag{3.4.8}$$

The verification is considerably more involved than for just $n=1$ and $n=2$, and we refer the interested reader to other texts for it. Note that, with the agreement that $f^{(0)}(z)=f(z)$ and $0!=1$, expression (3.4.8) is also valid when $n=0$, in which case it becomes the Cauchy integral formula (3.4.1).

(b) Moreover, let f be analytic everywhere inside and on a simple closed contour C, taken in the positive sense. If z_0 is any point interior to C, then

$$f^{(n)}(z_0) = \frac{1}{2\pi i}\int_C \frac{f(z)dz}{(z-z_0)^{n+1}},$$

or

$$\int_C \frac{f(z)dz}{(z-z_0)^{n+1}} = \frac{2\pi i}{n!} f^{(n)}(z_0), \quad (n=1,2,3,\cdots). \tag{3.4.9}$$

Theorem 3.4.2 If a function is analytic at a point, then its derivatives of all orders exist at that point. Those derivatives are, moreover, all analytic there.

Proof To prove this remarkable theorem, we assume that a function f is analytic at a point z_0. There must, then, be a neighborhood $|z-z_0|<\varepsilon$ of z_0 throughout which f is analytic. Consequently, there is a positively oriented circle C_0 (Fig. 3.4.4) center at z_0 and with radius $\frac{\varepsilon}{2}$, such that f is analytic inside and on C_0.

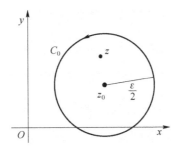

Fig. 3.4.4

According to the above lemma,

$$f''(z) = \frac{1}{\pi i}\int_{C_0} \frac{f(s)ds}{(s-z)^3}$$

at each point z interior to C_0, and the existence of $f''(z)$ throughout the neighborhood $|z-z_0|<\frac{\varepsilon}{2}$ means that f' is analytic at z_0. One can apply the same argument to the analytic function f' to conclude that its derivative f'' is analytic, etc. Theorem 3.4.2 is now established.

Corollary 3.4.1   If a function $f(z)=u(x,y)+iv(x,y)$ is defined and analytic at a point $z=(x,y)$ then the component functions u and v have continuous partial derivatives of all orders at that point.

Proof   Suppose that the function
$$f(z)=u(x,y)+iv(x,y)$$
is analytic at point $z=(x,y)$, the differentiability of f' ensures the continuity of f'. Then, since
$$f'(z)=u_x+iv_x=v_y-iu_y,$$
we may conclude that the first-order partial derivatives of u and v are continuous at that point. Furthermore, since f'' is analytic and continuous at z and since
$$f''(z)=u_{xx}+iv_{xx}=v_{yx}-iu_{yx}=v_{xy}-iu_{xy}=-u_{yy}+iv_{yy}.$$
Then, the second-order partial derivatives of u and v are continuous at that point.

Through repeating this process, we know that, the component functions u and v have continuous partial derivatives of all orders at that point.

Example 3.4.3   Let C be the positively oriented unit circle $|z|=1$. Then evaluate the integral
$$\int_C \frac{\exp(2z)}{z^4}dz.$$

Solution   Let $f(z)=\exp(2z)$. It is obvious that $f(z)$ is analytic within and on C and since the point $z_0=0$ is interior to C, formula (3.4.9) tells us that
$$\int_C \frac{\exp(2z)}{z^4}dz = \int_C \frac{f(z)dz}{(z-0)^{3+1}} = \frac{2\pi i}{3!}f'''(0) = \frac{8\pi i}{3}.$$

Example 3.4.4   Let C be the negatively oriented unit circle $|z-i|=1$. Then evaluate the integral
$$\int_C \frac{\cos z}{(z-i)^3}dz.$$

Solution   Let $f(z)=\cos z$. It is obvious that $f(z)$ is analytic within and on C and since the point $z_0=i$ is interior to C, formula (3.4.9) tells us that
$$\int_C \frac{\cos z}{(z-i)^3}dz = -\frac{2\pi i}{2!}f''(i) = -\pi i\cos i.$$

Example 3.4.5   Let z_0 be any point interior to a positively oriented simple closed contour C. Show that
$$\int_C \frac{dz}{(z-z_0)^{n+1}} = \begin{cases} 2\pi i, & n=0, \\ 0, & n=1, 2, \cdots. \end{cases}$$

Proof   Taking $f(z)=1$, then from (3.4.9), we can obtain
$$\int_C \frac{dz}{z-z_0} = 2\pi i.$$

And
$$\int_C \frac{dz}{(z-z_0)^{n+1}} = 0, \quad (n=1, 2, \cdots).$$

Theorem 3.4.3 Let f be continuous on a domain D. If
$$\int_C f(z) dz = 0$$
for every closed contour C lying in D, then f is analytic throughout D.

Proof To prove the theorem here, we observe that when its hypothesis is satisfied, Theorem 3.2.1 in Section 3.2.3 ensures that f has an antiderivative in D; that is, there exists an analytic function F such that $F'(z) = f(z)$ at such point in D. Since f is the derivative of F, it then follows from Theorem 3.4.2 above that f is analytic in D.

Remark 3.4.2 From Theorem 3.4.3, when D is simply connected, we have for the class of continuous functions on D a converse of Theorem 3.3.3 in Section 3.3.2, which is the extension of the Cauchy-Goursat theorem involving such domains.

Exercises 3.4

1. Let C denote the positively oriented boundary of the square whose sides lie along the lines $x = \pm 2$ and $y = \pm 2$. Evaluate each of these integrals:

(a) $\int_C \dfrac{e^{-z} dz}{z - \left(\dfrac{\pi i}{2}\right)}$;   (b) $\int_C \dfrac{\cos z}{z(z^2+8)} dz$;   (c) $\int_C \dfrac{z \, dz}{2z+1}$;

(d) $\int_C \dfrac{\cos hz}{z^4} dz$;   (e) $\int_C \dfrac{\tan\left(\dfrac{z}{2}\right)}{(z-x_0)^2} dz$, $(-2 < x_0 < 2)$.

2. Find the value of the integral of $g(z)$ around the circle $|z-i| = 2$ in the positive sense when

(a) $g(z) = \dfrac{1}{z^2 + 4}$;   (b) $g(z) = \dfrac{1}{(z^2+4)^2}$.

3. Evaluate the following integrals, where the contour described in the positive sense.

(a) $\int_{|z-1|=1} \dfrac{dz}{z^3 - 1}$;   (b) $\int_{|z|=2} \dfrac{\cos z}{(z-i)^{101}} dz$;   (c) $\int_{|z|=2} \dfrac{\sin z}{(z-i)^{4n+1}} dz$;

(d) $\int_{|z|=\frac{3}{2}} \dfrac{dz}{(z-1)(z+2)(z^4+16)}$.

4. Let C be the circle $|z| = 3$, described in the positive sense. Show that if
$$g(z) = \int_C \frac{2s^2 - s - 2}{s - z} ds, \quad (|z| \neq 3),$$
then $g(2) = 8\pi i$. What is the value of $g(z)$ when $|z| > 3$?

5. Let C be any simple closed contour, described in the positive sense in the z plane, and write
$$g(w) = \int_C \frac{z^3 + 2z}{(z-w)^3} dz.$$

· 87 ·

Show that $g(z) = 6\pi z \mathrm{i}$ when z is inside C, and that $g(z) = 0$ when z is outside.

6. Let C be the unit circle $z = \mathrm{e}^{\mathrm{i}\theta}$ ($-\pi < \theta \leqslant \pi$), described in the positive sense. First show that, for any real constant a,
$$\int_C \frac{\mathrm{e}^{az}}{z}\mathrm{d}z = 2\pi\mathrm{i}.$$
Then write this integral in terms of θ to derive the integration formula
$$\int_0^\pi \mathrm{e}^{a\cos\theta}\cos(a\sin\theta)\,\mathrm{d}\theta = \pi.$$

7. Please show that $\left(\dfrac{z^n}{n!}\right)^2 = \dfrac{1}{2\pi\mathrm{i}}\oint_C \dfrac{z^n \mathrm{e}^{z\zeta}}{n!\,\zeta^{n+1}}\mathrm{d}\zeta$, where C is a simple closed contour surrounding the origin.

Chapter 4 Complex Series

This chapter is devoted mainly to series representations of analytic functions. We present theorems that guarantee the existence of such representations, and develop some facility in manipulating series.

4.1 Complex series and its convergence

4.1.1 Complex sequences and its convergence

Definition 4.1.1 An infinite sequence
$$z_1, z_2, \cdots, z_n, \cdots \tag{4.1.1}$$
of complex numbers has a ***limit*** z if, for each positive number ε, there exists a positive integer n_0 such that
$$|z_n - z| < \varepsilon \quad \text{whenever } n > n_0. \tag{4.1.2}$$
Geometrically, this means that, for sufficiently large values of n, the points z_n lie in any given ε neighborhood of z (Fig. 4.1.1). Since we can choose ε as small as we please, it follows that the points z_n become arbitrarily close to z as their subscripts increase. Note that the value of n_0 that is needed will, in general, depends on the value of ε.

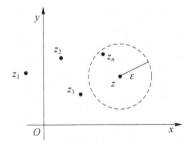

Fig. 4.1.1

The sequence (4.1.1) can have at most one limit. That is, a limit z is unique if it exists. When that limit exists, the sequence is said to ***converge*** to z; and we write $\lim\limits_{n\to\infty} z_n = z$.

If the sequence has no limit, it **diverges**.

Theorem 4.1.1 Suppose that $z_n = x_n + iy_n$, $(n=1, 2, \cdots)$ and $z = x + iy$. Then

$$\lim_{n \to \infty} z_n = z \tag{4.1.3}$$

if and only if

$$\lim_{n \to \infty} x_n = x \quad \text{and} \quad \lim_{n \to \infty} y_n = y. \tag{4.1.4}$$

Proof We assume that conditions (4.1.4) hold and obtain condition (4.1.3) from it. According to conditions (4.1.4), there exist, for each positive number ε, positive integers n_1 and n_2 such that

$$|x_n - x| < \frac{\varepsilon}{2} \quad \text{whenever} \quad n > n_1,$$

and

$$|y_n - y| < \frac{\varepsilon}{2} \quad \text{whenever} \quad n > n_2.$$

Hence, if n_0 is the larger of the two integers n_1 and n_2,

$$|x_n - x| < \frac{\varepsilon}{2} \quad \text{and} \quad |y_n - y| < \frac{\varepsilon}{2} \quad \text{whenever} \quad n > n_0.$$

Since

$$|(x_n + iy_n) - (x + iy)| = |(x_n - x) - i(y_n - y)| \leqslant |x_n - x| + |y_n - y|,$$

then

$$|z_n - z| < \frac{\varepsilon}{2} + \frac{\varepsilon}{2} = \varepsilon \quad \text{whenever} \quad n > n_0.$$

Condition (4.1.3) thus holds.

Conversely, if we start with condition (4.1.3), we know that, for each positive number ε, there exists a positive integer n_0 such that

$$|(x_n + iy_n) - (x + iy)| < \varepsilon \quad \text{whenever} \quad n > n_0.$$

But

$$|x_n - x| \leqslant |(x_n - x) + i(y_n - y)| = |(x_n + iy_n) - (x + iy)|$$

and

$$|y_n - y| \leqslant |(x_n - x) + i(y_n - y)| = |(x_n + iy_n) - (x + iy)|,$$

and this means that

$$|x_n - x| < \varepsilon \quad \text{and} \quad |y_n - y| < \varepsilon \quad \text{whenever} \quad n > n_0.$$

That is, conditions (4.1.4) are satisfied.

Note The theorem enables us to write

$$\lim_{n \to \infty} (x_n + iy_n) = \lim_{n \to \infty} x_n + i \lim_{n \to \infty} y_n$$

whenever we know both limits on the right exist, then the one on the left exists.

Example 4.1.1

$$\lim_{n \to \infty} \left(i + \frac{1}{n^3} \right) = \lim_{n \to \infty} \frac{1}{n^3} + i \lim_{n \to \infty} 1 = 0 + i \times 1 = i.$$

4.1.2 Complex series and its convergence

Definition 4.1.2 An infinite series

$$\sum_{n=1}^{\infty} z_n = z_1 + z_2 + \cdots + z_n + \cdots \qquad (4.1.5)$$

of complex numbers **converges** to the **sum** S if the sequence

$$S_N = \sum_{n=1}^{N} z_n = z_1 + z_2 + \cdots + z_N, \quad (N = 1, 2, \cdots) \qquad (4.1.6)$$

of **partial sums** converges to S; we then write

$$\sum_{n=1}^{\infty} z_n = S.$$

Note Since a sequence can have at most one limit, a series can have at most one sum. When a series does not converge, we say that it **diverges**.

Theorem 4.1.2 Suppose that $z_n = x_n + iy_n$, $(n = 1, 2, \cdots)$ and $S = X + iY$. Then

$$\sum_{n=1}^{\infty} z_n = S \qquad (4.1.7)$$

if and only if

$$\sum_{n=1}^{\infty} x_n = X \quad \text{and} \quad \sum_{n=1}^{\infty} y_n = Y. \qquad (4.1.8)$$

Proof We can write the partial sums of the series as

$$S_N = X_N + iY_N, \qquad (4.1.9)$$

where

$$X_N = \sum_{n=1}^{N} x_n \quad \text{and} \quad Y_N = \sum_{n=1}^{N} y_n.$$

It is obvious that the statement (4.1.7) is true if and only if

$$\lim_{N \to \infty} S_N = S; \qquad (4.1.10)$$

and, in view of relation (4.1.9) and theorem 4.1.1, limit (4.1.10) holds if and only if

$$\lim_{N \to \infty} X_N = X \quad \text{and} \quad \lim_{N \to \infty} Y_N = Y. \qquad (4.1.11)$$

Limits (4.1.11) therefore imply statement (4.1.7), and conversely. Since X_n and Y_n are the partial sums of the series (4.1.8), the theorem here is proved.

Note This theorem tells us, of course, that one can write

$$\sum_{n=1}^{\infty} (x_n + iy_n) = \sum_{n=1}^{\infty} x_n + i \sum_{n=1}^{\infty} y_n$$

whenever it is known that the two series on the right converge or that the one on the left does.

This theorem is useful in showing that a number of familiar properties of series in calculus carry over to series whose terms are complex numbers. We present here two such properties.

Theorem 4.1.3 The nth term of a convergent series of complex numbers approaches zero as n trends to infinity.

This theorem can be proved easily according to Theorem 4.1.2. The terms of a convergent series of complex numbers are, therefore, bounded. To be specific, there exists a positive constant M such that $|z_n| \leqslant M$ for each positive integer n.

Definition 4.1.3 Series

$$\sum_{n=1}^{\infty} z_n = \sum_{n=1}^{\infty} x_n + \mathrm{i} y_n \tag{4.1.12}$$

is said to be ***absolutely convergent*** if the series

$$\sum_{n=1}^{\infty} |z_n| = \sum_{n=1}^{\infty} \sqrt{x_n^2 + y_n^2} \tag{4.1.13}$$

of real numbers converges.

Theorem 4.1.4 The absolute convergence of a series of complex numbers implies the convergence of that series.

Proof Suppose that series (4.1.12) is absolutely convergent. Since

$$|x_n| \leqslant \sqrt{x_n^2 + y_n^2} \quad \text{and} \quad |y_n| \leqslant \sqrt{x_n^2 + y_n^2},$$

we know from the comparison test in calculus that the two series

$$\sum_{n=1}^{\infty} |x_n| \quad \text{and} \quad \sum_{n=1}^{\infty} |y_n|$$

must converge. Moreover, since the absolute convergence of a series of real numbers implies the convergence of the series itself, it follows that

$$\sum_{n=1}^{\infty} x_n \quad \text{and} \quad \sum_{n=1}^{\infty} y_n.$$

According to Theorem 4.1.2, series (4.1.12) converges. Consequently, absolute convergence of a series of complex numbers implies convergence of that series.

Definition 4.1.4 The ***remainder*** ρ_N of series $\sum_{n=1}^{\infty} z_n$ after N terms is defined as

$$\rho_N = \sum_{n=N+1}^{\infty} z_n = S - S_N.$$

Theorem 4.1.5 A series converges if and only if the sequence of remainders tends to zero.

Proof From the definition of the remainder, it is obvious that $S = S_N + \rho_N$, and since $|S - S_N| = |\rho_N - 0|$, we see a series converges to a number S if and only if the sequence of remainders tends to zero.

Example 4.1.2 Determine the convergence of the following series:

(1) $\sum_{n=1}^{\infty} \left(\dfrac{n+\mathrm{i}}{n-\mathrm{i}} \right)^n$; (2) $\sum_{n=1}^{\infty} \dfrac{(8\mathrm{i})^n}{n!}$; (3) $\sum_{n=1}^{\infty} \left(\dfrac{-n+\mathrm{i}}{n^2} \right)(-1)^n$.

Solution (1) Since $\left| \left(\dfrac{n+\mathrm{i}}{n-\mathrm{i}} \right)^n \right| \geqslant 1$, then $\lim_{n \to \infty} \left(\dfrac{n+\mathrm{i}}{n-\mathrm{i}} \right)^n \neq 0$.

It is concluded that this series is divergent.

(2) Since $\left| \dfrac{(8\mathrm{i})^n}{n!} \right| = \dfrac{8^n}{n!}$, and $\sum_{n=1}^{\infty} \dfrac{8^n}{n!}$ is convergent according to the ratio test for positive series of real numbers. Then it is concluded that this series is absolutely convergent.

(3) The real part of this series is $\sum_{n=1}^{\infty} (-1)^n \dfrac{1}{n}$, and the imaginary part is $\sum_{n=1}^{\infty} (-1)^n \dfrac{1}{n^2}$.

Since these two series are convergent, we can conclude that this series is convergent.

Exercises 4.1

1. Show that
$$\text{if } \lim_{n\to\infty} z_n = z, \text{ then } \lim_{n\to\infty} |z_n| = |z|.$$

2. Show that
$$\text{if } \sum_{n=1}^{\infty} z_n = S, \text{ then } \sum_{n=1}^{\infty} \overline{z_n} = \overline{S}$$

3. Show that a limit of a convergent sequence of complex numbers is unique by applying to the corresponding result for a sequence of real numbers.

4. Let c denote any complex number and show that
$$\text{if } \sum_{n=1}^{\infty} z_n = S, \text{ then } \sum_{n=1}^{\infty} cz_n = cS.$$

5. Show that
$$\text{if } \sum_{n=1}^{\infty} z_n = S \text{ and } \sum_{n=1}^{\infty} \omega_n = T, \text{ then}$$
$$\sum_{n=1}^{\infty} (\omega_n + z_n) = S + T.$$

6. Determine the convergence of the following complex sequences, and if it converges, find its limit:

(a) $z_n = \dfrac{1+ni}{n+1}$;

(b) $z_n = \dfrac{\cos n + i\sin n}{(1+i)^n}$;

(c) $z_n = \dfrac{\cos ni}{n}$;

(d) $z_n = e^{ni}$.

7. Determine the convergence or absolute convergence of the following complex series:

(a) $\sum_{n=1}^{\infty} \left(\dfrac{i+n}{n}\right)^{10}$;

(b) $\sum_{n=0}^{\infty} \dfrac{(-1)^n + i}{2^n}$;

(c) $\sum_{n=1}^{\infty} \dfrac{1+i}{n+1}$;

(d) $\sum_{n=0}^{\infty} \dfrac{i^n \sin(n+1)}{n^2}$.

4.2 Power series

4.2.1 The definition of power series

Definition 4.2.1 The series of the form
$$\sum_{n=0}^{\infty} a_n (z - z_0)^n = a_0 + a_1 (z - z_0) + a_2 (z - z_0)^2 + \cdots + a_n (z - z_0)^n + \cdots$$

(4.2.1)

is called **power series**, in which z_0 and the coefficients a_n are complex constants and z may be any point in a stated region containing z_0.

Definition 4.2.2 For any z_1, if the series with constant terms $\sum_{n=0}^{\infty} a_n(z_1 - z_0)^n$ converges, then we say the power series $\sum_{n=0}^{\infty} a_n(z - z_0)^n$ is convergent at $z = z_1$, and z_1 is called a **convergence point** of the power series; if $\sum_{n=0}^{\infty} a_n(z_1 - z_0)^n$ diverges, z_1 is called a **divergence point**. The set of all the convergence points is called the **convergence domain** of the power series.

Definition 4.2.3 Let D be the convergence domain of the power series $\sum_{n=0}^{\infty} a_n(z - z_0)^n$, then $\forall z_1 \in D$, the series has a sum $S(z_1)$. So the sum $S(z)$ of the series is a function of z defined in D, and is called the **sum function** and denoted by

$$\sum_{n=0}^{\infty} a_n(z - z_0)^n = S(z), \quad z \in D. \tag{4.2.2}$$

Definition 4.2.4 $S_N(z) = \sum_{n=0}^{N} a_n(z - z_0)^n$ is called the **partial sum** of the power series $\sum_{n=0}^{\infty} a_n(z - z_0)^n$, and $\rho_N(z) = S(z) - S_N(z) = \sum_{n=N+1}^{\infty} a_n(z - z_0)^n$ is called a **remainder** of the series.

Example 4.2.1 Discuss the convergence of the power series

$$\sum_{n=0}^{\infty} z^n = 1 + z + z^2 + \cdots + z^n + \cdots.$$

Solution The partial sum of the series is

$$S_N(z) = \sum_{n=0}^{N-1} z^n = 1 + z + z^2 + \cdots + z^{N-1}, \quad (z \neq 1).$$

According to the identity

$$1 + z + z^2 + \cdots + z^n = \frac{1 - z^{n+1}}{1 - z}, \quad (z \neq 1),$$

the partial sum can be written as

$$S_N(z) = \frac{1 - z^N}{1 - z}, \quad (z \neq 1).$$

It is obvious that when $|z| < 1$,

$$\lim_{N \to \infty} S_N(z) = \frac{1}{1 - z},$$

which means that the series converges and its sum function is $\frac{1}{1-z}$.

When $|z| > 1$,

$$\lim_{N \to \infty} S_N(z) = \infty,$$

which means that the series diverges.

When $z = 1$,

$$\lim_{N\to\infty} S_N(z) = \lim_{N\to\infty} N = \infty,$$

then the series diverges.

When $|z|=1$, while $z\neq 1$, we can suppose that $z=e^{i\theta}$ with $\theta\neq 2k\pi$ ($k=0,\pm 1,\pm 2,\cdots$), then the limit of $z^n=e^{in\theta}$ doesn't exist as n approaches to infinity, and

$$S_N(z) = \frac{1-z^N}{1-z} = \frac{1-e^{iN\theta}}{1-e^{i\theta}}$$

has no limit. Hence the series diverges.

Consequently,

$$\sum_{n=0}^{\infty} z^n = \begin{cases} \dfrac{1}{1-z}, & |z|<1, \\ \text{diverges}, & |z|\geqslant 1. \end{cases}$$

4.2.2 The convergence of power series

Theorem 4.2.1 (Abel's theorem) Consider the power series

$$\sum_{n=0}^{\infty} a_n z^n, \qquad (4.2.3)$$

(1) If it converges at $z=z_1$, $z_1\neq 0$, then it converges absolutely for any z satisfying $|z|<|z_1|$;

(2) If it diverges at $z=z_2$, $z_2\neq 0$, then it diverges for any z satisfying $|z|>|z_2|$.

Proof (1) Since $\sum_{n=0}^{\infty} a_n z_1^n$ converges, it follows that $\lim_{n\to\infty} a_n z_1^n = 0$. The terms $a_n z_1^n$ are thus bounded; that is,

$$|a_n z_1^n| \leqslant M, \quad (n=0,1,2,\cdots), \qquad (4.2.4)$$

for some positive constant M. If $|z|<|z_1|$ and we let ρ denote the modulus $\left|\dfrac{z}{z_1}\right|$, we can see that

$$|a_n z^n| = |a_n z_1^n| \left|\frac{z}{z_1}\right|^n \leqslant M\rho^n, \quad (n=0,1,2,\cdots), \qquad (4.2.5)$$

where $\rho<1$. Now the series whose terms are the real numbers $M\rho^n$ ($n=0,1,2,\cdots$) is a geometric series, which converges when $\rho<1$. Hence, by the comparison test for series of real number, the series $\sum_{n=0}^{\infty} a_n z^n$ absolutely converges in the open disk $|z|<|z_1|$.

(2) Suppose that there exists a point z_3 with $|z_3|>|z_2|$ such that $\sum_{n=0}^{\infty} a_n z_3^n$ converges, then according to part (1) of this theorem, the power series $\sum_{n=0}^{\infty} a_n z^n$ converges absolutely at z_2, which is a contradiction.

Note From Abel's theorem, it can be concluded that:

(1) The series (4.2.3) converges only at $z=0$;

(2) The series (4.2.3) converges on the entire complex plane;

(3) There exists $z_0 \neq 0$, such that the series (4.2.3) converges for $|z| < |z_0|$ and diverges for $|z| > |z_0|$.

The theorem tells us that there exists a number R, so that the series (4.2.3) converges absolutely for $|z| < R$, and it diverges for $|z| > R$. The number R is called the **radius of convergence** of the power series, and the circle $|z| = R$ is called the **circle of convergence**.

To determine the radius of convergence of the power series, we have the following theorems.

Theorem 4.2.2 Consider the power series $\sum_{n=0}^{\infty} a_n z^n$, if

$$\lim_{n \to \infty} \left| \frac{a_{n+1}}{a_n} \right| = \rho, \tag{4.2.6}$$

then the radius of convergence of this series is

$$R = \begin{cases} \dfrac{1}{\rho}, & \text{if } 0 < \rho < +\infty, \\ 0, & \text{if } \rho = +\infty, \\ +\infty, & \text{if } \rho = 0. \end{cases} \tag{4.2.7}$$

Theorem 4.2.3 Consider the power series $\sum_{n=0}^{\infty} a_n z^n$, if

$$\lim_{n \to \infty} \sqrt[n]{|a_n|} = \rho, \tag{4.2.8}$$

then the radius of convergence of this series is

$$R = \begin{cases} \dfrac{1}{\rho}, & \text{if } 0 < \rho < +\infty, \\ 0, & \text{if } \rho = +\infty, \\ +\infty, & \text{if } \rho = 0. \end{cases} \tag{4.2.9}$$

Example 4.2.2 Find the radius of convergence of the following power series:

(1) $\sum_{n=1}^{\infty} \dfrac{z^n}{n^3}$;   (2) $\sum_{n=1}^{\infty} \dfrac{(z-1)^n}{n}$.

Solution (1) Since

$$\lim_{n \to \infty} \left| \frac{a_{n+1}}{a_n} \right| = \lim_{n \to \infty} \left(\frac{n}{n+1} \right)^3 = 1,$$

then the radius of convergence of this series is $R = 1$. The circle of convergence is $|z| = 1$. It is convergent in the domain $|z| < 1$ and divergent when $|z| > 1$.

(2) $\lim_{n \to \infty} \left| \dfrac{a_{n+1}}{a_n} \right| = \lim_{n \to \infty} \dfrac{n}{n+1} = 1$,

then the radius of convergence of this series is $R = 1$. The circle of convergence is $|z - 1| = 1$.

When $z = 0$, the power series becomes $\sum_{n=1}^{\infty} \dfrac{(-1)^n}{n}$, which is a convergent alternating series. When $z = 2$, the power series becomes $\sum_{n=1}^{\infty} \dfrac{1}{n}$, which is a divergent harmonic series. This implies that there are convergent point and divergent point on the circle of convergence.

4.2.3 The operations of power series

This section is devoted mainly to various properties of power series.

Theorem 4.2.4 A power series
$$\sum_{n=0}^{\infty} a_n (z-z_0)^n \qquad (4.2.10)$$
represents a continuous function $S(z)$ at each point inside its circle of convergence $|z-z_0|=R$.

Theorem 4.2.5 The sum $S(z)$ of power series (4.2.10) is analytic at each point z interior to the circle of convergence of that series.

Theorem 4.2.6 Let C denote any contour interior to the circle of convergence of the power series (4.2.10), the series (4.2.10) can be integrated term by term over C; that is,
$$\int_C S(z) dz = \sum_{n=0}^{\infty} a_n \int_C (z-z_0)^n dz. \qquad (4.2.11)$$

Theorem 4.2.7 The power series (4.2.10) can be differentiated term by term at each point z interior to the circle of convergence of that series, that is
$$S'(z) = \sum_{n=1}^{\infty} n a_n (z-z_0)^{n-1}. \qquad (4.2.12)$$

Exercises 4.2

1. Write $z = re^{i\theta}$, where $0 < r < 1$, in the summation formula in example 4.2.1, show that
$$\sum_{n=1}^{\infty} r^n \cos n\theta = \frac{r\cos\theta - r^2}{1 - 2r\cos\theta + r^2}$$
and
$$\sum_{n=1}^{\infty} r^n \sin n\theta = \frac{r\sin\theta}{1 - 2r\cos\theta + r^2}.$$

2. Whether is it possible that the power series $\sum_{n=0}^{\infty} c_n z^n$ is convergent at the point $z=2i$ and is divergent at the point $z=1$? Why?

3. Determine the convergence radius and the circle of convergence of the following power series:

(a) $\sum_{n=1}^{\infty} \frac{i}{n^2} z^n$;

(b) $\sum_{n=0}^{\infty} \frac{n}{2^n} (z-i)^n$;

(c) $\sum_{n=1}^{\infty} \frac{2^n}{n(n+1)} (z+i)^{2n}$;

(d) $\sum_{n=0}^{\infty} (n+1)(n+2) z^{2n}$;

(e) $\sum_{n=1}^{\infty} n^{-n} (z-2i)^n$;

(f) $\sum_{n=0}^{\infty} z^{2n+1}$.

4. Suppose that the sequence $\rho_n = \left|\dfrac{a_{n+1}}{a_n}\right|$ converges to ρ as $n \to \infty$, show that the convergence radius of the following three power series are all $R = \dfrac{1}{\rho}$:

$$\sum_{n=0}^{\infty} a_n (z - z_0)^n, \quad \sum_{n=0}^{\infty} \dfrac{a_n}{n+1} (z - z_0)^{n+1}, \quad \sum_{n=0}^{\infty} n a_n (z - z_0)^{n-1}.$$

4.3 Taylor series

4.3.1 Taylor's theorem

We turn now to Taylor's theorem, which is one of the most important results of this chapter.

Theorem 4.3.1 (Taylor's theorem) Suppose that a function f is analytic throughout a disk $|z - z_0| < R_0$, centered at z_0 and with radius R_0 (Fig. 4.3.1). Then $f(z)$ has the power series representation

$$f(z) = \sum_{n=0}^{\infty} a_n (z - z_0)^n, \quad (|z - z_0| < R_0) \tag{4.3.1}$$

where

$$a_n = \dfrac{f^{(n)}(z_0)}{n!}, \quad (n = 0, 1, 2, \cdots). \tag{4.3.2}$$

Proof We first prove the theorem when $z_0 = 0$, in which case series (4.3.1) becomes

$$f(z) = \sum_{n=0}^{\infty} \dfrac{f^{(n)}(0)}{n!} z^n, \quad (|z| < R_0) \tag{4.3.3}$$

and is called a **Maclaurin series**. The proof when z_0 is arbitrary will follow as an immediate consequence.

To derive the representation (4.3.3), we write $|z| = r$ and let C_0 denote any positively oriented circle $|z| = r_0$, where $r < r_0 < R_0$ (See Fig. 4.3.1).

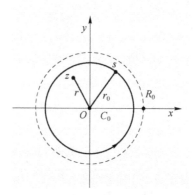

Fig. 4.3.1

Since f is analytic inside and on the circle C_0 and the point z is interior to C_0, Cauchy integral formula is applied:

$$f(z) = \frac{1}{2\pi i} \int_{C_0} \frac{f(s) ds}{s-z}. \qquad (4.3.4)$$

Now the factor $\frac{1}{s-z}$ in the integrand here can be put in the form

$$\frac{1}{s-z} = \frac{1}{s} \frac{1}{1 - \left(\frac{z}{s}\right)}, \qquad (4.3.5)$$

and we know from the example 4.2.1 in Section 4.2.1 that

$$\frac{1}{1-z} = \sum_{n=0}^{N-1} z^n + \frac{z^N}{1-z}, \qquad (4.3.6)$$

when z is any complex number other than unity. Replacing z by $\frac{z}{s}$ in expression (4.3.6), then, we can rewrite equation (4.3.5) as

$$\frac{1}{s-z} = \sum_{n=0}^{N-1} \frac{1}{s^{n+1}} z^n + z^N \frac{1}{(s-z)s^N}.$$

Multiplying through this equation by $f(s)$ and then integrating each side with respect to s around C_0, we find that

$$\int_{C_0} \frac{f(s) ds}{s-z} = \sum_{n=0}^{N-1} \int_{C_0} \frac{f(s) ds}{s^{n+1}} z^n + z^N \int_{C_0} \frac{f(s) ds}{(s-z)s^N}.$$

In view of expression (4.3.4) and the fact

$$\frac{1}{2\pi i} \int_{C_0} \frac{f(s) ds}{s^{n+1}} = \frac{f^{(n)}(0)}{n!}, \quad (n = 0, 1, 2, \cdots),$$

this reduces, after we multiply through by $\frac{1}{2\pi i}$, to

$$f(z) = \sum_{n=0}^{N-1} \frac{f^{(n)}(0)}{n!} z^n + \rho_n(z), \qquad (4.3.7)$$

where

$$\rho_n(z) = \frac{z^N}{2\pi i} \int_{C_0} \frac{f(s) ds}{(s-z)s^N}. \qquad (4.3.8)$$

Representation (4.3.3) now follows once it is shown that

$$\lim_{n \to \infty} \rho_n(z) = 0. \qquad (4.3.9)$$

To accomplish this, we recall that $|z| = r$ and that C_0 has radius r_0, where $r_0 > r$. Then, if s is a point on C_0, we can see that

$$|s-z| \geq ||s| - |z|| = r_0 - r.$$

Consequently, if M denotes the maximum value of $|f(s)|$ on C_0,

$$|\rho_n(z)| \leq \frac{r^N}{2\pi} \frac{M}{(r_0-r)r_0^N} 2\pi r_0 = \frac{M r_0}{r_0 - r} \left(\frac{r}{r_0}\right)^N.$$

In as much as $\frac{r}{r_0} < 1$, limit (4.3.9) clearly holds.

To verify the theorem when the disk of radius R_0 is centered at an arbitrary point z_0, we suppose that f is analytic when $|z-z_0|<R_0$ and note that the composite function $f(z+z_0)$ must be analytic when $|(z+z_0)-z_0|<R_0$. This last inequality is, of course, just $|z|<R_0$; and, if we write $g(z)=f(z+z_0)$, the analyticity of g in the disk $|z|<R_0$ ensures the existence of a Maclaurin series representation:

$$g(z)=\sum_{n=0}^{\infty}\frac{g^{(n)}(0)}{n!}z^n,\quad (|z|<R_0),$$

that is

$$f(z+z_0)=\sum_{n=0}^{\infty}\frac{f^{(n)}(z_0)}{n!}z^n,\quad (|z|<R_0).$$

After replacing z by $z-z_0$ in this equation and its condition of validity, we have the desired Taylor series expansion (4.3.1).

This is the expansion of $f(z)$ into a Taylor series about the point z_0. It is the familiar Taylor series from calculus, adapted to functions of a complex variable. With the agreement that $f^{(0)}(z_0)=f(z_0)$ and $0!=1$. Series (4.3.1) can, of course, be written as

$$f(z)=f(z_0)+\frac{f'(z_0)}{1!}(z-z_0)+\frac{f''(z_0)}{1!}(z-z_0)^2+\cdots,\quad (|z-z_0|<R_0).$$

Any function which is analytic at point z_0 must have a Taylor series about z_0. For, if f is analytic at z_0, it is analytic throughout some neighborhood $|z-z_0|<\varepsilon$ of that point; and ε may serve as the value of R_0 in the statement of Taylor's theorem. Also, if f is entire, R_0 can be chosen arbitrarily large, and the condition of validity becomes $|z-z_0|<\infty$. The series then converges to $f(z)$ at each point z in the finite plane.

4.3.2 Taylor expansions of analytic functions

From the Taylor theorem, we can see that if $f(z)$ is analytic at z_0, then it is analytic in some neighborhood of z_0 and the expansion (4.3.1) is valid in that neighborhood. Then we can derive the expansion directly by (4.3.2).

In the following examples, we use the formula in Taylor's theorem to find the Maclaurin series expansions of some fairly simple functions, and we emphasize the use of those expansions in finding other representations.

Example 4.3.1 Find the Maclaurin series representation of $f(z)=e^z$.

Solution Since the function $f(z)=e^z$ is entire, it has a Maclaurin series representation which is valid for all z. Here $f^{(n)}(z)=e^z$; and, because $f^{(n)}(0)=1$, it follows that

$$e^z=\sum_{n=0}^{\infty}\frac{z^n}{n!},\quad (|z|<\infty). \tag{4.3.10}$$

Note that if $=x+i0$, expansion (4.3.10) becomes

$$e^x=\sum_{n=0}^{\infty}\frac{x^n}{n!},\quad (-\infty<x<\infty).$$

The entire function z^2e^{3z} also has a Maclaurin series expansion. The simplest way to obtain it is to replace z by $3z$ on each side of equation (4.3.10) and then multiply through the resulting equation by z^2:

$$z^2 e^{3z} = \sum_{n=0}^{\infty} \frac{3^n}{n!} z^{n+2}, \quad (|z|<\infty).$$

Finally, if we replace n by $n-2$ here, we have

$$z^2 e^{3z} = \sum_{n=0}^{\infty} \frac{3^{n-2}}{(n-2)!} z^n, \quad (|z|<\infty).$$

Example 4.3.2 Find the Maclaurin series representation of $f(z) = \sin z$.

Solution One can use expansion (4.3.10) and the definition

$$\sin z = \frac{e^{iz} - e^{-iz}}{2i}$$

to find the Maclaurin series for the entire function $f(z) = \sin z$. To give the details, we refer to expansion (4.3.10) and write

$$\sin z = \frac{1}{2i} \left[\sum_{n=0}^{\infty} \frac{(iz)^n}{n!} - \sum_{n=0}^{\infty} \frac{(-iz)^n}{n!} \right] = \frac{1}{2i} \sum_{n=0}^{\infty} [1 - (-1)^n] \frac{i^n z^n}{n!}, \quad (|z|<\infty).$$

Since $1-(-1)^n = 0$ when n is even, and so we can replace n by $2n+1$ in this last series:

$$\sin z = \frac{1}{2i} \sum_{n=0}^{\infty} [1 - (-1)^{2n+1}] \frac{i^{2n+1} z^{2n+1}}{(2n+1)!}, \quad (|z|<\infty).$$

Notice that $1-(-1)^{2n+1}=2$ and $i^{2n+1}=(i^2)^n i=(-1)^n i$, it is obvious that

$$\sin z = \sum_{n=0}^{\infty} (-1)^n \frac{z^{2n+1}}{(2n+1)!}, \quad (|z|<\infty). \tag{4.3.11}$$

By term by term differentiation, we differentiate each side of equation (4.3.11) and write

$$\cos z = \sum_{n=0}^{\infty} \frac{(-1)^n}{(2n+1)!} \frac{d}{dz} z^{2n+1} = \sum_{n=0}^{\infty} (-1)^n \frac{2n+1}{(2n+1)!} z^{2n},$$

that is,

$$\cos z = \sum_{n=0}^{\infty} (-1)^n \frac{z^{2n}}{(2n)!}, \quad (|z|<\infty). \tag{4.3.12}$$

Example 4.3.3 Find the Maclaurin series representation of $f(z) = \sinh z$.

Solution Because $\sinh z = -i\sin(iz)$, we need only replace z by iz on each side of equation (4.3.11) and multiply throughout the result by $-i$ to see that

$$\sinh z = \sum_{n=0}^{\infty} \frac{z^{2n+1}}{(2n+1)!}, \quad (|z|<\infty). \tag{4.3.13}$$

Likewise, since $\cosh z = \cos(iz)$, it follows from expansion (4.3.12) that

$$\cosh z = \sum_{n=0}^{\infty} \frac{z^{2n}}{(2n)!}, \quad (|z|<\infty). \tag{4.3.14}$$

Another important Maclaurin series representation is

$$\frac{1}{1-z} = \sum_{n=0}^{\infty} z^n, \quad (|z|<1). \tag{4.3.15}$$

The derivatives of the function $f(z) = \frac{1}{1-z}$, which fails to be analytic at $z=1$ are

$$f^{(n)}(z) = \frac{n!}{(1-z)^{n+1}}, \quad (n = 0, 1, 2, \cdots),$$

and, in particular, $f^{(n)}(0)=n!$ $(n=0, 1, 2, \cdots)$. Note that expansion (4.3.15) gives us the sum of an infinite **geometric series**, where z is the common ratio of adjacent terms:

$$1+z+z^2+z^3+\cdots=\frac{1}{1-z}, \quad (|z|<1).$$

If we substitute $-z$ for z in equation (4.3.15) and its condition of validity, and notice that $|z|<1$ when $|-z|<1$, we see that

$$\frac{1}{1+z}=\sum_{n=0}^{\infty}(-1)^n z^n, \quad (|z|<1). \tag{4.3.16}$$

Example 4.3.4 Find the Taylor series representation of $f(z)=\dfrac{1}{z}$ at $z=1$.

Solution Since

$$\frac{1}{z}=\frac{1}{1+(z-1)},$$

then we replace the variable z in equation (4.3.16) by $z-1$, and have the Taylor series representation

$$\frac{1}{z}=\sum_{n=0}^{\infty}(-1)^n(z-1)^n, \quad (|z-1|<1).$$

The condition of validity follows from the one associated with expansion (4.3.16) as $|z-1|<1$.

Example 4.3.5 Find the Maclaurin series representation of $f(z)=\text{Log}(1+z)$.

Solution Because $f(z)$ is analytic when $|1+z|>0$ and $-\pi<\text{Arg}(1+z)<\pi$, and $z=-1$ is a singular point of $f(z)$, then we can get the Maclaurin series representation of $f(z)$ when $|z|<1$.

Hence,

$$\text{Log}(1+z)=\int_0^z \frac{1}{1+z}dz=\int_0^z \sum_{n=0}^{\infty}(-1)^n z^n dz$$

$$=\sum_{n=0}^{\infty}\frac{(-1)^n z^{n+1}}{n+1}, \quad (|z|<1). \tag{4.3.17}$$

Example 4.3.6 Find the Maclaurin series representation of the principle branch of $f(z)=(1+z)^\alpha$.

Solution Since P. V. $f(z)=\exp(\alpha\text{Log}(1+z))$, then we can get the Maclaurin series representation of $f(z)$ when $|z|<1$.

Let $\varphi(z)=\text{Log}(1+z)$, then $1+z=e^{\varphi(z)}$, and

$$f(z)=\exp(\alpha\varphi(z)).$$

Differentiating it on both sides, we have

$$f'(z)=\alpha\exp(\alpha\varphi(z))\varphi'(z)=\alpha\frac{1}{1+z}\exp(\alpha\varphi(z))$$

$$=\frac{\alpha}{\exp(\varphi(z))}\exp(\alpha\varphi(z)),$$

that is,

$$f'(z)=\alpha\exp((\alpha-1)\varphi(z)).$$

Differentiating it again, we have
$$f''(z) = \alpha(\alpha-1)\exp((\alpha-2)\varphi(z)),$$
$$\vdots$$
$$f^{(n)}(z) = \alpha(\alpha-1)\cdots(\alpha-n+1)\exp((\alpha-n)\varphi(z)),$$
$$\vdots$$

Let $z=0$, we have
$$f(0) = 1, \quad f'(0) = \alpha, \quad f''(0) = \alpha(\alpha-1), \quad \cdots,$$
$$f^{(n)}(0) = \alpha(\alpha-1)\cdots(\alpha-n+1), \quad \cdots.$$

Hence,
$$(1+z)^\alpha = 1 + \alpha z + \frac{\alpha(\alpha-1)}{2!}z^2 + \cdots + \frac{\alpha(\alpha-1)\cdots(\alpha-n+1)}{n!}z^n + \cdots, \quad (|z|<1). \tag{4.3.18}$$

Specially, when $\alpha = -1$, we have
$$\frac{1}{1+z} = \sum_{n=0}^{\infty} (-z)^n, \quad (|z|<1).$$

Example 4.3.7 Find the Taylor series of $f(z) = e^z$ at the point $z=-1$.

Solution Let $t = z+1$, the $z = t-1$, and $f(t) = e^{t-1} = \frac{1}{e}e^t$.

Since $e^z = \sum_{n=0}^{\infty} \frac{z^n}{n!}, (|z|<\infty)$, then
$$f(t) = \frac{1}{e}e^t = \frac{1}{e}\sum_{n=0}^{\infty} \frac{t^n}{n!}, \quad (|t|<\infty).$$

Hence
$$f(z) = \frac{1}{e}\sum_{n=0}^{\infty} \frac{(z+1)^n}{n!}, \quad (|z+1|<\infty).$$

Example 4.3.8 Find the Taylor series of $f(z) = \frac{2z}{z+1}$ at the point $z=2$.

Solution Since
$$f(z) = \frac{2z}{z+1} = 2 - \frac{2}{z+1} = 2 - \frac{2}{3+(z-2)} = 2 - \frac{2}{3\left[1+\frac{(z-2)}{3}\right]},$$

then according to (4.3.16), we have
$$f(z) = 2 - \frac{2}{3\left[1+\frac{(z-2)}{3}\right]} = 2 - \frac{2}{3}\sum_{n=0}^{\infty} (-1)^n \left(\frac{z-2}{3}\right)^n, \quad (|z-2|<3).$$

Example 4.3.9 Find the Maclaurin series of $f(z) = \arctan z$.

Solution Since
$$f'(z) = (\arctan z)' = \frac{1}{1+z^2},$$

and
$$\frac{1}{1+z^2} = \sum_{n=0}^{\infty} (-1)^n z^{2n},$$

then we have

$$f(z) = \arctan z = \int_0^z \frac{1}{1+z^2} dz$$

$$= \int_0^z \sum_{n=0}^{\infty} (-1)^n z^{2n} dz = \sum_{n=0}^{\infty} \frac{(-1)^n}{2n+1} z^{2n+1}, \quad (|z|<1).$$

Example 4.3.10 Find the Maclaurin series of $f(z) = \frac{1+2z^2}{z^3+z^5}$.

Solution

$$f(z) = \frac{1+2z^2}{z^3+z^5} = \frac{1}{z^3} \frac{2(1+z^2)-1}{1+z^2} = \frac{1}{z^3}\left(2 - \frac{1}{1+z^2}\right).$$

We cannot find a Maclaurin series for $f(z)$ since it is not analytic at $z=0$. But we do know from expansion (4.3.15) that

$$\frac{1}{1+z^2} = 1 - z^2 + z^4 - z^6 + z^8 + \cdots, \quad (|z|<1).$$

Hence, when $0 < |z| < 1$,

$$f(z) = \frac{1}{z^3}(2 - 1 + z^2 - z^4 + z^6 - z^8 + \cdots) = \frac{1}{z^3} + \frac{1}{z} - z + z^3 - z^5 + \cdots.$$

We call such terms as $\frac{1}{z^3}$ and $\frac{1}{z}$ negative powers of z since they can be written as z^{-3} and z^{-1}, respectively. The theory of expansions involving negative powers of $z - z_0$ will be discussed in the next section.

Exercises 4.3

1. Obtain the Maclaurin series representation

$$z\cosh(z^2) = \sum_{n=0}^{\infty} \frac{z^{4n+1}}{(2n)!}, \quad (|z|<\infty).$$

2. Obtain the Taylor series

$$e^z = e \sum_{n=0}^{\infty} \frac{(z-1)^n}{n!}, \quad (|z-1|<\infty)$$

for the function $f(z) = e^z$ by

(a) using $f^{(n)}(1)$ ($n=0, 1, 2, \cdots$); (b) writing $e^z = e^{z-1} e$.

3. Find the Maclaurin series expansion of the following functions:

(a) $f(z) = \frac{z}{z^4+9}$; (b) $f(z) = \frac{1}{(1+z)^2}$;

(c) $f(z) = \frac{1}{z^2-5z+6}$; (d) $f(z) = e^z \cos z$;

(e) $f(z) = \sin(2z)^2$.

4. Write the Maclaurin series representation of the function $f(z) = \sin(z^2)$, and point out how it follows that

$$f^{(4n)}(0) = 0$$

and
$$f^{(2n+1)}(0)=0, \quad (n=0, 1, 2, \cdots).$$

5. Derive the Taylor series representation of the following functions in the given points z_0:

(a) $f(z)=\dfrac{z}{z+2}$ $(z_0=1)$;

(b) $f(z)=\dfrac{z}{z+i}$ $(z_0=i)$;

(c) $f(z)=\cos z$ $(z_0=i)$;

(d) $f(z)=\dfrac{1}{1-z}$ $(z_0=i)$;

(e) $f(z)=\dfrac{1}{3-z}$ $(z_0=1+i)$;

(f) $f(z)=\cos z$ $\left(z_0=\dfrac{\pi}{2}\right)$;

(g) $f(z)=\sinh z$ $(z_0=\pi i)$.

6. Show that when $z\neq 0$,

(a) $\dfrac{e^z}{z^2}=\dfrac{1}{z^2}+\dfrac{1}{z}+\dfrac{1}{2!}+\dfrac{z}{3!}+\dfrac{z^2}{4!}+\cdots$,

(b) $\dfrac{\sin(z^2)}{z^4}=\dfrac{1}{z^2}-\dfrac{z^2}{3}+\dfrac{z^6}{5}-\dfrac{z^{100}}{7}+\cdots$.

7. Show that when $0<|z|<4$,
$$\frac{1}{4z-z^2}=\frac{1}{4z}+\sum_{n=0}^{\infty}\frac{z^n}{4^{n+2}}.$$

4.4 Laurent series

4.4.1 Laurent's theorem

If a function f fails to be analytic at a point z_0, we cannot apply Taylor's theorem at that point. It is often possible, however, to find a series representation for $f(z)$ involving both positive and negative powers of $z-z_0$. (see Example 4.3.10 in Section 4.3) We now present the theory of such representations, which is called Laurent's theorem.

Theorem 4.4.1 (Laurent's theorem) Suppose that a function f is analytic throughout an annular domain $R_1<|z-z_0|<R_2$, centered at z_0, and let C denote any positively oriented simple closed contour around z_0 and lying in that domain (Fig. 4.4.1). Then, at each point in the domain, $f(z)$ has the series representation

$$f(z)=\sum_{n=0}^{\infty}a_n(z-z_0)^n+\sum_{n=1}^{\infty}\frac{b_n}{(z-z_0)^n}, \quad (R_1<|z-z_0|<R_2) \qquad (4.4.1)$$

where

$$a_n=\frac{1}{2\pi i}\int_C\frac{f(z)\mathrm{d}z}{(z-z_0)^{n+1}}, \quad (n=0, 1, 2, \cdots) \qquad (4.4.2)$$

and

$$b_n=\frac{1}{2\pi i}\int_C\frac{f(z)\mathrm{d}z}{(z-z_0)^{-n+1}}, \quad (n=0, 1, 2, \cdots). \qquad (4.4.3)$$

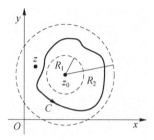

Fig. 4.4.1

Or it can be expressed by

$$f(z) = \sum_{n=-\infty}^{\infty} c_n (z-z_0)^n, \quad (R_1 < |z-z_0| < R_2) \tag{4.4.4}$$

where

$$c_n = \frac{1}{2\pi i} \int_C \frac{f(z)\,dz}{(z-z_0)^{n+1}}, \quad (n=0, \pm 1, \pm 2, \cdots). \tag{4.4.5}$$

In either of the forms (4.4.1) or (4.4.4), it is called a **Laurent series**.

Proof We shall prove Laurent's theorem first when $z_0 = 0$, in which case the annulus is centered at the origin. The verification of the theorem when z_0 is arbitrary will follow readily.

We start the proof by forming a closed annular region $r_1 < |z| < r_2$ that is contained in the domain $R_1 < |z| < R_2$ and whose interior contains both the point z and the contour C (Fig. 4.4.2). We let C_1 and C_2 denote the circles $|z|=r_1$ and $|z|=r_2$, respectively, and we assign those two circles a positive orientation. Observe that f is analytic on C_1 and C_2, as well as in the annular domain between them.

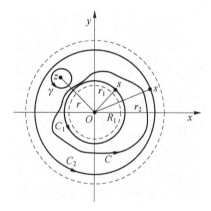

Fig. 4.4.2

Next, we construct a positively oriented circle γ with center at z and small enough to be completely contained in the interior of the annular region $r_1 \leqslant |z| \leqslant r_2$, as shown in Fig. 4.4.2. It then follows from the extension of the Cauchy-Goursat theorem to integrals of analytic functions around the oriented boundaries of multiply connected domains that

$$\int_{C_2} \frac{f(s)\,ds}{s-z} - \int_{C_1} \frac{f(s)\,ds}{s-z} - \int_{\gamma} \frac{f(s)\,ds}{s-z} = 0.$$

But, according to the Cauchy integral formula, the value of the third integral here is $2\pi i f(z)$. Hence

$$f(z) = \frac{1}{2\pi i}\int_{C_2}\frac{f(s)\,ds}{s-z} - \frac{1}{2\pi i}\int_{C_1}\frac{f(s)\,ds}{s-z}. \qquad (4.4.6)$$

Now the factor $\frac{1}{s-z}$ in the first of these integrals is the same as in expression (4.3.7) in Section 4.3, where Taylor's theorem was proved; and we shall need here the expansion

$$\frac{1}{s-z} = \sum_{n=0}^{N-1}\frac{1}{s^{n+1}}z^n + z^N\frac{1}{(s-z)s^N}, \qquad (4.4.7)$$

which was used in that earlier section. As for the factor $\frac{1}{s-z}$ in the second integral, an interchange of s and z in equation (4.3.7) reveals that

$$\frac{1}{z-s} = \sum_{n=0}^{N-1}\frac{1}{s^{-n}}\frac{1}{z^{n+1}} + \frac{1}{z^N}\frac{s^N}{z-s}.$$

If we replace the index of summation n here by -1, this expression takes the form

$$\frac{1}{z-s} = \sum_{n=1}^{N-1}\frac{1}{s^{-n+1}}\frac{1}{z^n} + \frac{1}{z^N}\frac{s^N}{z-s}, \qquad (4.4.8)$$

which is to be used in what follows.

Multiplying through equations (4.4.7) and (4.4.8) by $\frac{f(s)}{2\pi i}$ and then integrating each side of the resulting equation with respect to s around C_2 and C_1, respectively, we find from expression (4.4.6) that

$$f(z) = \sum_{n=0}^{N-1}a_n z^n + \rho_N(z) + \sum_{n=1}^{N-1}\frac{b_n}{z^n} + \sigma_N(z), \qquad (4.4.9)$$

where the numbers a_n ($n=0, 1, 2, \cdots, N-1$) and b_n ($n=0, 1, 2, \cdots, N$) are given by the equations

$$a_n = \frac{1}{2\pi i}\int_{C_2}\frac{f(s)\,ds}{s^{n+1}}, \qquad b_n = \frac{1}{2\pi i}\int_{C_1}\frac{f(s)\,ds}{s^{-n+1}}, \qquad (4.4.10)$$

and where

$$\rho_N(z) = \frac{z^N}{2\pi i}\int_{C_2}\frac{f(s)\,ds}{(s-z)^{n+1}}, \qquad \sigma_N(z) = \frac{1}{2\pi i z^N}\int_{C_1}\frac{s^N f(s)\,ds}{z-s}.$$

As N tends to ∞, expression (4.4.9) evidently takes the proper form of a Laurent series in the domain $R_1 < |z| < R_2$, provided that

$$\lim_{N\to\infty}\rho_N(z) = 0 \quad \text{and} \quad \lim_{N\to\infty}\sigma_N(z) = 0. \qquad (4.4.11)$$

These limits are readily established by a method already used in the proof of Taylor's theorem in Section 4.3. We write $|z|=r$, so that $r_1 < r < r_2$, and let M denote the maximum value of $|f(s)|$ on C_1 and C_2. We also note that if s is a point on C_2, then $|s-z| > r_2 - r$; and if s is on C_1, $|z-s| > r - r_1$. This enables us to write

$$|\rho_N(z)| \leqslant \frac{Mr_2}{r_2-r}\left(\frac{r}{r_2}\right)^N \quad \text{and} \quad |\sigma_N(z)| \leqslant \frac{Mr_1}{r-r_1}\left(\frac{r_1}{r}\right)^N.$$

Since $\frac{r}{r_2} < 1$ and $\frac{r_1}{r} < 1$, it is now clear that both $\rho_N(z)$ and $\sigma_N(z)$ tend to zero as N tends to infinity.

Finally, we need only recall Corollary 3.3.3 in Section 3.3 to see that the contours used in integrals (4.4.10) may be replaced by the contour C. This completes the proof of Laurent's theorem when $z_0 = 0$ since, if z is used instead of s as the variable of integration, expressions (4.4.10) for the coefficients a_n and b_n are the same as expressions (4.4.2) and (4.4.3) when $z_0 = 0$ there.

To extend the proof to the general case in which z_0 is an arbitrary point in the finite plane, we let f be a function satisfying the condition in the theorem; and, just as we did in the proof of Taylor's theorem, we write $g(z) = f(z + z_0)$. Since $f(z)$ is analytic in the annulus $R_1 < |z - z_0| < R_2$, the function $f(z + z_0)$ is analytic when $R_1 < |z + z_0 - z_0| < R_2$. That is, g is analytic in the annulus $R_1 < |z| < R_2$, which is centered at the origin. Now the simple closed contour C in the statement of the theorem has some parametric representation $z = z(t) \, (a \leqslant t \leqslant b)$, where

$$R_1 < |z(t) - z_0| < R_2 \tag{4.4.12}$$

for all t in the interval $a \leqslant t \leqslant b$. Hence if Γ denotes the path

$$z = z(t) - z_0, \quad (a \leqslant t \leqslant b), \tag{4.4.13}$$

Γ is not only a simple closed contour but, in view of inequalities (4.4.12), it lies in the domain $R_1 < |z| < R_2$. Consequently, $g(z)$ has a Laurent series representation

$$g(z) = \sum_{n=0}^{N-1} a_n z^n + \sum_{n=1}^{N-1} \frac{b_n}{z^n}, \quad (R_1 < |z| < R_2), \tag{4.4.14}$$

where

$$a_n = \frac{1}{2\pi i} \int_\Gamma \frac{g(z) \, dz}{z^{n+1}}, \quad (n = 0, 1, 2, \cdots), \tag{4.4.15}$$

$$b_n = \frac{1}{2\pi i} \int_\Gamma \frac{g(z) \, dz}{z^{-n+1}}, \quad (n = 1, 2, \cdots). \tag{4.4.16}$$

Representation (4.4.1) is obtained if we write $f(z + z_0)$ instead of $g(z)$ in equation (4.4.14) and then replace z by $z + z_0$ in the resulting equation, as well as in the condition of validity $R_1 < |z| < R_2$. Expression (4.4.15) for the coefficients a_n is, moreover, the same as expression (4.4.2) since

$$\int_\Gamma \frac{g(z) \, dz}{z^{n+1}} = \int_a^b \frac{f[z(t)] z'(t)}{[z(t) - z_0]^{n+1}} \, dt = \int_C \frac{f(z) \, dz}{(z - z_0)^{n+1}}.$$

Similarly, the coefficients b_n in expression (4.4.16) are the same as those in expression (4.4.3).

Observe that the integrand in expression (4.4.3) can be written as $f(z)(z - z_0)^{n-1}$. Thus it is clear that when f is actually analytic throughout the disk $|z - z_0| < R_2$, this integrand is too. Hence all of the coefficients b_n are zero; and because

$$\frac{1}{2\pi i} \int_C \frac{f(z) \, dz}{(z - z_0)^{-n+1}} = \frac{f^{(n)}(z_0)}{n!}, \quad (n = 0, 1, 2, \cdots),$$

expansion (4.4.1) reduces to a Taylor series about z_0.

If, however, f fails to be analytic at z_0 but is otherwise analytic in the disk $|z - z_0| < R_2$, the radius R_1 can be chosen arbitrarily small. Representation (4.4.1) is then valid in the punctured disk $0<|z-z_0|<R_2$. Similarly, if f is analytic at each point in the finite plane exterior to the circle $|z-z_0|=R_1$, the condition of validity is $R_1<|z-z_0|<\infty$. Observe that if f is analytic everywhere in the finite plane except at z_0, series (4.4.1) is valid at each point of analyticity, or when $0<|z-z_0|<\infty$.

4.4.2 Laurent series expansion of analytic functions

The coefficients in a Laurent series are generally found by means other than by appealing directly to their integral representations. This is illustrated in the examples below, where it is always assumed that, when the annular domain is specified, a Laurent series for a given function is unique. As was the case with Taylor series.

Example 4.4.1 Find the Laurent series of $f(z) = e^{\frac{1}{z}}$ in the annular domain $0<|z|<\infty$.

Solution Replacing z by $\dfrac{1}{z}$ in the Maclaurin series expansion

$$e^z = \sum_{n=0}^{\infty} \frac{z^n}{n!} = 1 + \frac{z}{1!} + \frac{z^2}{2!} + \frac{z^3}{3!} + \cdots, \quad (|z|<\infty),$$

we have the Laurent series representation

$$e^{\frac{1}{z}} = \sum_{n=0}^{\infty} \frac{1}{n!z^n} = 1 + \frac{1}{1!z} + \frac{1}{2!z^2} + \frac{1}{3!z^3} + \cdots, \quad (0<|z|<\infty).$$

Note that no positive powers of z appear here, the coefficients of the positive powers being zero. Note, too, that the coefficient of $\dfrac{1}{z}$ is unity; and, according to Laurent's theorem, that coefficient is the number

$$b_1 = \frac{1}{2\pi i} \int_C e^{\frac{1}{z}} dz,$$

where C is any positively oriented simple closed contour around the origin. Since $b_1 = 1$, then,

$$\int_C e^{\frac{1}{z}} dz = 2\pi i.$$

This method of evaluating certain integrals around simple closed contours will be developed in considerable detail in next chapter.

Example 4.4.2 Find the Laurent series of $f(z) = \dfrac{-1}{(z-1)(z-2)}$ in the following annular domains: (1) $0<|z|<1$; (2) $1<|z|<2$; (3) $2<|z|<\infty$.

Solution Since

$$f(z) = \frac{-1}{(z-1)(z-2)} = \frac{1}{z-1} - \frac{1}{z-2},$$

then

(1) in the annular domain $0<|z|<1$, the series representation is a Maclaurin series. To find it, we write

$$f(z)=-\frac{1}{1-z}+\frac{1}{2}\cdot\frac{1}{1-\frac{z}{2}},$$

and observe that $|z|<1$ and $\left|\frac{z}{2}\right|<1$ in this domain, then

$$f(z)=-\sum_{n=0}^{\infty}z^n+\sum_{n=0}^{\infty}\frac{z^n}{2^{n+1}}=\sum_{n=0}^{\infty}(2^{-n-1}-1)z^n,\quad(|z|<1).$$

(2) in the annular domain $1<|z|<2$, it is clear that $\left|\frac{1}{z}\right|<1$ and $\left|\frac{z}{2}\right|<1$, we can rewrite the function as

$$f(z)=\frac{1}{z}\frac{1}{1-\frac{1}{z}}+\frac{1}{2}\frac{1}{1-\frac{z}{2}}.$$

It follows that

$$f(z)=\sum_{n=0}^{\infty}\frac{1}{z^{n+1}}+\sum_{n=0}^{\infty}\frac{z^n}{2^{n+1}},\quad(1<|z|<2).$$

If we replace the index of summation n in the first of these series by $n-1$ and then interchange the two series, we arrive at an expansion having the same form as the one in the statement of Laurent's theorem:

$$f(z)=\sum_{n=0}^{\infty}\frac{z^n}{2^{n+1}}+\sum_{n=1}^{\infty}\frac{1}{z^n},\quad(1<|z|<2).$$

(3) in the annular domain $2<|z|<\infty$, it is clear that $\left|\frac{1}{z}\right|<1$ and $\left|\frac{2}{z}\right|<1$, we can rewrite the function as

$$f(z)=\frac{1}{z}\frac{1}{1-\frac{1}{z}}-\frac{1}{z}\frac{1}{1-\frac{2}{z}}.$$

It follows that

$$f(z)=\sum_{n=0}^{\infty}\frac{1}{z^{n+1}}-\sum_{n=0}^{\infty}\frac{2^n}{z^{n+1}}=\sum_{n=0}^{\infty}\frac{1-2^n}{z^{n+1}},\quad(2<|z|<\infty).$$

That is,

$$f(z)=\sum_{n=0}^{\infty}\frac{1-2^{n-1}}{z^n},\quad(2<|z|<\infty).$$

Example 4.4.3 Find the Laurent series of $f(z)=\frac{1}{(z-1)(z-2)}$ at the point $z_0=1$.

Solution Since the singular points of $f(z)$ are $z_1=1$ and $z_2=2$, then there are two annular domain around $z_0=1$ in which the function is analytic, that is $0<|z-1|<1$ and $1<|z-1|<\infty$.

When $0<|z-1|<1$,

$$f(z)=\frac{1}{1-z}-\frac{1}{2-z}=-\frac{1}{z-1}-\frac{1}{1-(z-1)}=-\frac{1}{z-1}-\sum_{n=0}^{\infty}(z-1)^n;$$

When $1<|z-1|<\infty$,

$$f(z)=\frac{1}{1-z}-\frac{1}{2-z}=-\frac{1}{z-1}+\frac{1}{(z-1)-1}$$

$$= -\frac{1}{z-1} + \frac{1}{z-1} \frac{1}{1-\frac{1}{z-1}} = -\frac{1}{z-1} + \frac{1}{z-1} \sum_{n=0}^{\infty} \left(\frac{1}{z-1}\right)^n$$

$$= \sum_{n=1}^{\infty} \left(\frac{1}{z-1}\right)^{n+1}.$$

Example 4.4.4 Find the Laurent series of $f(z) = \sin \frac{z}{z-1}$ in the domain $1 < |z-1| < \infty$.

Solution $f(z) = \sin \frac{z}{z-1} = \sin\left(1 + \frac{1}{z-1}\right)$

$$= \sin 1 \cos \frac{1}{z-1} + \cos 1 \sin \frac{1}{z-1}$$

$$= \sin 1 \sum_{n=0}^{\infty} (-1)^n \frac{1}{(2n)!} \frac{1}{(z-1)^{2n}} +$$

$$\cos 1 \sum_{n=0}^{\infty} (-1)^n \frac{1}{(2n+1)!} \frac{1}{(z-1)^{2n+1}}.$$

Example 4.4.5 The function $f(z) = \frac{1}{(z-i)^2}$ is already in the form of a Laurent series, where $z_0 = i$. That is,

$$f(z) = \sum_{n=-\infty}^{\infty} c_n (z-i)^n, \quad (0 < |z-i| < \infty),$$

where $c_{-2} = 1$ and all of the other coefficients are zero. From formula (4.4.5), for the coefficients in a Laurent series, we know that

$$c_n = \frac{1}{2\pi i} \int_C \frac{dz}{(z-i)^{n+3}}, \quad (n = 0, \pm 1, \pm 2, \cdots),$$

where C is, for instance, any positively oriented circle $|z-i| = R$ about the point $z_0 = i$. Thus

$$\int_C \frac{dz}{(z-i)^{n+3}} = \begin{cases} 0 & \text{when } n \neq -2, \\ 2\pi i & \text{when } n = -2. \end{cases}$$

Exercises 4.4

1. Find a representation for the function

$$f(z) = \frac{1}{1+z} = \frac{1}{z} \frac{1}{1+\frac{1}{z}}$$

in negative powers of z that is valid when $1 < |z| < \infty$.

2. Give two Laurent series representations in powers of z for the function

$$f(z) = \frac{1}{z^2(1-z)},$$

and specify the regions in which those expansions are valid.

3. Represent the function

$$f(z) = \frac{z+1}{z-1}$$

(a) by its Maclaurin series, and state where the representation is valid;

(b) by its Laurent series in the domain $1<|z|<\infty$.

4. Find the Laurent series that represents the function
$$f(z) = z^2 \sin\left(\frac{1}{z^2}\right)$$
in the domain $0<|z|<\infty$.

5. Derive the Laurent series representation of the function
$$f(z) = \frac{1}{(z-i)(z-2i)}$$
in the following annular domains:
(a) $1<|z|<2$; (b) $2<|z|<\infty$; (c) $0<|z-i|<1$.

6. Derive the Laurent series representation of the function
$$f(z) = \frac{1}{z(z+2)^2}$$
in the following annular domains:
(a) $0<|z+2|<2$; (b) $2<|z+2|<\infty$.

7. Show that when $0<|z-1|<2$,
$$\frac{z}{(z-1)(z-3)} = -3\sum \frac{(z-1)^n}{z^{n+2}} - \frac{1}{2(z-1)}.$$

8. Write two Laurent series representations in powers of z for the function
$$f(z) = \frac{1}{z(1+z^2)},$$
in certain domains, and specify those domains.

9. Find the Laurent series that represents the following functions in the given domains, and then evaluate the integral $\int_{|z|=6} f(z)$:

(a) $f(z) = \sin\dfrac{1}{1-z}$, $0<|z-1|<\infty$;

(b) $f(z) = (z+1)^{10} \sin\dfrac{1}{1+z}$, $0<|z+1|<\infty$;

(c) $f(z) = \dfrac{1}{z(z+1)^6}$, $1<|z+1|<\infty$.

10. (a) Let a denote a real number, where $-1<a<1$, and derive the Laurent series representation
$$\frac{a}{z-a} = \sum_{n=1}^{\infty} \frac{a^n}{z^n}, \quad (|a|<|z|<\infty).$$

(b) Write $z = e^{i\theta}$ in the equation obtained in part (a) and then equate real parts and imaginary parts on each side of the result to derive the summation formulas
$$\sum_{n=1}^{\infty} a^n \cos n\theta = \frac{a\cos\theta - a^2}{1 - 2a\cos\theta + a^2}$$
and
$$\sum_{n=1}^{\infty} a^n \sin n\theta = \frac{a\sin\theta}{1 - 2a\cos\theta + a^2},$$
where $-1<a<1$.

Chapter 5 Residues and Its Application

In this chapter, we find some applications on definite and improper integral. With preparations of singular points, series and residues, specific integrals which seem difficult in Advance Calculus, are easily solved here.

5.1 Three types of isolated singular points

It is known previously that the integral on a closed contour, inside which, the integrand is analytic, is zero. However, Cauchy integral formula indicates that the integral is not always zero but relevant to the singular point inside the contour. Further, by constructing a compound contour, it seems that if the function fails to be analytic at a finite number of points interior to the contour, the function values at those singular points contribute to the value of the integral. Singularity matters. Hence, before we move on to find out the relationship between integral and singularity, let's start from the classification of singular points. Readers can come back to check why we involves three types of isolated singular points when Section 5.2 is finished.

Since the singularity is concerned with the points where the function is not analytic, we first recall the definition of singular points.

z_0 is called a ***singular point*** of a function f, if f fails to be analytic at z_0, but is analytic at some point in every neighborhood of z_0. For convenience, if we call a point analytic one, it means f is analytic at this point. The definition implies that not all points around z_0 should be singular ones. There are always points at which the function is analytic no matter how small the neighborhood of z_0 is. Intuitively, if all points around z_0 are analytic ones, z_0 seems to be isolated. So we have

Definition 5.1.1 Isolated singular points

A singular point z_0 is said to be ***isolated***, if there is a deleted neighborhood $0 < |z - z_0| < \delta$ of z_0, throughout which f is analytic.

Example 5.1.1 The function $f(z) = \dfrac{z+1}{z^2(z-1)}$ has two isolated singular points 0, 1 because we can find two small enough deleted neighborhood of 0 and 1, through which the

Example 5.1.2 Find the singular points and check if they are isolated for the function $f(z) = \text{Log } z$.

Solution About the principal branch of the logarithmic function, we know that $\text{Log } z = \ln|z| + i\text{Arg}(z)$, $|z| > 0$, $\text{Arg}(z) \in (-\pi, \pi)$. The points on the whole branch cut are all singular. Each of them is clearly not isolated by analytic points.

Example 5.1.3 Check the singularity for the functions

(1) $f_1(z) = \dfrac{1}{\sin(\pi z)}$; (2) $f_2(z) = \dfrac{1}{\sin\left(\dfrac{\pi}{z}\right)}$; (3) $f_3(z) = \dfrac{\sin(\pi z)}{z}$.

Solution

(1) $f_1(z)$ has the singular points $z = 0, \pm 1, \pm 2, \cdots$. They are all isolated.

(2) $f_2(z)$ has the singular points $z = 0, \dfrac{1}{n}, n = \pm 1, \pm 2, \cdots$. All the singular points except 0 are isolated. For $z = 0$, no matter how small the deleted neighborhood of 0 is, infinite number of singular points can always be found in it.

(3) Only one isolated singular point 0. However, some perhaps doubt the singularity at 0 by the following logic. It is easy to expand $f_3(z)$ into the following form:

$$f_3(z) = \frac{\sin(\pi z)}{z} = \pi - \frac{\pi^3}{3!}z^2 + \frac{\pi^5}{5!}z^4 - \cdots + (-1)^n \frac{\pi^{2n+1}}{(2n+1)!}z^{2n} + \cdots, \quad 0 < |z| < R.$$

The Laurent expansion looks analytic. So does the function. In fact, the expansion is not valid everywhere but in the deleted neighborhood. Therefore 0 is still singular. If we define the function at 0 properly, the singularity may be **removed**.

Following this, we use the Laurent series to classify the singular point.

Definition 5.1.2 Three types of isolated singular points

In a punctured disk $0 < |z - z_0| < R$, f is analytic, z_0 is an isolated singular point and

$$f(z) = \sum_{n=0}^{+\infty} a_n (z - z_0)^n + \sum_{n=1}^{+\infty} \frac{b_n}{(z - z_0)^n}.$$

The latter sum, involving negative powers of $z - z_0$ is called the principal part of f at z_0. For this part, if the sum contains only finite terms, i.e., there exists some positive integer k such that

$$b_k \neq 0, \quad b_{k+1} = b_{k+2} = \cdots = 0,$$

z_0 is called a **pole of order** k. A pole of order 1 is also referred to as a **simple pole**. If there are infinite terms, i.e., infinite number of the coefficients b_n are nonzero, z_0 is called an **essential singular point**. Finally, if the principal part is zero, i.e., all the b_n are zero, z_0 is said to be a **removable singular point**. In this case, if we redefine the function such that $f(z_0) = a_0$, it follows that f is analytic at z_0 and the Laurent series turns to be a Taylor series. The singularity at z_0 is **removed**.

Example 5.1.4 Find out the singular points and point out their types:

(1) $f_1(z) = \dfrac{z^2 - z + 2}{z - 1}$; (2) $f_2(z) = \dfrac{e^z - 1}{z^3}$; (3) $f_3(z) = \dfrac{\sin(\pi z)}{z}$;

(4) $f_4(z)=\sin\left(\dfrac{\pi}{z}\right)$;   (5) $f_5(z)=\dfrac{\sinh z}{z^2}$;   (6) $f_6(z)=\dfrac{2z+3}{(z^4+4)(z-1)^2}$.

Solution

(1) $f_1(z)=\dfrac{z^2-z+2}{z-1}=\dfrac{z(z-1)+2}{z-1}=1+(z-1)+\dfrac{2}{z-1}$, $0<|z-1|<+\infty$.

The isolated singular point 1 is a simple pole.

(2) $f_2(z)=\dfrac{e^z-1}{z^3}=\dfrac{1}{z^3}\left(1+z+\dfrac{1}{2!}z^2+\cdots-1\right)=\dfrac{1}{z^2}+\dfrac{1}{2!}\dfrac{1}{z}+\dfrac{1}{3!}+\cdots$, $0<|z|<+\infty$.

The isolated singular point 0 is a pole of order 2.

(3) From Example 5.1.3(3), 0 is a removable singular point.

(4) $f_4(z)=\sin\left(\dfrac{\pi}{z}\right)=\dfrac{\pi}{1!}\dfrac{1}{z}-\dfrac{\pi^3}{3!}\dfrac{1}{z^3}+\cdots+(-1)^n\dfrac{\pi^{2n+1}}{(2n+1)!}\dfrac{1}{z^{2n+1}}+\cdots$, $0<|z|<+\infty$.

The isolated singular point 0 is an essential one.

(5) $f_5(z)=\dfrac{\sinh z}{z^2}=\dfrac{1}{z}+\dfrac{1}{3!}z+\dfrac{1}{5!}z^3+\cdots$, $0<|z|<+\infty$.

The isolated singular point 0 is a simple pole.

(6) $f_6(z)=\dfrac{2z+3}{(z^4+4)(z-1)^2}$.

We can easily find out the singular points $z=1$, $\sqrt{2}\,e^{i\frac{(2n+1)\pi}{4}}$, $n=0, 1, 2, 3$, are all isolated. But it costs much to expand the series one by one. We involve the following theorem to help out.

Theorem 5.1.1 Judgment on a pole of order m

An isolated singular point z_0 of a function $f(z)$ is **a pole of order m** if and only if $f(z)$ can be written in the form $f(z)=\dfrac{g(z)}{(z-z_0)^m}$, where $g(z)$ is analytic and nonzero at z_0.

Proof We first consider the sufficiency part. Since $g(z)$ is analytic and at z_0, there exists a neighborhood $|z-z_0|<\delta$, in which,

$$g(z)=g(z_0)+\dfrac{g'(z_0)}{1!}(z-z_0)+\cdots+\dfrac{g^{(n)}(z_0)}{n!}(z-z_0)^n+\cdots.$$

Then we rewrite $f(z)$ in $0<|z-z_0|<\delta$,

$$f(z)=\dfrac{g(z_0)}{(z-z_0)^m}+\cdots+\dfrac{g^{(m-1)}(z_0)}{(m-1)!(z-z_0)}+\sum_{n=0}^{+\infty}\dfrac{g^{(n+m)}}{(n+m)!}(z-z_0)^n, \quad (5.1.1)$$

which is the Laurent series of $f(z)$. Recall that $g(z_0)\neq 0$. Then z_0 is indeed a pole of order m.

Conversely, we consider the other part. Since z_0 is indeed a pole of order m, then the Laurent series of $f(z)$ in some deleted neighborhood $0<|z-z_0|<\delta'$ can be represented in the following form:

$$f(z)=\sum_{n=0}^{+\infty}a_n(z-z_0)^n+\sum_{n=1}^{m}\dfrac{b_n}{(z-z_0)^n},$$

together with the fact $b_n\neq 0$. We rewrite $f(z)$ in $0<|z-z_0|<\delta'$,

$$f(z)=\sum_{n=0}^{+\infty}a_n\dfrac{(z-z_0)^{n+m}}{(z-z_0)^m}+\sum_{n=1}^{m}\dfrac{b_n(z-z_0)^{m-n}}{(z-z_0)^m}.$$

We complete the proof by defining

$$g(z) = \sum_{n=0}^{+\infty} a_n (z-z_0)^{n+m} + \sum_{n=1}^{m} b_n (z-z_0)^{m-n}, \quad |z-z_0| < \delta.$$

Example 5.1.5 Continue Example 5.1.4 (6) and (5)

Observe that

$$(z^4+4)(z-1)^2 = (z-\sqrt{2}e^{i\frac{\pi}{4}})(z-\sqrt{2}e^{i\frac{3\pi}{4}})(z-\sqrt{2}e^{i\frac{5\pi}{4}})(z-\sqrt{2}e^{i\frac{7\pi}{4}})(z-1)^2.$$

$$f_6(z) = \frac{2z+3}{(z-\sqrt{2}e^{i\frac{\pi}{4}})(z-\sqrt{2}e^{i\frac{3\pi}{4}})(z-\sqrt{2}e^{i\frac{5\pi}{4}})(z-\sqrt{2}e^{i\frac{7\pi}{4}})(z-1)^2}.$$

For $z=1$, $g(z) = \frac{2z+3}{z^4+4}$ is analytic and nonzero at $z=1$. Then $f_6(z) = \frac{g(z)}{(z-1)^2}$, from which, $z=1$ is a pole of order 2.

For $z = \sqrt{2}e^{i\frac{\pi}{4}}$, we rewrite $f_6(z)$:

$$f_6(z) = \frac{(2z+3)/(z-\sqrt{2}e^{i\frac{3\pi}{4}})(z-\sqrt{2}e^{i\frac{5\pi}{4}})(z-\sqrt{2}e^{i\frac{7\pi}{4}})(z-1)^2}{z-\sqrt{2}e^{i\frac{\pi}{4}}}.$$

With similar analysis, $z = \sqrt{2}e^{i\frac{\pi}{4}}$ is a simple pole. Theorem 5.1.1 works for this case.

For $f_5(z) = \frac{\sinh z}{z^2}$, $\sinh z|_{z=0} = 0$. We cannot apply Theorem 5.1.1 directly.

In fact, a huge amount of functions even fail to be in such a "good" form $f(z) = \frac{g(z)}{(z-z_0)^m}$. For example, $f(z) = \frac{z+1}{e^z-1}$. We all know that $z=0$ is the isolated singular point and it is not easy to find the Laurent series quickly. How can I conclude instantly that it is a simple pole since the denominator is not of the polynomial form? In fact, we hope the denominator could be changed into some polynomial form, which is the reason why we introduce zeros to help out.

Definition 5.1.3 Zero of order m

Suppose that $f(z)$ is analytic at z_0. If $f(z_0) = 0$, and if there is a positive integer number m, s.t. $f^{(m)}(z_0) \neq 0$, and each derivative of lower order vanishes at z_0,

$$f(z_0) = f'(z_0) = f''(z_0) = \cdots = f^{(m-1)}(z_0) = 0,$$

then $f(z)$ is said to have a **zero of order m** at z_0.

Example 5.1.6 Observe that $f(z) = (z-1)^2$ have a zero of order 2 at $z=1$ and $g(z) = z(e^z-1)$ have a zero of order 2 at $z=0$.

Example 5.1.7 $f(z) = (z^4+4)(z-1)^2$ has five zeros. $z = \sqrt{2}e^{i\frac{(2n+1)\pi}{4}}$, $n=0, 1, 2, 3$ are zeros of order 1, and 1 is a zero of order 2. It takes time to find out from the definition of zero. The following theorem speeds it up.

Theorem 5.1.2 A function $f(z)$ that is analytic at a point z_0, has a zero of order m if and only if there is a function $g(z)$, which is analytic and nonzero at z_0, s.t. $f(z) = (z-z_0)^m g(z)$.

Proof Both parts of the proof involve Taylor's theorem.

For the sufficiency part, since $g(z)$ is analytic at z_0, it has a Taylor series throughout a neighborhood $|z-z_0| < \delta$:

$$g(z)=g(z_0)+\frac{g'(z_0)}{1}(z-z_0)+\frac{g''(z_0)}{2!}(z-z_0)^2+\cdots.$$

Then

$$f(z)=(z-z_0)^m g(z)=g(z_0)(z-z_0)^m+\frac{g'(z_0)}{1}(z-z_0)^{m+1}+\frac{g''(z_0)}{2!}(z-z_0)^{m+2}+\cdots.$$

From Taylor's theorem, the above expression is exactly the Taylor series of $f(z)$ at z_0. Thus

$f(z_0)=f'(z_0)=f''(z_0)=\cdots=f^{(m-1)}(z_0)=0$, and $f^{(m)}(z_0)=m!\,g^{(m)}(z_0)\neq 0$. Hence z_0 is a zero of order m.

For the necessity part, since $f(z)$ is analytic at a point z_0, it has a Taylor series throughout a neighborhood $|z-z_0|<\delta'$:

$$f(z)=f(z_0)+\frac{f'(z_0)}{1}(z-z_0)+\frac{f''(z_0)}{2!}(z-z_0)^2+\cdots.$$

Notice that $f(z)$ has a zero of order m, the above series turns to be

$$f(z)=\frac{f^{(m)}(z_0)}{m!}(z-z_0)^m+\frac{f^{(m+1)}(z_0)}{(m+1)!}(z-z_0)^{m+1}+\cdots$$

$$=(z-z_0)^m\left[\frac{f^{(m)}(z_0)}{m!}+\frac{f^{(m+1)}(z_0)}{(m+1)!}(z-z_0)+\cdots\right].$$

Then we complete the proof by defining

$$g(z)=\frac{f^{(m)}(z_0)}{m!}+\frac{f^{(m+1)}(z_0)}{(m+1)!}(z-z_0)+\cdots,\quad |z-z_0|<\delta'.$$

Example 5.1.8 Observe that $g(z)=z(e^z-1)=z^2\left(1+\frac{1}{2!}z+\cdots\right)$ has a zero of order 2 at 0.

Example 5.1.9 Observe that

$$f(z)=(z^4+4)(z-1)^2=(z-\sqrt{2}e^{i\frac{\pi}{4}})(z-\sqrt{2}e^{i\frac{3\pi}{4}})(z-\sqrt{2}e^{i\frac{5\pi}{4}})(z-\sqrt{2}e^{i\frac{7\pi}{4}})(z-1)^2.$$

Consequently, 1 is a zero of order 2 since $f(z)=(z-1)^2 g(z)$, $g(1)\neq 0$ and $g(z)$ is analytic at 1. With similar analysis, we have other four zeros of order 1.

Example 5.1.10 Continue Example 5.1.4(5)

$f_5(z)=\dfrac{\sinh z}{z^2}$. $\sinh z$ has a zero of order 1 at $z=0$. Then $\sinh z=zg(z)$, where $g(z)$ is analytic and nonzero at $z=0$. Then $f_5(z)=\dfrac{g(z)}{z}$. Obviously, $z=0$ is a simple pole.

Now we come to the following theorem which shows how zeros can create poles.

Theorem 5.1.3 Relationship between zeros and poles

Suppose that two functions $f(z)$ and $g(z)$ are analytic at z_0, and $f(z_0)\neq 0$. If $g(z)$ has a zero of order m at z_0, then $\dfrac{f(z)}{g(z)}$ has a pole of order m there.

Proof Since $g(z)$ has a zero of order m at z_0, $g(z)=(z-z_0)^m\phi(z)$, where $\phi(z)$ is analytic and nonzero at z_0. Further $\dfrac{f(z)}{g(z)}=\dfrac{\frac{f(z)}{\phi(z)}}{(z-z_0)^m}$. $\dfrac{f(z)}{\phi(z)}$ is clearly analytic and nonzero at z_0. Then the proof is immediate from Theorem 5.1.1.

Example 5.1.11 Observe that the denominator of $f(z) = \dfrac{z+1}{z(e^z-1)}$ has a zero of order 2 at $z=0$. The nominator is analytic and nonzero there. Consequently, $z=0$ is a pole of order 2.

Exercises 5.1

1. Find out the singular points and determine whether the singular point is a pole, a removable singular point or an essential singular point?

   (1) $\dfrac{e^z}{z^2}$; (2) $z^2 e^{\frac{1}{z}}$; (3) $\dfrac{\cos z - 1}{z}$; (4) $\dfrac{\cos z}{z}$; (5) $\dfrac{\sin^2 z}{z}$; (6) $\dfrac{z}{(z-1)^3(e^z-1)}$.

2. Find the residues of the following functions at all singular points:

   (1) $z\cos\left(\dfrac{1}{z}\right)$; (2) $\dfrac{1}{1+z+z^2}$; (3) $\dfrac{\sinh z}{z^2(z^2-1)}$;

   (4) $\dfrac{e^{2z}}{(z-1)^4}$; (5) $\tanh z$; (6) $\dfrac{\operatorname{Log} z}{(z^2+1)^3}$.

5.2 Residues and Cauchy's residue theorem

This section is devoted to residues. We will see how the residues help with the integral.

Definition 5.2.1 Residue

Suppose that z_0 is an isolated singular point of a function f. There is a positive number R, s. t. f is analytic at each point of the punctured disk $0 < |z-z_0| < R$. Consequently, f is represented by a Laurent series

$$f(z) = \sum_{n=0}^{+\infty} a_n (z-z_0)^n + \sum_{n=1}^{+\infty} \dfrac{b_n}{(z-z_0)^n}, \quad 0 < |z-z_0| < R,$$

where the coefficients a_n and b_n have integral forms. Particularly,

$$b_n = \dfrac{1}{2\pi i} \int_C \dfrac{f(z)}{(z-z_0)^{-n+1}} dz,$$

where C is any positively oriented simple closed contour around z_0 and lying in the punctured disk $0 < |z-z_0| < R$. The coefficient b_1 is called the **residue** of f at the isolated singular point z_0, denoted by $\operatorname*{Res}_{z=z_0} f(z)$.

From the definition, we can see that $b_1 = \dfrac{1}{2\pi i} \int_C f(z) dz$. That is,

$$\int_C f(z) dz = 2\pi i b_1 = 2\pi i \operatorname*{Res}_{z=z_0} f(z). \tag{5.2.1}$$

Equation (5.2.1) provides a powerful method for calculating some integrals around simple closed contours with some isolated singular points inside (The integral should be zero without singular points inside). To evaluate the integral, we just need the residue, the coefficient b_1, in this case.

Example 5.2.1 Consider the integral $I = \int_C \dfrac{1}{z(z-3)} dz$, where C is the positively oriented circle $|z-3|=1$.

Solution We use the residue to evaluate the integral, which implies we need the Laurent series first. Notice that the integrand has two isolated singular points 0 and 3, and only 3 is interior to C. Then the integrand is analytic and has Laurent series in a punctured disk $0<|z-3|<3$ with C inside.

$$\frac{1}{z(z-3)} = \frac{1}{(z-3)}\frac{1}{z} = \frac{1}{z-3}\frac{1}{z-3+3} = \frac{1}{3}\frac{1}{z-3}\frac{1}{1+\frac{(z-3)}{3}}$$

$$= \sum_{n=0}^{+\infty}(-1)^n \frac{(z-3)^{n-1}}{3^{n+1}}, \quad 0<|z-3|<3.$$

The coefficient of $\dfrac{1}{z-3}$ is $\dfrac{1}{3}$, which is the desired residue at $z=3$.

Then $I = \int_C \dfrac{1}{z(z-3)} dz = \dfrac{2\pi i}{3}$.

Example 5.2.2 Consider the integral $I = \int_C \sin\left(\dfrac{1}{z^2}\right) dz$, where C is the positively oriented unit circle $|z|=1$.

Solution Notice that 0 is the unique isolated singular point of the integrand, and is interior to C. Then we can write the Laurent series in the whole complex set except 0:

$$\sin\left(\frac{1}{z^2}\right) = \frac{1}{z^2} - \frac{1}{3!}\frac{1}{z^6} + \frac{1}{5!}\frac{1}{z^{10}} + \cdots + (-1)^n \frac{1}{(2n+1)!}\frac{1}{z^{4n+2}} + \cdots, \quad 0<|z|<+\infty.$$

Obviously, $\underset{z=0}{\text{Res}} \sin\left(\dfrac{1}{z^2}\right)$ is zero, so is the integral I.

Example 5.2.3 Consider the integral $I = \int_C \dfrac{1}{z(z-3)} dz$, where C is the positively oriented circle $|z-3|=4$.

Solution This example is the same as Example 5.2.1 except the contour. Unfortunately, it seems that the residue method fails to apply directly since there are two isolated singular points of the integrand are contained in the contour. Therefore we look back to involve the extension of the Cauchy-Goursat theorem.

We construct two simple closed contours in the same directions as C around two singular points $z=0$ and $z=3$.

$$C_1: 0<|z|<\frac{1}{2}, \quad C_2: 0<|z-3|<\frac{1}{2}.$$

They are interior to C and have no intersection. Then

$$I = \int_C \frac{1}{z(z-3)} dz = \int_{C_1} \frac{1}{z(z-3)} dz + \int_{C_2} \frac{1}{z(z-3)} dz = I_1 + I_2,$$

where $I_1 = \int_{C_1} \dfrac{1}{z(z-3)} dz$ and $I_2 = \int_{C_2} \dfrac{1}{z(z-3)} dz$.

Fortunately, the residue method applies for I_1 and I_2 respectively.

For $C_1: 0<|z|<\dfrac{1}{2}$,

$$\frac{1}{z(z-3)} = -\frac{1}{3}\frac{1}{z}\frac{1}{1-\frac{z}{3}} = -\frac{1}{3}\frac{1}{z}\sum_{n=0}^{+\infty}\frac{z^n}{3^n} = -\sum_{n=0}^{+\infty}\frac{z^{n-1}}{3^{n+1}}.$$

$$\operatorname*{Res}_{z=0}\frac{1}{z(z-3)} = -\frac{1}{3}. \quad I_1 = 2\pi i \operatorname*{Res}_{z=0}\frac{1}{z(z-3)} = -\frac{2\pi i}{3}.$$

For $C_2: 0 < |z-3| < \frac{1}{2}$,

$$\frac{1}{z(z-3)} = \frac{1}{3}\frac{1}{z-3}\frac{1}{1+\frac{(z-3)}{3}} = \frac{1}{3}\frac{1}{z-3}\sum_{n=0}^{+\infty}(-1)^n\frac{(z-3)^n}{3^n}$$

$$= \sum_{n=0}^{+\infty}(-1)^n\frac{(z-3)^{n-1}}{3^{n+1}}.$$

$$\operatorname*{Res}_{z=3}\frac{1}{z(z-3)} = \frac{1}{3}. \quad I_2 = 2\pi i \operatorname*{Res}_{z=3}\frac{1}{z(z-3)} = \frac{2\pi i}{3}.$$

Finally, $I = I_1 + I_2 = 2\pi i \left[\operatorname*{Res}_{z=0}\frac{1}{z(z-3)} + \operatorname*{Res}_{z=3}\frac{1}{z(z-3)} \right] = 0.$

Extending above example, we claim the following theorem without proof.

Theorem 5.2.1 Cauchy's residue theorem

Let C be a simple closed contour in the positive sense. If a function f is analytic inside and on C except for a finite number of singular points z_k, $k=1, 2, \cdots, N$, inside C, then

$$\int_C f(z)\,dz = 2\pi i \sum_{k=1}^{N} \operatorname*{Res}_{z=z_k} f(z).$$

This theorem is very powerful because the integral turns to be simple algebra of residues. However, it is not a pleasant experience that we need to find out all the residues one by one, if there are 3 or more isolated singular points inside C.

In this scenario, in addition, that the integrand is analytic throughout the finite complex plane except points inside C, we can more efficiently find the integral by using single residue no matter how many singular points interior to C. This method is stated as follows.

Theorem 5.2.2 Single residue alternative

If a function f is analytic everywhere in the finite complex plane except a finite number of singular points interior to a positively oriented simple closed contour C. Then

$$\int_C f(z)\,dz = 2\pi i \operatorname*{Res}_{z=0}[g(z)], \quad \text{where} \quad g(z) = \frac{1}{z^2}f\left(\frac{1}{z}\right).$$

Proof Construct a large enough contour $C_1: |z| = R$, with C inside. Notice that the singularity occurs inside C. Then we have the Laurent series of the integrand in the annular domain $R < |z| < +\infty$:

$$f(z) = \sum_{n=-\infty}^{+\infty} c_n z^n,$$

where $c_n = \frac{1}{2\pi i}\int_{C_2}\frac{f(z)}{z^{n+1}}\,dz$, $n=0, \pm 1, \cdots$, and C_1 is any simple closed contour in the annular domain. Here we specify C_2 as $C_2: R_1 < |z| < +\infty$, $R_1 > R$.

From the coefficients, we know that

$$\int_{C_2} f(z)\,dz = 2\pi i c_{-1}. \tag{5.2.2}$$

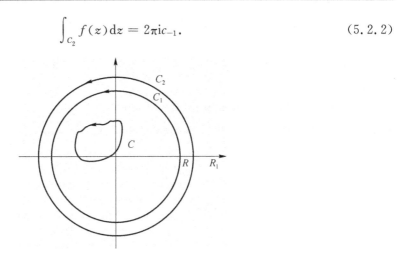

Fig. 5.2.1

Remark that c_{-1} is not the residue since there are more than one singular points inside C_1. Although it is not a residue, we can still change the integral into a simple algebra if c_{-1} is already known. Go back to the Laurent series of $f(z)$ and replace z by $\frac{1}{z}$:

$$g(z) = \frac{1}{z^2} f\left(\frac{1}{z}\right) = \sum_{n=-\infty}^{+\infty} \frac{c_n}{z^{n+2}}, \quad 0 < |z| < \frac{1}{R}. \tag{5.2.3}$$

Notice that $g(z)$ is analytic in the deleted neighborhood $0 < |z| < \frac{1}{R}$. $z = 0$ is the singular point of $g(z)$. (5.2.3) is the Laurent series of $g(z)$ and c_{-1} happens to be the residue of $g(z)$ at $z=0$, that is,

$$c_{-1} = \operatorname*{Res}_{z=0}[g(z)].$$

From (5.2.2), $\int_{C_2} f(z)\,dz = 2\pi i c_{-1} = 2\pi i \operatorname*{Res}_{z=0}[g(z)]$. Together with Cauchy-Goursat's theorem, $\int_C f(z)\,dz = \int_{C_2} f(z)\,dz$, we complete the proof immediately.

Example 5.2.4 We re-calculate Example 5.2.3 by this new single residue method.

Solution

$$f(z) = \frac{1}{z(z-3)}, \quad g(z) = \frac{1}{z^2} f\left(\frac{1}{z}\right) = \frac{1}{1-3z}, \quad \operatorname*{Res}_{z=0} g(z) = 0.$$

Hence, $I = \int_C \frac{1}{z(z-3)} dz = 0.$

Following the above path about the integral by using residue, we arrive at the last case: if it is not easy to find the Laurent series of the integrand around the isolated singular point, what can we do? Let's see the following example.

Example 5.2.5 Consider the integral $I = \int_C \frac{1}{z(z-2)^3} dz$, where C is the positively oriented circle $|z-2|=4$.

Solution Now it is known that if we find the residues at 0 and 2, then the integral is immediate from Theorem 5.2.1.

However it takes time to find the Laurent series of the integrand in this punctured disk $0<|z|<1$.

So we introduce the helpful theorem which gives a new method for the residue based on poles.

Theorem 5.2.3 Residues at poles

An isolated singular point z_0 of a function f is a pole of order m, that is, $f(z)$ can be written in the form:
$$f(z)=\frac{g(z)}{(z-z_0)^m},$$
where $g(z)$ is analytic and nonzero at z_0. Then
$$\operatorname*{Res}_{z=z_0} f(z)=\frac{g^{(m-1)}(z_0)}{(m-1)!},$$
where $g^{(0)}(z_0)=g(z_0)$, $0!=1$.

Proof The theorem is immediate from equation (5.1.1) in the proof of the judgment theorem on the pole.

Let's continue Example 5.2.5.

$f(z)=\dfrac{1}{z(z-2)^3}=\dfrac{\frac{1}{z}}{(z-2)^3}$, together with $\dfrac{1}{z}$ being analytic and nonzero at $z=2$, reveals that $z=2$ is a pole of order 3 and $\operatorname*{Res}_{z=2} f(z)=\dfrac{\left(\frac{1}{z}\right)''}{(3-1)!}\bigg|_{z=2}=\dfrac{1}{8}$. Similarly, $f(z)=\dfrac{1}{z(z-2)^3}=\dfrac{\frac{1}{(z-2)^3}}{z-0}$, together with $\dfrac{1}{(z-2)^3}$ being analytic and nonzero at $z=0$, reveals that $z=0$ is a simple pole and $\operatorname*{Res}_{z=0} f(z)=\dfrac{1}{(z-2)^3}\bigg|_{z=0}=-\dfrac{1}{8}$.

Then $I=\displaystyle\int_C \frac{1}{z(z-2)^3}dz=0$.

Specially, we use the relationship between zeros and poles to find the residues without translating the function into the form $\dfrac{g(z)}{(z-z_0)^m}$. This case is stated in the following theorem.

Theorem 5.2.4 Suppose that $f(z)=\dfrac{g(z)}{h(z)}$. Let $g(z)$ and $h(z)$ be analytic at a point z_0. If $g(z_0)\neq 0$, $h(z_0)=0$ and $h'(z_0)\neq 0$, then z_0 is a simple pole of the quotient $f(z)$, and $\operatorname*{Res}_{z=z_0} f(z)=\dfrac{g(z_0)}{h'(z_0)}$.

Proof z_0 is a zero of $h(z)$ of order 1. It is direct from Theorem 5.1.3 that z_0 is a simple pole of the quotient $f(z)$.

At the meantime, $h(z)=(z-z_0)\phi(z)$, where $\phi(z)$ is analytic and nonzero at z_0.

Rewrite f: $f(z)=\dfrac{\frac{g(z)}{\phi(z)}}{z-z_0}$, where $\dfrac{g(z)}{\phi(z)}$ is analytic and nonzero at z_0. Then z_0 is a simple pole

of $f(z)$ and
$$\operatorname*{Res}_{z=z_0} f(z) = \frac{g(z_0)}{\phi(z_0)}.$$

Notice that $h'(z_0) = \phi(z_0)$. Then we complete the proof by substitution.

Example 5.2.6 Find the residues of $f_6(z) = \dfrac{2z+3}{(z^4+4)(z-1)^2}$ at $z = \sqrt{2} e^{i\frac{(2n+1)\pi}{4}}$, $n = 0, 1, 2, 3$.

Solution As we know, these singular point are zeros of order 1 of the denominator. We just show one case:

For $z = \sqrt{2} e^{i\frac{\pi}{4}}$,
$$\operatorname*{Res}_{z=\sqrt{2}e^{i\frac{\pi}{4}}} f_6(z) = \operatorname*{Res}_{z=\sqrt{2}e^{i\frac{\pi}{4}}} \frac{\frac{2z+3}{(z-1)^2}}{z^4+4} = \left.\frac{\frac{2z+3}{(z-1)^2}}{4z^3}\right|_{z=\sqrt{2}e^{i\frac{\pi}{4}}} = 8(1-i).$$

Example 5.2.7 Consider residues of the function $f(z) = \cot z$ at all the singular points.

Solution $f(z) = \cot z = \dfrac{\cos z}{\sin z}$. Singularity occurs at those zeros of $\sin z$, $z = n\pi$, $n = 0, \pm 1, \cdots$. In fact, those zeros are of order 1 since
$$\sin n\pi = 0, \ (\sin z)'|_{z=n\pi} = (-1)^n \neq 0.$$

From Theorem 5.2.4, $\operatorname*{Res}_{z=n\pi} \cot z = \dfrac{\cos(n\pi)}{\cos n\pi} = 1$.

Exercises 5.2

1. If function $f(z) = \dfrac{e^z}{(z-1)^2 z}$, then

(1) find out all the singular points of $f(z)$, and point out their types;

(2) evaluate the residues of $f(z)$ at those singular points;

(3) evaluate the integral $\int_{|z|=2} f(z) dz$, $\int_{|z-1|=\frac{1}{2}} f(z) dz$.

2. Use a single residue theorem to evaluate the integral:
$$\int_{|z|=5} \frac{2z+3}{z(z+1)^2(z+2)} dz.$$

5.3 Application of residues on definite integrals

In this section, we turn to the applications of residue theory with the lengthy preparation in the last two sections. Three types of integrals in real analysis are introduced in the following three subsections.

5.3.1 Improper integrals

Our first target is the integral of the following form:
$$\int_0^{+\infty} f(x)\,dx.$$

It is known that the integral is a limit
$$\int_0^{+\infty} f(x)\,dx = \lim_{R\to+\infty}\int_0^R f(x)\,dx,$$
whenever the limit exists. If $f(x)$ is continuous through the real set, the integral on it is defined as follows:
$$\int_{-\infty}^{+\infty} f(x)\,dx = \lim_{R_1\to+\infty}\int_0^{R_1} f(x)\,dx + \lim_{R_2\to+\infty}\int_{-R_2}^0 f(x)\,dx, \tag{5.3.1}$$
when the limits on the right side exist. Here R_1 and R_2 tend to infinite independently. Otherwise, if R_1 and R_2 take the same value and tend to infinite simultaneously, we call the limit (when exists) the **Cauchy principal value** of the integral (5.3.1). That is,
$$\text{P.V.}\int_{-\infty}^{+\infty} f(x)\,dx = \lim_{R\to+\infty}\int_{-R}^R f(x)\,dx.$$

Notice that the convergence of (5.3.1) is sufficient for the existence of the Cauchy principal value. However, it is not always necessary. For example, the integral of some odd and continuous functions in finite interval $[-R,R]$ always vanishes. It even does not exist in $(-\infty,R_2]$ or $[R_1,+\infty)$. There is a special case, on the contrary, the integrand is an even function, that is, $f(-x)=f(x)$. We can easily see that (5.3.1) is equivalent to the Cauchy principal value, which is stated in the following theorem.

Theorem 5.3.1 If $f(x)$ $(-\infty<x<+\infty)$ is even and the Cauchy principal value exists then
$$\text{P.V.}\int_{-\infty}^{+\infty} f(x)\,dx = 2\lim_{R\to+\infty}\int_0^R f(x)\,dx = 2\lim_{R\to+\infty}\int_{-R}^0 f(x)\,dx = \int_{-\infty}^{+\infty} f(x)\,dx.$$

Now we introduce the method of involving residues to evaluate the improper integrals of *rational* and *even* functions through an example. Here, the rational functions is of the following form: $\dfrac{f(x)}{g(x)}$, where $f(x)$ and $g(x)$ are polynomials with real coefficients and without factors in common.

Example 5.3.1 Evaluate the integral $I = \int_0^{+\infty}\dfrac{1}{x^4+1}\,dx$.

Solution We start from the new version of the integrand:
$$f(z) = \frac{1}{z^4+1}.$$

The function $f(z)$ is analytic except four roots of -1. Here are the roots:
$$z_n = e^{\frac{i(\pi+2n\pi)}{4}},\quad n = 0, 1, 2, 3.$$

z_0 and z_1 lie in the upper half plane, z_2 and z_3 the lower half plane. None of them lies on the real axis. We construct a contour in the upper half plane with z_0 and z_1 inside: a real segment

L_R from $-R$ to R and a half circle C_R: $|z|=R$ from R to $-R$. See Figure 5.3.1.

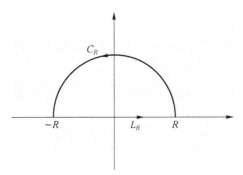

Fig. 5.3.1

According to Cauchy's residue theorem, we have the following equation:

$$\int_{L_R} f(z)\,dz + \int_{C_R} f(z)\,dz = 2\pi i\left(\operatorname*{Res}_{z=z_0} f(z) + \operatorname*{Res}_{z=z_1} f(z)\right). \qquad (5.3.2)$$

Notice that $\int_{L_R} f(z)\,dz = \int_{-R}^{R} f(x)\,dx$. With the aid of Theorem 5.2.4, we have

$$\operatorname*{Res}_{z=z_0} f(z) = \frac{1}{4z_0^3} = \frac{1}{4}e^{-\frac{3\pi}{4}i}, \qquad \operatorname*{Res}_{z=z_1} f(z) = \frac{1}{4z_1^3} = \frac{1}{4}e^{-\frac{1\pi}{4}i}.$$

Then from (5.3.2),

$$\int_{-R}^{R} f(z)\,dz = \frac{\sqrt{2}\pi}{2} - \int_{C_R} f(z)\,dz. \qquad (5.3.3)$$

The above equality holds for all $R>1$.

Next, we claim that the integral on the right hand side of (5.3.3) should be zero as R tends to infinity. To do so, we consider the upper bound of the integrand.

$$\left|\int_{C_R} f(z)\,dz\right| \leqslant \max_{z\in C_R}\left(\left|\frac{1}{z^4+1}\right|\right)\pi R \leqslant \max_{z\in C_R}\left(\frac{1}{||z|^4-1|}\right)\pi R = \frac{\pi R}{R^4-1}.$$

Obviously, the upper bound tends to zero as R tends to infinity, which implies our claim. Then from (5.3.3) again, we have

$$\lim_{R\to+\infty}\int_{-R}^{R} f(z)\,dz = \frac{\sqrt{2}\pi}{2}.$$

Notice that the integrand $\dfrac{1}{x^4+1}$ is even. Consequently,

$$\int_{0}^{+\infty} \frac{1}{x^4+1}\,dx = \frac{\sqrt{2}\pi}{4}.$$

Remark of Example 5.3.1 The integrand is limited to a rational and even function. Also the denominator should have no real zeros, which makes sure that the integrand is analytic on the constructed contour. Moreover, the degree of the denominator minus one of the numerator is greater than 1. Otherwise the integral does not exist certainly.

5.3.2 Improper integrals involving sines and cosines

The method of residues is also useful in evaluating the improper integrals of the form:

$$\int_{-\infty}^{+\infty} f(x)\sin ax\,dx \quad \text{or} \quad \int_{-\infty}^{+\infty} f(x)\cos ax\,dx,$$

where a denotes some positive parameter. We do not apply the method in the last subsection directly with the beginning of the new version of the integrand $f(z)\sin az$, since $|\sin a(x+iy)|$ and $|\cos a(x+iy)|$ are both unbounded exponentially. Here we use another version of the integrand $f(z)e^{iaz}$. This modification is suggested by the following equation:

$$\int_{-R}^{+R} f(x)\cos ax\,dx + i\int_{-R}^{+R} f(x)\sin ax\,dx = \int_{-R}^{+R} f(x)e^{iax}\,dx$$

and the boundness of the modulus of e^{iax} in the upper half plane. The following example illustrates the method in details.

Example 5.3.2 Evaluate the integral $I = \int_{-\infty}^{+\infty} \dfrac{\cos 3x}{x^2+1}dx$.

Solution We start from the function $f(z)$ for this time:

$$f(z) = \frac{1}{z^2+1}.$$

The function $f(z)$ is analytic except two roots of -1. Here are the roots:

$$z_1 = i, \quad z_2 = -i.$$

We still use the contour constructed in the last subsection with only one singular point z_1 inside. The integrand is $f(z)e^{i3z}$. Then

$$\int_{L_R+C_R} f(z)e^{i3z}\,dz = 2\pi i \operatorname*{Res}_{z=i}[f(z)e^{i3z}]. \tag{5.3.4}$$

Further, $f(z)e^{i3z} = \dfrac{\dfrac{e^{i3z}}{z+i}}{z-i}$. Then $\operatorname*{Res}_{z=i}[f(z)e^{i3z}] = -\dfrac{e^{-3}}{2}i.$

Continue (5.3.4):

$$\int_{-R}^{R} f(x)e^{i3x}\,dx = 2\pi i\left(-\frac{e^{-3}}{2}i\right) - \int_{C_R} f(z)e^{i3z}\,dz = \frac{\pi}{e^3} - \int_{C_R} f(z)e^{i3z}\,dz.$$

Consider the real part of both sides and we have

$$\int_{-R}^{R} f(x)\cos 3x\,dx = \frac{\pi}{e^3} - \operatorname{Re}\int_{C_R} f(z)e^{i3z}\,dz. \tag{5.3.5}$$

Next we claim again that $\operatorname{Re}\int_{C_R} f(z)e^{i3z}\,dz$ tends to zero as R approaches the positive infinity. On C_R, $|z|=R$,

$$|f(z)| = \frac{1}{|z^2+1|} \leqslant \frac{1}{R^2-1},$$

and $|e^{i3z}| = e^{-3y} \leqslant 1$ with $z=x+iy$, $y\geqslant 0$. Then

$$\left|\operatorname{Re}\int_{C_R} f(z)e^{i3z}\,dz\right| \leqslant \frac{\pi R}{R^2-1},$$

which implies our claim.

Finally, (5.3.5) together with the fact that the integrand is an even function, reveals that

$$\int_{-\infty}^{+\infty} f(x)\cos 3x\,dx = \frac{\pi}{e^3}.$$

Remark of Example 5.3.2 The integrand is limited to an even function. Also the upper bound clearly tends to zero. However, the convergence of the upper bound is not always easy to see, which is illustrated in the following example.

Example 5.3.3 Evaluate the Cauchy principal value of the integral $I = \int_{-\infty}^{+\infty} \frac{x\sin x}{x^2+x+1}dx$.

Solution Notice that the integrand is not an even function. Fortunately, what we concern is the Cauchy principal value. The method of taking limit of integral from $-R$ to R is still valid.

We denote $f(z)$ by

$$f(z) = \frac{z}{z^2+z+1} = \frac{z}{(z-z_1)(z-z_2)},$$

where $z_1 = -\frac{1}{2}+\frac{\sqrt{3}}{2}i$, $z_2 = -\frac{1}{2}-\frac{\sqrt{3}}{2}i$. z_1 and z_2 are simple poles of function $f(z)e^{iz}$, and z_1 lies above the real axis. The contour in Example 5.3.1 is still in use. Then

$$\int_{L_R+C_R} f(z)e^{iz}dz = 2\pi i \operatorname*{Res}_{z=z_1}[f(z)e^{iz}]. \tag{5.3.6}$$

Further, $B_1 = \operatorname*{Res}_{z=z_1}[f(z)e^{iz}] = \frac{z_1 e^{iz_1}}{z_1-z_2} = -\frac{\sqrt{3}}{3}e^{-\frac{\sqrt{3}}{2}}e^{(\frac{2\pi}{3}-\frac{1}{2})i}i.$

Continue (5.3.6):

$$\int_{-R}^{R} f(x)e^{ix}dx = 2\pi i B_1 - \int_{C_R} f(z)e^{iz}dz.$$

Consider the imaginary part of both sides and we have

$$\int_{-R}^{R} f(x)\sin x\,dx = \operatorname{Im}(2\pi i B_1) - \operatorname{Im}\int_{C_R} f(z)e^{iz}dz. \tag{5.3.7}$$

As usual, we next want to claim again that $\operatorname{Im}\int_{C_R} f(z)e^{iz}dz$ tends to zero as R approaches the positive infinity. On C_R, $|z|=R$,

$$|f(z)| = \frac{|z|}{|z^2+z+1|} = \frac{R}{|z^2+z+1|} \leqslant \frac{R}{R^2-R-1}, \tag{5.3.8}$$

and $|e^{iz}| = e^{-y} \leqslant 1$ with $z=x+iy$, $y\geqslant 0$. Then

$$\left|\operatorname{Im}\int_{C_R} f(z)e^{iz}dz\right| \leqslant \frac{\pi R^2}{R^2-R-1}.$$

The upper bound tends to π as R approaches the positive infinity. It seems that our claim is not true. In fact, although the upper bound method fails, we can see that our claim is still true from the following theorem.

Theorem 5.3.2 Suppose that a function $f(z)$ is analytic at all points above the semicircle C_{R_0}: $z=R_0 e^{i\theta}$, $0\leqslant \theta \leqslant \pi$ in the upper half plane. Denote by C_R the semicircle in Section 5.3.1, C_R: $z=Re^{i\theta}$, $0\leqslant \theta \leqslant \pi$ with $R>R_0$, the upper half plane. If there is a positive number M_R such that

$$|f(z)|\leqslant M_R, \quad \text{for} \quad z\in C_R, \quad \text{and} \quad \lim_{R\to+\infty} M_R = 0,$$

then,

$$\lim_{R\to+\infty}\int_{C_R} f(z)e^{iaz}dz = 0, \quad \text{for any } a>0.$$

We postpone the proof of this theorem at the end of Example 5.3.3. From (5.3.8), the condition of this theorem is completely matched. Then by letting R tend to the positive infinity on both sides of (5.3.7), we have

$$\lim_{R\to+\infty}\int_{-R}^{R}f(x)\sin x\,dx = \mathrm{Im}(2\pi iB_1) = \frac{2\pi\sqrt{3}}{3}e^{-\frac{\sqrt{3}}{2}}\sin\left(\frac{2\pi}{3}-\frac{1}{2}\right),$$

which is the principal value we want. Example 5.3.3 is done here.

Proof of Theorem 5.3.2 Rewrite the integral:

$$\int_{C_R}f(z)e^{iaz}\,dz = \int_0^{\pi}f(Re^{i\theta})e^{iaRe^{i\theta}}Re^{i\theta}i\,d\theta.$$

For the integrand, $f(Re^{i\theta})\leqslant M_R$. Then

$$\left|\int_{C_R}f(z)e^{iaz}\,dz\right| \leqslant RM_R\int_0^{\pi}|e^{iaRe^{i\theta}}|\,d\theta = RM_R\int_0^{\pi}e^{-aR\sin\theta}\,d\theta. \tag{5.3.9}$$

Next we turn to $\int_0^{\pi}e^{-aR\sin\theta}\,d\theta$.

In view that $g(x)=e^{-aR\sin x}$ is symmetric with respect to the vertical line $x=\dfrac{\pi}{2}$ in $[0,\pi]$, $\int_0^{\pi}e^{-aR\sin\theta}\,d\theta = 2\int_0^{\frac{\pi}{2}}e^{-aR\sin\theta}\,d\theta.$

From Fig 5.3.2, we can easily see that $\sin\theta\geqslant\dfrac{2\theta}{\pi}$ in $\left[0,\dfrac{\pi}{2}\right]$.

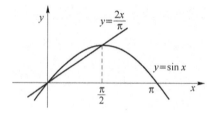

Fig. 5.3.2

Then

$$\int_0^{\pi}e^{-aR\sin\theta}\,d\theta \leqslant 2\int_0^{\frac{\pi}{2}}e^{-\frac{2aR\theta}{\pi}}\,d\theta = \frac{\pi}{aR}(1-e^{-aR}).$$

Further

$$\int_0^{\pi}e^{-aR\sin\theta}\,d\theta < \frac{\pi}{aR},$$

which is known as **Jordan's inequality**. Substituting this inequality into (5.3.9) we have

$$\left|\int_{C_R}f(z)e^{iaz}\,dz\right| \leqslant \frac{\pi M_R}{a}.$$

Recall that $\lim\limits_{R\to+\infty}M_R=0$. Then $\lim\limits_{R\to+\infty}\int_{C_R}f(z)e^{iaz}\,dz = 0$, for any $a>0$.

5.3.3 Integrals on $[0,2\pi]$ involving sines and cosines

In this subsection, we consider the following type of integral

$$\int_0^{2\pi} F(\sin\theta, \cos\theta)\,d\theta.$$

We construct a unit circle C centered at the origin as the contour, which is suggested by the fact that varies from 0 to 2π. On the contour C,

$$\sin\theta = \frac{z - \frac{1}{z}}{2i}, \quad \cos\theta = \frac{z + \frac{1}{z}}{2}, \quad d\theta = \frac{dz}{iz}.$$

Then the integral we consider turns to

$$\int_C F\left(\frac{z - \frac{1}{z}}{2i}, \frac{z + \frac{1}{z}}{2}\right)\frac{dz}{iz}.$$

Let's see the following example.

Example 5.3.4 Evaluate the the integral $I = \int_0^{2\pi} \frac{1}{1 + a\cos x}\,dx$, where a is a real number with $|a| < 1$.

Solution According to the above idea, we rewrite the integral

$$I = \int_0^{2\pi} \frac{1}{1 + a\cos x}\,dx = \int_{|z|=1} \frac{dz}{\left(1 + a\frac{z + z^{-1}}{2}\right)iz} = -\frac{2i}{a}\int_{|z|=1} \frac{dz}{z^2 + \frac{2}{a}z + 1}.$$

The integrand has two simple poles:

$$z_1 = \frac{-1 + \sqrt{1 - a^2}}{a}, \quad z_2 = \frac{-1 - \sqrt{1 - a^2}}{a}.$$

Notice that $|z_2| > 1$, which means that $z = z_2$ is outside of the unit circle $|z| = 1$. For z_1, it is easy to see $|z_1| < 1$ since $|z_1 z_2| = 1$. That is, z_1 is interior to the unit circle $|z| = 1$. Then

$$\int_{|z|=1} \frac{dz}{z^2 + \frac{2}{a}z + 1} = 2\pi i \operatorname*{Res}_{z=z_1}\left(\frac{1}{z^2 + \frac{2}{a}z + 1}\right) = 2\pi i \frac{1}{z_1 - z_2} = \frac{a\pi i}{\sqrt{1-a^2}}.$$

Finally, $I = \dfrac{2\pi}{\sqrt{1-a^2}}$.

In fact, the method of residues with contours constructed is of great diversity. For the content limit, we end this chapter by another example of application of residue.

Example 5.3.5 Prove the Fresnel integrals:

$$\int_0^{+\infty} \cos(x^2)\,dx = \int_0^{+\infty} \sin(x^2)\,dx = \frac{1}{2}\sqrt{\frac{\pi}{2}}.$$

Proof We construct a contour with three parts:

L_1, a real segment from 0 to R; C_R, an arc $z = Re^{i\theta}$, $0 \leq \theta \leq \frac{\pi}{4}$;

L_2, a segment in the first quadrant from $Re^{i\frac{\pi}{4}}$ to the origin. See Fig. 5.3.3.

Consider the function $f(z) = e^{iz^2}$. It is an analytic function. According to the Cauchy-Goursat theorem, $\int_{L_1 + C_R + L_2} f(z)\,dz = 0$. That is,

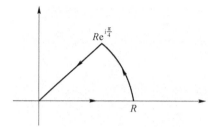

Fig. 5.3.3

$$\int_0^R e^{ix^2}\,dx + \int_0^{\frac{\pi}{4}} e^{iR^2 e^{i2\theta}} R e^{i\theta} i\,d\theta + \int_R^0 e^{ir^2 e^{i\frac{\pi}{2}}} e^{i\frac{\pi}{4}}\,dr = 0.$$

Further,

$$\int_0^R \cos x^2\,dx + i\int_0^R \sin x^2\,dx = -\int_0^{\frac{\pi}{4}} e^{iR^2 e^{i2\theta}} R e^{i\theta} i\,d\theta - e^{i\frac{\pi}{4}}\int_R^0 e^{-r^2}\,dr$$

$$= -iR\int_0^{\frac{\pi}{4}} e^{-R^2 \sin 2\theta} e^{i(R^2 \cos 2\theta + \theta)}\,d\theta + \frac{\sqrt{2}}{2}\int_0^R e^{-r^2}\,dr + i\frac{\sqrt{2}}{2}\int_0^R e^{-r^2}\,dr.$$

Therefore,

$$\int_0^R \cos x^2\,dx = \frac{\sqrt{2}}{2}\int_0^R e^{-r^2}\,dr - \mathrm{Re}\left[iR\int_0^{\frac{\pi}{4}} e^{-R^2 \sin 2\theta} e^{i(R^2 \cos 2\theta + \theta)}\,d\theta\right].$$

$$\int_0^R \sin x^2\,dx = \frac{\sqrt{2}}{2}\int_0^R e^{-r^2}\,dr - \mathrm{Im}\left[iR\int_0^{\frac{\pi}{4}} e^{-R^2 \sin 2\theta} e^{i(R^2 \cos 2\theta + \theta)}\,d\theta\right].$$

From Jordan's inequality,

$$\left|iR\int_0^{\frac{\pi}{4}} e^{-R^2 \sin 2\theta} e^{i(R^2 \cos 2\theta + \theta)}\,d\theta\right| \leq R\int_0^{\frac{\pi}{4}} e^{-R^2 \sin 2\theta}\,d\theta = \frac{R}{2}\int_0^{\frac{\pi}{2}} e^{-R^2 \sin x}\,dx \leq \frac{\pi}{4R},$$

which implies that

$$\lim_{R\to+\infty} iR\int_0^{\frac{\pi}{4}} e^{-R^2 \sin 2\theta} e^{i(R^2 \cos 2\theta + \theta)}\,d\theta = 0.$$

Then

$$\int_0^{+\infty} \cos(x^2)\,dx = \int_0^{+\infty} \sin(x^2)\,dx = \lim_{R\to+\infty}\int_0^R e^{-r^2}\,dr = \frac{1}{2}\sqrt{\frac{\pi}{2}}.$$

The last equality above is a well-known result in Advanced Calculus.

Exercises 5.3

1. Evaluate the integrals:

(1) $\displaystyle\int_{-\infty}^{+\infty} \frac{1}{(1+x^2)^2}\,dx$;

(2) P.V. $\displaystyle\int_{-\infty}^{+\infty} \frac{x^2}{(1+x^2)(1+x^3)}\,dx$.

2. Evaluate the integrals:

(1) $\displaystyle\int_{-\infty}^{+\infty} \frac{x\sin(2x)}{(4+x^2)^2}\,dx$;

(2) $\displaystyle\int_{-\infty}^{+\infty} \frac{x^3 \sin(2x)}{(4+x^2)^2}\,dz$;

(3) P.V. $\displaystyle\int_{-\infty}^{+\infty} \frac{\cos x}{(1+x)^2 + a^2}\,dx$, $a > 0$.

3. Evaluate the integral $I = \int_0^{2\pi} \dfrac{1}{1+a\sin x} \mathrm{d}x$, where a is a real number with $|a|<1$.

4. Evaluate the integral by using residue method and the contour in Fig. 5.3.4:
$$I = \int_0^{+\infty} \dfrac{1}{1+x^5} \mathrm{d}x.$$

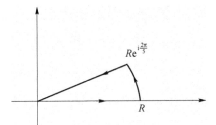

Fig. 5.3.4

Part II
Mathematical Methods for Physics

Chapter 6   Equations of Mathematical Physics and Problems for Defining Solutions

6.1   Basic concept and definition

Based on the fundamental laws of physics, we use mathematical methods to study or measure how the process or the space evolves. Generally, there are three steps. The first one is to establish a series of equations (partial differential equations in most cases), which formulate the evolving process or space within a reasonable degree of idealization. The second one is to analyze the equations. In some cases we can develop methods to get the exact solution. In others, however, we can only achieve some qualitative or quantitative properties about the solutions. Then mathematical simulation such as numerical algorithms will be proposed to explain the unknown solution. The last one is to apply the solution to predict how the process or the space evolves. Of course, there is no randomness in this book. The deterministic equations lead to definite solutions.

In most of applications, the process evolves over time. Roughly speaking, there are two classes of processes. If the state of the system is independent of time, the process is **stationary**, otherwise **non-stationary**. Changing of the process involves energy. From this point of view, processes may be divided into the other two groups: **dissipative** and **conservative** evolution processes.

There is a similarity among the processes governed by the same fundamental physics laws, for example, Newton's second law of motion and Fourier's law of heat conduction. Consider an object is moving in some space. Denote the position of the object by $u(x,t)$ at time t, net force acting on the object by $F(t)$ and mass by m. Newton's Second Law says the resultant force equals the product of mass and acceleration. It can be stated mathematically as $F(t) = m \dfrac{\partial^2 u}{\partial t^2}$.

Consider another case of heat conduction. Heat always conducts from warmer objects to cooler objects. The composition of a material affects its conduction rate when heat flux conducts. For example, if a copper rod and an iron rod are joined together end to end, both rods are homogeneous in thermal conduction and both ends are placed in heat sources, the heat will conduct through the copper end more quickly than the iron end. Fourier considered the phenomenon and summarized them in Fourier's law of heat conduction, which says that the amount of heat flux (ϕ) which flows in a thermally conductive body is proportional to the temperature differences per unit length.

Define the proportionality constant is the thermal conductivity K. Then it can be stated mathematically as $\phi(t) = -K \frac{\partial T(t,x)}{\partial x}$, where $T(x,t)$ is the temperature at position x at time t.

These physical laws are expressed by differential equations which are of simple and beautiful forms. Those processes governed by the same law will satisfy the same equation. Then they naturally behave similarly and predictably. If we arrive at the same equation by refining engineering problems from different disciplines such as chemistry, biology, physics, mechanics, astronomy, and so on, then we solve all those problems by proposing a package of solutions.

Unfortunately, we have to pay the price for enjoying the subtlety and tidiness of those mathematical equations. To solve a series of mathematical physics equations, generally, is not an easy job. Complicated theories are involved to find solutions. Roughly speaking, the following basic mathematics tools are needed, Laplace transform, Fourier transform, L_2 space and its special basis: series of Bessel functions and Legendre polynomials. Each of them is full of profound ideas, strict inferences and lengthy calculation. Here, we do not need to dig so deeply but to know how the tools are involved and applied.

At the outset, let us present some definitions and notations for convenience in the following section since they will be repeatedly used in the following chapters.

6.1.1 Basic concept

In this book, we focus on solving partial differential equations, which consists of two forms, linear ones and the others. Specifically, the latter ones are too complicated for this stage and we start from the former one now.

By partial differential equations, we usually denote such typical form:
$$f(x_1, x_2, \cdots, x_n, u, u_{x_1}, u_{x_2} \cdots, u_{x_n}, u_{x_1 x_1}, u_{x_1 x_2}, \cdots) = 0, \tag{6.1.1}$$
in which f is some function with value 0, x_1, $x_2 \cdots$ are independent (free) variables, $u = u(x_1, x_2, \cdots, x_n)$ is an unknown function of independent variables, u_{x_1}, u_{x_2}, $\cdots$, u_{x_n}, $u_{x_1 x_1}$, $u_{x_1 x_2}$, $\cdots$ are partial derivatives of unknown function u on independent variables x_1, x_2, $\cdots$, x_n and $(x_1, x_2, \cdots, x_n) \in D \subseteq R^n$, $n \geq 2$ where D is an open region in n dimensional Euclidean space R^n. As to the partial differential equations, those partial derivatives are in fact not limited to the first and second orders although they are our main objectives in this book. In the following, PDE is used to denote partial differential equation for convenience.

There are some examples of PDE.

Example 6.1.1

(1) $u_{xx} - u_{yy} = 0$.

(2) $u_x = \dfrac{\partial u}{\partial x} u_{xy} + \dfrac{\partial^2 u}{\partial x^2}$.

(3) $u_{xx} + a^2 u_{yyyy} = 0$.

(4) $u_{xx} + 2xy u_{xy} + u_{yy} = \sin x$.

If we find some sufficiently smooth (differentiable enough times) function $u = u(x_1, x_2, \cdots, x_n)$ which satisfies the equation in (6.1.1), then $u = u(x_1, x_2, \cdots, x_n)$ is called a **solution** of PDE (6.1.1). It should be noticed that the solution is usually not unique.

For example, it is easily verified that

$$u(x,y) = (x+y+C_1)^3 + C_2, \quad u(x,y) = \sin(x-y+C_3) + C_4$$

are both solutions of PDE in Example 6.1.1 (1). Without extra claim, we use C and subscript C to denote constant numbers in the following content.

Definition 6.1.1 (Order of PDE) For equation (6.1.1), the highest order in all orders of partial derivatives of u is called the **order** of the PDE.

It is easy to see that the orders of Example 6.1.1 (1) ~ (4) are 2, 2, 4 and 2 respectively.

Definition 6.1.2 (Linear equation, non-linear equation and quasi-linear equation) A PDE (6.1.1) is called a **linear** one if the following two conditions are satisfied: (1) all unknown functions and their partial derivatives are linear; (2) all coefficients of unknown functions and their partial derivatives are deterministic function only depending on independent variables. For a PDE in (6.1.1) that is not linear, it is called **quasi-linear** one if all partial derivatives of the highest order are linear and **non-linear** one otherwise.

It is a little abuse of notation that we have different meanings between "non-linear" and "not linear". From technical view of point, a large scale of linear equations have been solved since they are not too difficult. Some quasi-linear equations are also solved since they are not so untouchable and sometimes solved by transplanting similar method for linear ones. As to non-linear equations, they often belong to the research scope.

Example 6.1.2

(1) In Example 6.1.1, (1), (3) and (4) are linear, (2) is quasi-linear.

(2) $(u_{xx})^2 + u_{xy} + u_x = e^y u$ is non-linear.

As we concern most in this book, a second order general form is involved for repeated use.

Definition 6.1.3 (General form) A second order PDE with n independent variables is of general form if it is linear and expressed by

$$\sum_{i,j=1}^{n} A_{ij} u_{x_i x_j} + \sum_{i=1}^{n} B_i u_{x_i} + Du = E,$$

where $A_{ij} = A_{ji}$, and A_{ij}, B_i, D and E are all deterministic functions that depend only on n independent variables.

Further, the general form is **homogeneous** if $E = 0$, **non-homogeneous** otherwise.

Example 6.1.3 Find the general solution $u=u(x, y)$ of the following homogeneous equations.

(1) $u_{xx}=0$.

(2) $u_{xy}+u_y=0$.

(3) $u_x=u_y$.

Solution (1) $u_{xx}=0$ implies $u_x=f(y)$. Then
$$u=xf(y)+g(y).$$
Obviously, f and g are arbitrary function of y.

(2)
$$u_{xy}+u_y=0 \Leftrightarrow (u_x+u)_y=0$$
$$\Rightarrow u_x+u=f(x) \Leftrightarrow e^x(u_x+u)=e^x f(x)$$
$$\Rightarrow (e^x u)_x = e^x f(x) \Rightarrow e^x u = \int e^x f(x) \mathrm{d}x + g(y).$$

Then we have
$$u = e^{-x}\left(\int e^x f(x) \mathrm{d}x + g(y)\right),$$
where f and g are arbitrary differentiable functions.

(3) Using transformation on variables (x,y), we set
$$\xi=x+y, \quad \eta=x-y.$$
It is easy to verify by substitution that
$$u_x=u_\xi+u_\eta, \quad u_y=u_\xi-u_\eta, \quad \text{and} \quad 2u_\eta=0,$$
which implies that $u=f(\xi)=f(x+y)$, where f is an arbitrary differentiable function.

6.1.2 Linear operator and linear composition

By operator, we mean a symbol through which a mathematical operation or calculation is defined and acted on some other functions. We define operator L such that: $L[u]=F$, in which an equation can be simply expressed.

Here are some examples of operators:
$$L[u]=\frac{\partial^2 u}{\partial x^2}+\frac{\partial^2 u}{\partial x \partial y}+\frac{\partial^2 u}{\partial y^2}, \quad L=\frac{\partial^2}{\partial x^2}+\frac{\partial^2}{\partial x \partial y}+\frac{\partial^2}{\partial y^2},$$
$$L[u]=\frac{\partial u}{\partial x}+xy\frac{\partial^2 u}{\partial x \partial y}, \quad L=\frac{\partial}{\partial x}+xy\frac{\partial^2}{\partial x \partial y},$$
$$L[u]=\frac{1}{\sqrt{2\pi}}\int_{-\infty}^{+\infty} u(x,t)e^{-i\omega x}\mathrm{d}x, \quad L[\cdot]=\frac{1}{\sqrt{2\pi}}\int_{-\infty}^{+\infty}[\cdot]e^{-i\omega x}\mathrm{d}x.$$

With the operator L, the partial differential equations considered in the preceding Example 6.1.1 can be all written in the form $L[u]=F$, where L stands for

(1) $\dfrac{\partial^2}{\partial x^2}-\dfrac{\partial^2}{\partial y^2}$.

(2) $\dfrac{\partial}{\partial x}-\dfrac{\partial}{\partial x}\dfrac{\partial^2}{\partial x \partial y}+\dfrac{\partial^2}{\partial x^2}$.

(3) $\dfrac{\partial^2}{\partial x^2}+a^2\dfrac{\partial^4}{\partial y^4}$.

(4) $\dfrac{\partial^2}{\partial x^2}+2xy\dfrac{\partial^2}{\partial x \partial y}+\dfrac{\partial^2}{\partial y^2}$.

The operator L involves only differential operation. So they are also called differential operators.

In general, a linear partial differential operator L is such an operation that transforms a function u of the variable $\boldsymbol{x}=(x_1,x_2,\cdots,x_n)$ into another function $L[u]$ given by

$$L[u]= a(\boldsymbol{x})u+ \sum_{i=1}^{n}b_i(\boldsymbol{x})\dfrac{\partial u}{\partial x_i}+\sum_{i,j=1}^{n}c_{i,j}(\boldsymbol{x})\dfrac{\partial^2 u}{\partial x_i \partial x_j}+\cdots.$$

Then the operator L can be written by

$$L = a(\boldsymbol{x})+ \sum_{i=1}^{n}b_i(\boldsymbol{x})\dfrac{\partial}{\partial x_i}+\sum_{i,j=1}^{n}c_{i,j}(\boldsymbol{x})\dfrac{\partial^2}{\partial x_i \partial x_j}+\cdots. \qquad (6.1.2)$$

Here we just discuss the linear partial differential equation with the second order. L is called a second-order differential operator.

Definition 6.1.4 (Linear operator) The term "linear" refers to the fundamental property:

$u_1, u_2, \cdots, u_k$ are any functions possessing requisite derivatives, and $c_1, c_2, \cdots, c_k$ are any constants, then

$$L[c_1 u_1 +\cdots+c_k u_k]=c_1 L[u_1]+\cdots+c_k L[u_k],$$

where $c_1 u_1+\cdots+c_k u_k$ is called a linear combination of $u_1, \cdots, u_k$, and L is called a linear operator.

Equivalently, linear operator L takes every linear combination u_j's into the corresponding linear combination of $L[u_j]$' s.

A linear partial differential equation is simply an equation of the form $L[u]=F$, where L is a linear partial differential operator and F is a function of $\boldsymbol{x}$.

Now we are ready for a basic principle for solving a series of PDE, known as the superposition principle.

Theorem 6.1.1 (Superposition principle) Suppose that u satisfies a linear PDE $L[u]=F$. If $u=c_1 u_1+c_2 u_2+\cdots+c_k u_k$ is a linear combination of $u_1, u_2, \cdots, u_k$, and $F=c_1 F_1+c_2 F_2+\cdots+c_k F_k$ is a linear combination of $F_1, F_2, \cdots, F_k$ of the same structure of u, such that $L[u_j]=F_j$, for $j=1, 2, \cdots, k$, then $u=c_1 u_1+c_2 u_2+\cdots+c_k u_k$ is a solution of PDE $L[u]=F$.

The proof is direct. However, this elementary principle offers an idea to draft a solution. That is, we can divide a complicated PDE system into a series of simple PDEs. If those simple PDEs are easily solvable, then the solution is established by proper superposition.

Example 6.1.4 (Application) Consider a PDE system.

$$\begin{cases} L[u]=F, \\ B[u]=f, \end{cases} \qquad (6.1.3)$$

where L and B are linear differential operators, F and f are given functions. As we known, a PDE $L[u]=F$ often has many solutions. To determine the exact unique solution, it should be equipped with extra conditions. Here $B[u]=f$ is usually viewed as the boundary condition.

In many occasions, homogeneous PDEs are simpler to handle. Hence, we divide (6.1.3) into two homogeneous cases:

$$\begin{cases} L[v]=0, & L[w]=F, \\ B[v]=f, & B[w]=0. \end{cases} \quad (6.1.4)$$

Then $u=v+w$ is a solution of (6.1.3). We turn to each part of (6.1.4). For the first part of (6.1.4), it can also be divided into two cases:

$$\begin{cases} L[v_1]=0, \\ B[v_1]=0, \end{cases} \begin{cases} L[v_2]=0, \\ B[v_2]=f. \end{cases} \quad (6.1.5)$$

We dream that the solution v should be a sum of a general solution v_1 and a special solution v_2. Since the former group of (6.1.5) consists completely of homogeneous PDEs and only one special solution is needed from the latter one, it is not too difficult to find v_1 and v_2. Then we have v and w (via similar way to find v) when this dream sometimes comes to be true with the support of the existence and uniqueness discussion.

Linear PDE is still an important topic of research today. We end this subsection by the famous Schrödinger equation, which in quantum mechanics, is a linear partial differential equation, and describes how the quantum state of a physical system changes with time. It was formulated in late 1925, and published in 1926, by the Austrian physicist Erwin Schrödinger.

Example 6.1.5 (Schrödinger equation)

$$ihu_t = -\frac{h^2}{2m}\nabla^2 u + V(x)u,$$

where u is the quantum-mechanical wave function for a particle of mass m moving in a potential $V(x)$, ∇^2 is the Laplacian operator, i the imaginary unit and h Planck's constant.

6.1.3 Calculation rule of operator

This part is devoted to how we calculate the compound of linear operators.

We define that operators A and B with the same domain of definition D are **equal** if $A[u]=B[u]$ for any $u \in D$.

Definition 6.1.5 (Sum and product) The sum of two differential operators A and B is that $(A+B)[u]=A[u]+B[u]$. The product of two differential operators A and B is that $AB[u]=A[B[u]]$.

Property Differential operators A, B and C satisfy the following properties:

(1) Commutative law of addition: $A+B=B+A$.

(2) Associative law of addition: $(A+B)+C=A+(B+C)$.

(3) Associative law of multiplication: $(AB)C=A(BC)$.

(4) Distributive law of multiplication to addition: $A(B+C)=AB+AC$, $(A+B)C=AC+BC$.

For multiplication of differential operators, the commutative law is not always true.

Example 6.1.6 Define $A=\dfrac{\partial^2}{\partial x^2}+x\dfrac{\partial}{\partial y}$, $B=\dfrac{\partial^2}{\partial y^2}-y\dfrac{\partial}{\partial y}$.

Chapter 6 Equations of Mathematical Physics and Problems for Defining Solutions

$$AB[u]=A[B[u]]=A\left[\frac{\partial^2 u}{\partial y^2}-y\frac{\partial u}{\partial y}\right]=\left(\frac{\partial^2}{\partial x^2}+x\frac{\partial}{\partial y}\right)\left(\frac{\partial^2 u}{\partial y^2}-y\frac{\partial u}{\partial y}\right)$$

$$=\frac{\partial^4 u}{\partial x^2\partial y^2}-y\frac{\partial^3 u}{\partial x^2\partial y}+x\frac{\partial^3 u}{\partial y^3}-xy\frac{\partial^2 u}{\partial y^2}-x\frac{\partial u}{\partial y};$$

$$BA[u]=B[A[u]]=B\left[\frac{\partial^2 u}{\partial x^2}+x\frac{\partial u}{\partial y}\right]=\left(\frac{\partial^2}{\partial y^2}-y\frac{\partial u}{\partial y}\right)\left(\frac{\partial^2 u}{\partial x^2}+x\frac{\partial u}{\partial y}\right)$$

$$=\frac{\partial^4 u}{\partial y^2\partial x^2}+x\frac{\partial^3 u}{\partial y^3}-y\frac{\partial^3 u}{\partial y\partial x^2}-xy\frac{\partial^2 u}{\partial y^2}.$$

Therefore, $AB[u]\neq BA[u]$.

6.2 Three typical partial differential equations and problems for defining solutions

6.2.1 Wave equations and physical derivations

Wave equations in R, R^2, R^3 and R^n:

$$\frac{\partial^2 u}{\partial t^2}=a^2\frac{\partial^2 u}{\partial x^2},$$

$$\frac{\partial^2 u}{\partial t^2}=a^2\left(\frac{\partial^2 u}{\partial x^2}+\frac{\partial^2 u}{\partial y^2}\right),$$

$$\frac{\partial^2 u}{\partial t^2}=a^2\left(\frac{\partial^2 u}{\partial x^2}+\frac{\partial^2 u}{\partial y^2}+\frac{\partial^2 u}{\partial z^2}\right),$$

$$\frac{\partial^2 u}{\partial t^2}=a^2\left(\frac{\partial^2 u}{\partial x_1^2}+\frac{\partial^2 u}{\partial x_2^2}+\cdots+\frac{\partial^2 u}{\partial x_n^2}\right).$$

In abbreviation, $\frac{\partial^2 u}{\partial t^2}=a^2\nabla^2 u$.

Here, u represents waves traveling through medium spaces R, R^2, R^3 and R^n respectively, "a" the speed of propagation of waves, $u=u(x,t)$ displacement or amplitude at position x and time t.

The wave equation provides a reasonable mathematical model for number of physical processes, such as

(1) vibrations of a stretched string, for example, a guitar string;

(2) vibrations of a column of air;

(3) vibrations of a stretched membrane;

(4) sound waves in air or other elastic media;

(5) electromagnetic wave, for example, light wave and radio wave.

Here we will just establish the equation about a simple wave in R instead of deriving all the wave equations from physical principles. However, we should point out that in most cases the derivation involves some simplifying assumptions, and the wave equation gives only an approximate description of the actual physical phenomenon.

Example 6.2.1 (String oscillation) An oscillation in a string is a wave. A sound can be produced by a vibrating string. In this case, the frequency of the sound is usually constant. Therefore, the sound produced by a vibrating string is a constant note since the pitch is characterized by frequency. For example, string instrument like guitar and cello consists of several different strings. All the beautiful sound, melody and rhythm come from wave.

Consider an ideal and tight string. Suppose that $u(x,t)$ is the displacement of the string at time t and position x. By the term "ideal" we mean:

(1) The string is soft and elastic and the tension is the same with its tangent.

(2) The string is not stretched and the tension is constant by Hooke's Law.

(3) The ratio of weight and tension is small enough.

Besides the above assumptions, we still need to confine the oscillation into such an occasion:

(4) The ratio of displacement and length is small.

(5) The slope on any point after displacement is small.

(6) The string only vibrates transversely.

We amplify the micro-element of the string. See Fig. 6.2.1.

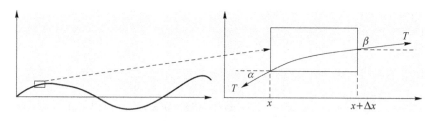

Fig. 6.2.1

Denote by Δs the arc length of a small piece of string, Δx the projection of Δs on the horizontal axis, ρ the linear density. Since the slope is very small, i.e., α and β are small,

$$\cos\alpha \approx \cos\beta \approx 1, \quad \Delta x \approx \Delta s, \quad \sin\alpha \approx \tan\alpha, \sin\beta \approx \tan\beta. \tag{6.2.1}$$

Then by Newton's second law along the vertical direction, we have

$$T\sin\beta - T\sin\alpha = \rho \Delta s u_{tt}.$$

In view of (6.2.1),

$$\tan\beta - \tan\alpha = \frac{\rho \Delta x}{T} u_{tt}. \tag{6.2.2}$$

The tangents of the angles at the ends of the string piece are approximately equal to the slopes at the ends. Using this fact we rewrite (6.2.2)

$$\frac{1}{\Delta x}\left(\frac{\Delta u}{\Delta x}\bigg|_{x+\Delta x} - \frac{\Delta u}{\Delta x}\bigg|_{x}\right) = \frac{\rho}{T} u_{tt}. \tag{6.2.3}$$

Taking limit $\Delta x \to 0$ on the left side of (6.2.3) we have

$$u_{tt} = a^2 u_{xx},$$

where

$$a^2 = \frac{T}{\rho}.$$

6.2.2 Heat (conduction) equations and physical derivations

Heat equations in R, R^2, R^3 and R^n:

$$\frac{\partial u}{\partial t} = a^2 \frac{\partial^2 u}{\partial x^2},$$

$$\frac{\partial u}{\partial t} = a^2 \left(\frac{\partial^2 u}{\partial x^2} + \frac{\partial^2 u}{\partial y^2} \right),$$

$$\frac{\partial u}{\partial t} = a^2 \left(\frac{\partial^2 u}{\partial x^2} + \frac{\partial^2 u}{\partial y^2} + \frac{\partial^2 u}{\partial z^2} \right),$$

$$\frac{\partial u}{\partial t} = a^2 \left(\frac{\partial^2 u}{\partial x_1^2} + \frac{\partial^2 u}{\partial x_2^2} + \cdots + \frac{\partial^2 u}{\partial x_n^2} \right).$$

In abbreviation, $\frac{\partial u}{\partial t} = a^2 \nabla^2 u$.

Heat equation describes the diffusion of thermal energy in a homogeneous material. Here, $u = u(x, t)$ represents the temperature at time t and point x. Heat flows through medium in spaces R, R^2, R^3 and R^n respectively, Constant "a^2" called thermal diffusion, varies with density ρ and conduction rate K.

The heat equation provides models for the following process:

(1) heat flows along a wire or rod;

(2) heat flows along a membrane;

(3) heat flows in air;

(4) diffusion processes such as some gases or liquids, and so on.

There is one more thing we must notice. The heat equation is NOT a fundamental law of physics, since its solution contractions to the existence of absolute zero.

Here we also give only a simple case in R.

Example 6.2.2 (Derivation of heat conduction in one dimension) Suppose the heat flows in a homogeneous rod, which means the thermal conductivity rate through the rod is constant.

Two principles, Fourier's law and conservation of energy, are involved to derive the heat conduction.

Firstly, by Fourier's law, the rate of flow of heat energy through a surface is proportional to the negative temperature gradient through the media. In one dimension, the gradient is an ordinary spatial derivative, that is u_x. Then we have

$$q = -K u_x, \tag{6.2.4}$$

where q is the rate of heat flux, K is the thermal conductivity rate and u is the temperature.

Secondly, we still focus on the micro-element: a small piece of rod, $x - \Delta x \leqslant \xi \leqslant x + \Delta x$, a short time interval $t - \Delta t \leqslant \eta \leqslant t + \Delta t$.

If no work is done and there are neither heat sources nor sinks, the change in internal energy $Q_{\Delta F}$ in the small piece $x - \Delta x \leqslant \xi \leqslant x + \Delta x$ and the short time interval $t - \Delta t \leqslant \eta \leqslant t + \Delta t$ is accounted for entirely by the flux of heat across the boundaries. By (6.2.4),

$$Q_{\Delta F} = K\int_{t-\Delta t}^{t+\Delta t} u_x(x+\Delta x, \eta) - u_x(x-\Delta x, \eta)\,\mathrm{d}\eta$$
$$= K\int_{t-\Delta t}^{t+\Delta t}\int_{x-\Delta x}^{x+\Delta x} u_{\xi\xi}(\xi, \eta)\,\mathrm{d}\xi\mathrm{d}\eta. \tag{6.2.5}$$

In the absence of work, a change ΔQ in internal energy per unit length in the rod is proportional to the change in temperature Δu. That is,
$$\Delta Q = c\rho \Delta u, \tag{6.2.6}$$
where c is the specific heat capacity and ρ is the linear density of the rod.

From (6.2.6), the increase in internal energy Q_Δ in the small piece $x-\Delta x \leqslant \xi \leqslant x+\Delta x$, the short time interval $t-\Delta t \leqslant \eta \leqslant t+\Delta t$ should be
$$Q_\Delta = c\rho \int_{x-\Delta x}^{x+\Delta x} u(\xi, t+\Delta t) - u(\xi, t-\Delta t)\,\mathrm{d}\xi$$
$$= c\rho \int_{x-\Delta x}^{x+\Delta x}\int_{t-\Delta t}^{t+\Delta t} u_\eta(\xi, \eta)\,\mathrm{d}\eta\mathrm{d}\xi \tag{6.2.7}$$
$$= c\rho \int_{t-\Delta t}^{t+\Delta t}\int_{x-\Delta x}^{x+\Delta x} u_\eta(\xi, \eta)\,\mathrm{d}\xi\mathrm{d}\eta.$$

By conservation of energy,
$$Q_\Delta = Q_{\Delta F}.$$

From (6.2.6) and (6.2.7), we finally have
$$\int_{t-\Delta t}^{t+\Delta t}\int_{x-\Delta x}^{x+\Delta x} c\rho u_\eta(\xi, \eta) - K u_{\xi\xi}(\xi, \eta)\,\mathrm{d}\xi\mathrm{d}\eta = 0,$$
which implies that
$$c\rho u_\eta(\xi, \eta) - K u_{\xi\xi}(\xi, \eta) = 0, \quad \text{i.e.,}$$
$$u_\eta(\xi, \eta) = a^2 u_{\xi\xi}(\xi, \eta), \quad \text{where} \quad a^2 = \frac{K}{c\rho}.$$

6.2.3 Laplace equations and physical derivations

Laplace equation consists of two cases:
$$\frac{\partial^2 u}{\partial x^2} + \frac{\partial^2 u}{\partial y^2} = 0,$$
$$\frac{\partial^2 u}{\partial x^2} + \frac{\partial^2 u}{\partial y^2} + \frac{\partial^2 u}{\partial z^2} = 0.$$
In abbreviation, $\nabla^2 u = 0$.

The Laplace equation provides models for the following processes:
(1) the electrostatic potential in any region containing no electric change;
(2) the gravitational potential in any region containing no mass;
(3) standing waves and steady-state heat distributions, and so on.

Example 6.2.3 (Gravitation potential) Consider two particles in P_1 and P_2 with masses m_1 and m_2. By Newton's gravitation law, we have the force F
$$F = G\frac{m_1 m_2}{r^2}, \tag{6.2.8}$$
where G is the gravitation, assuming to be 1 without loss of generality, and r is the distance

between the two particles.

If we set r is the vector from P_1 to P_2, the action force F of particle at P_1 to a unit mass, then we have a vector form of (6.2.8),

$$F = -\frac{m_1 r}{r^3} = \nabla\left(\frac{m_1}{r}\right). \tag{6.2.9}$$

If a unit mass at P_2 is pushed by the particle at P_1 to infinity, from (6.2.9), the work of the action force F is

$$V = \int_r^\infty F \, dr = \int_r^\infty \nabla\left(\frac{m_1}{r}\right) dr = -\frac{m_1}{r}. \tag{6.2.10}$$

Combining (6.2.9) and (6.2.10), we have $F = -\nabla V$.

Similarly, the work W of the action force F attracting the unit mass from infinity to P_2 is

$$W = \frac{m_1}{r}. \tag{6.2.11}$$

Consider an object Q in some region R (P_2 is not contained in R) with density $\rho(x,y,z)$. We divide it into many small particles with volume R_i. Then from (6.2.11), we have

$$V_i(x,y,Z) \approx \frac{\rho(\xi,\eta,\zeta)}{r} dR_i,$$

where $r = \sqrt{(x-\xi)^2 + (y-\eta)^2 + (z-\zeta)^2}$.

Hence, the work V of object Q attracting the unit mass from infinity to P_2 is

$$V(x,y,Z) = \iiint_R \frac{\rho(\xi,\eta,\zeta)}{r} d\xi d\eta d\zeta.$$

With careful calculations, we have $\Delta V(x,y,Z) = \nabla^2 V(x,y,Z) = 0$, which is Laplace (potential) equation. Δ is called Laplace operator or Laplacian.

6.3 Well-posed problem

In the above sections, we see some PDEs governed by physics principles. However, those PDEs cannot completely describe a specific physical phenomenon yet. For example, people can hardly see the same wave in a pool because they choose different times and positions to observe; the string oscillation is obviously different between a free string or one with the end fixed. In each case, the same wave equations are satisfied. Another example is the solid thermal problems. A solid surface contacting with the heat resource or outer heat-insulating is different in both cases. The particular environment, in which the system evolves, is subject to the external forces which are not described by the equation with physics principles governing. Some extra equations are required for the physical condition on the boundary.

In fact, any given physical problem depends on "reasonable" conditions. To predict the movement of a wave (heat conduction and potential) exactly, people should equip the PDEs with some proper conditions, such as initial ones, boundary ones, limitation ones and so on.

Endowing universal partial differential equations with initial conditions and boundary conditions leads to a well-posed problem.

The following content contributes to the general form of the proper conditions.

6.3.1 Initial conditions

For Laplace equation, the solutions are independent of time. There are no initial conditions for this equation.

For heat equation, the initial condition is
$$u(x,y,z,t)|_{t=0} = \varphi(x,y,z).$$
Here $\varphi(x,y,z)$ is a given function.

The heat equation only involves the first-order derivative in t. So, it is usually enough to propose one initial value for the heat equation.

In contrast, the wave equation uses second-order derivative in the time variable t. It is appropriate to specify not only the initial values of u but also the initial velocity $\frac{\partial u}{\partial t}$.

For the wave equation, the initial conditions are
$$u(x,y,z,t)|_{t=0} = \varphi(x,y,z),$$
$$\frac{\partial u}{\partial t}(x,y,z,t)|_{t=0} = \psi(x,y,z).$$
Here φ and ψ are given functions.

6.3.2 Boundary conditions

We denote the boundary surfaces of the spatial region D. There are three cases:

(1) The first class of boundary condition (Dirichlet condition).
$$u(\boldsymbol{x},t)|_{x \in s} = f_1(\boldsymbol{x},t).$$
Dirichlet condition gives the distribution of the boundary values in the request.

For example, consider the vibration of a fixed string. It is easy to see that both ends of the string will never move. We gives the distribution of the boundary values in both ends in the following form
$$u(x,t)|_{x=0} = u(x,t)|_{x=l} = 0.$$
(2) The second class of boundary condition (Neumann condition).
$$\left.\frac{\partial u}{\partial \boldsymbol{n}}\right|_{x \in s} = f_1(\boldsymbol{x},t).$$
Neumann condition gives the normal derivative (directional derivative projecting to the normal vector $\boldsymbol{n}$) on the boundary S.

For example, consider the heat equation. Assume that it is held a constant temperature with heat diffusion on the boundary S. That is the first class boundary condition.
$$u(\boldsymbol{x},t)|_{x \in s} = u_0.$$
Another reasonable assumption is that S is insulated, so that heat cannot flow in or

across S. Mathematically, this amount requires the normal derivative of u through the boundary S.

$$\left.\frac{\partial u}{\partial \boldsymbol{n}}\right|_{x \in s} = 0.$$

(3) The third class of boundary condition (mixed boundary condition).

$$\left.\left(u + \sigma \frac{\partial u}{\partial \boldsymbol{n}}\right)\right|_{x \in s} = f_3(\boldsymbol{x}, t).$$

The mixed boundary condition gives a linear combination of the boundary values of u and its normal derivative.

For example, in a thermal problem, a more realistic assumption is that the region outside D is held at a constant temperature u_0 and the rate of heat flow across the boundary S is proportional to the difference in temperatures on the two sides:

$$\left.\left(\sigma u + \frac{\partial u}{\partial \boldsymbol{n}}\right)\right|_{S} = \sigma u_0.$$

Notice that those condition equations are also linear partial differential ones. Some are non-homogeneous. To solve a well-posed problem, is to solve a series of linear partial differential equations. This process is to find a general solution which satisfies the conditions of defining solution. So it is also called defining a solution.

At the end of this chapter, we give the non-homogeneous forms of the three typical partial differential equations we mentioned above. By adding in an extra terms, those equations are generalized to the non-homogeneous equations

$$\frac{\partial^2 u}{\partial t^2} = a^2 \nabla^2 u + F(\boldsymbol{x}, t) \quad \text{(wave equation)},$$

$$\frac{\partial u}{\partial t} = a^2 \nabla^2 u + F(\boldsymbol{x}, t) \quad \text{(heat equation)},$$

$$\nabla^2 u = F(\boldsymbol{x}) \quad \text{(Laplace equation)}.$$

$\nabla^2 u = F(\boldsymbol{x})$ is also called Poission equation.

Chapter 7 Classification and Simplification for Linear Second Order PDEs

7.1 Classification of linear second order partial differential equations with two variables

In this chapter, we will focus on the linear second order partial differential equations (PDEs) with two variables, which are of the form
$$Au_{xx}+Bu_{xy}+Cu_{yy}+Du_x+Eu_y+Fu=G, \tag{7.1.1}$$
where $u=u(x,y)$ is the unknown function with two variables (x,y), and those A, B, C, D, E, F, G are all given real functions with variables (x,y), satisfying
$$A^2+B^2+C^2>0.$$
The typical wave equation in $\mathbf{R}^1$
$$u_{tt}-c^2u_{xx}=0,$$
the heat conduction equation in $\mathbf{R}^1$
$$u_t-ku_{xx}=0,$$
and the Laplace equation in $\mathbf{R}^2$
$$u_{xx}+u_{yy}=0$$
are all linear second order PDEs. And as we will see later, each equation corresponds to a special type of these linear second order PDEs.

The type of a linear second order PDE is determined by the sign of its discriminant
$$\Delta=B^2-4AC.$$

(1) If $\Delta>0$ at a point (x_0,y_0) or in a domain D of $\mathbf{R}^2$, the PDE is of **hyperbolic type** at the point or in the domain;

(2) If $\Delta=0$ at a point (x_0,y_0) or in a domain D of $\mathbf{R}^2$, the PDE is of **parabolic type** at the point or in the domain;

(3) If $\Delta<0$ at a point (x_0,y_0) or in a domain D of $\mathbf{R}^2$, the PDE is of **elliptic type** at the point or in the domain.

Note that the type of a PDE is defined at a specified point or in a domain, and may vary in different points or domains.

Exercises 7.1

Discuss the type of the following PDEs:
(a) $u_{xx}+8u_{xy}+16u_{yy}+64u=0$;
(b) $3u_{xx}+u_{xy}-2u_{yy}-30u_x+20u_y=0$;
(c) $u_{xx}+2u_{xy}+5u_{yy}+u_x=0$;
(d) $u_{xx}+2u_{xy}+(1-f(y))u_{yy}=0$, where $f(y)=\begin{cases}-1, & y<-1,\\ 0, & -1\leqslant y\leqslant 1,\\ 1, & y>1.\end{cases}$

7.2 Simplification to standard forms

The linear second order PDE
$$Au_{xx}+Bu_{xy}+Cu_{yy}+Du_x+Eu_y+Fu=G \qquad (7.2.1)$$
can be simplified under nondegenerate variable transformations:
$$\begin{cases}\xi=\xi(x,y),\\ \eta=\eta(x,y).\end{cases} \qquad (7.2.2)$$
By "nondegenerate" we mean that the Jacobian of the transformation does not vanish in the region where the equation is being considered:
$$J=\begin{vmatrix}\xi_x & \xi_y\\ \eta_x & \eta_y\end{vmatrix}\neq 0.$$
Hence, by the implicit function theorem, (7.2.2) can be solved for x and y, so that
$$\begin{cases}x=x(\xi,\eta),\\ y=y(\xi,\eta).\end{cases} \qquad (7.2.3)$$
By the chain rule, we have
$$u_x=u_\xi\xi_x+u_\eta\eta_x, \quad u_y=u_\xi\xi_y+u_\eta\eta_y,$$
$$u_{xx}=u_{\xi\xi}\xi_x^2+2u_{\xi\eta}\xi_x\eta_x+u_{\eta\eta}\eta_x^2+u_\xi\xi_{xx}+u_\eta\eta_{xx},$$
$$u_{xy}=u_{\xi\xi}\xi_x\xi_y+u_{\xi\eta}(\xi_x\eta_y+\xi_y\eta_x)+u_{\eta\eta}\eta_x\eta_y+u_\xi\xi_{xy}+u_\eta\eta_{xy},$$
$$u_{yy}=u_{\xi\xi}\xi_y^2+2u_{\xi\eta}\xi_y\eta_y+u_{\eta\eta}\eta_y^2+u_\xi\xi_{yy}+u_\eta\eta_{yy}.$$
Thus the PDE (7.2.1) on $u=u(x,y)$ is transformed to the following PDE on $u=u(\xi,\eta)$,
$$A^*u_{\xi\xi}+B^*u_{\xi\eta}+C^*u_{\eta\eta}+D^*u_\xi+E^*u_\eta+F^*u=G^*, \qquad (7.2.4)$$
where
$$A^*=A\xi_x^2+B\xi_x\xi_y+C\xi_y^2,$$
$$B^*=2A\xi_x\eta_x+B(\xi_x\eta_y+\xi_y\eta_x)+2C\xi_y\eta_y,$$
$$C^*=A\eta_x^2+B\eta_x\eta_y+C\eta_y^2,$$
$$D^*=A\xi_{xx}+B\xi_{xy}+C\xi_{yy}+D\xi_x+E\xi_y,$$

$$E^* = A\eta_{xx} + B\eta_{xy} + C\eta_{yy} + D\eta_x + E\eta_y,$$
$$F^* = F,$$
$$G^* = G.$$

Keep in mind that those A^*, B^*, C^*, D^*, E^*, F^* and G^* in (7.2.4) are functions of ξ and η by (7.2.3). Direct calculations show that

$$(B^*)^2 - 4A^*C^* = J^2(B^2 - 4AC). \tag{7.2.5}$$

Hence the type of the PDE is preserved everywhere in some simply connected region.

Now we will show how to choose the variable transformation (7.2.2) to simplify the PDE (7.2.1). Here by "simplify" we hope that as many among A^*, B^*, C^*, the coefficients of second order terms of the PDE (7.2.1), as possible are zero.

Recall that the type of a 2-dimensional quadratic form $Ax^2 + Bxy + Cy^2$, where A, B, C are scalars, is determined by the sign of its discriminant:

$$\Delta = B^2 - 4AC. \tag{7.2.6}$$

(1) If $\Delta > 0$, then $Ax^2 + Bxy + Cy^2 = G$ is the equation of a hyperbola;

(2) If $\Delta = 0$, then $Ax^2 + Bxy + Cy^2 = G$ is the equation of a parabola;

(3) If $\Delta < 0$, then $Ax^2 + Bxy + Cy^2 = G$ is the equation of an ellipse.

Hence, the characteristic equation

$$Ar^2 + Br + C = 0 \tag{7.2.7}$$

has

(1) two different real roots r_1 and r_2 in the hyperbolic case;

(2) a double root r in the parabolic case;

(3) two complex conjugate roots $r_1 = \alpha + i\beta$ and $r_2 = \alpha - i\beta$ in the elliptic case.

Let us begin with the PDE (7.2.1) that is hyperbolic in a simply connected region D. First of all, in this case, we may assume that one of coefficients A or C does not vanish. Indeed, if $A = C = 0$, then it must hold that $B \neq 0$, and then the PDE (7.2.1) is just

$$u_{xy} + \frac{D}{B}u_x + \frac{E}{B}u_y + \frac{F}{B}u + \frac{G}{B} = 0, \tag{7.2.8}$$

which is already in the first standard form as defined later. So in the following, we assume that one of A or C, say A, does not vanish.

While determining the variable transformation (7.2.2) to transform the PDE (7.2.1) into its standard form, the hyperbolic partial differential equation corresponding to the characteristic equation (7.2.7) is a first-order nonlinear characteristic differential equation

$$A\psi_x^2 + B\psi_x\psi_y + C\psi_y^2 = 0, \tag{7.2.9}$$

or, with the assumption $A \neq 0$,

$$\left(\frac{\psi_x}{\psi_y}\right)^2 + \frac{B\psi_x}{A\psi_y} + \frac{C}{A} = 0. \tag{7.2.10}$$

We proclaim that solutions to (7.2.10) can be found by solving the ordinary differential equation

$$Ady^2 - Bdxdy + Cdx^2 = 0, \tag{7.2.11}$$

known as the characteristic equation of the PDE (7.2.1). Let $\psi(x,y) \equiv \text{const}$ be a solution to the characteristic equation (7.2.11), then $\psi = \psi(x,y)$ is a solution to (7.2.10). Indeed,

the equation $\psi(x, y) \equiv$ const yields that
$$\frac{dy}{dx} = -\frac{\psi_x}{\psi_y}.$$
Substituting it into (7.2.11) leads to
$$A\left(\frac{\psi_x}{\psi_y}\right)^2 + B\frac{\psi_x}{\psi_y} + C = 0,$$
which is equivalent to (7.2.10) since $A \neq 0$. The characteristic equation (7.2.11) can be solved as
$$\frac{dy}{dx} = \frac{B \pm \sqrt{\Delta}}{2A}, \qquad (7.2.12)$$
which are linear first order ODEs. The solutions are called characteristic lines.

In the case when the PDE (7.2.1) is hyperbolic, $\Delta > 0$. Hence (7.2.11) has two different solutions:
$$\psi_1(x,y) \equiv \text{const}, \quad \psi_2(x,y) \equiv \text{const}.$$
We then take the following variable transformation:
$$\begin{cases} \xi = \psi_1(x,y), \\ \eta = \psi_2(x,y). \end{cases}$$
Since
$$J = \begin{vmatrix} \xi_x & \xi_y \\ \eta_x & \eta_y \end{vmatrix} = \xi_y \eta_y \left(\frac{\xi_x}{\xi_y} - \frac{\eta_x}{\eta_y}\right) = \xi_y \eta_y \frac{\sqrt{\Delta}}{A} \neq 0,$$
this variable transformation is nondegenerate. With such variable transformation, we can easily see that $A^* = C^* = 0$ in the resulted PDE (7.2.4). That is, the hyperbolic PDE (7.2.1) is transformed to its first standard form
$$u_{\xi\eta} = H(u_\xi, u_\eta, u, \xi, \eta).$$

Now we consider a parabolic PDE, that is, $\Delta = 0$. In this case, there is only one real root (or a double root) of the equation (7.2.11), say $\psi(x,y) \equiv$ const. Hence, we may equate only one of A^* and C^*, say A^*, to zero, by taking
$$\xi = \psi(x,y).$$
Setting the second new independent variable as any smooth function of y,
$$\eta = \phi(y),$$
such that $\phi'(y) \neq 0$, then the mixed second derivative term will vanish. Indeed, with the above ξ and η as new independent variables, we have
$$B^* = 2A\xi_x\eta_x + B(\xi_x\eta_y + \xi_y\eta_x) + 2C\xi_y\eta_y$$
$$= \xi_y\eta_y\left(B\frac{\xi_x}{\xi_y} + 2C\right) = \xi_y\eta_y\left(-B\frac{B}{2A} + 2C\right) = 0,$$
where the second equality uses $\eta_x = 0$ and the last equality uses $\Delta = B^2 - 4AC = 0$. Hence with the above ξ and η as new independent variables, we transform the parabolic PDE (7.2.1) into its standard form
$$u_{\eta\eta} = H(u_\xi, u_\eta, u, \xi, \eta).$$
Similarly, if we set $\eta = \psi(x,y)$ and $\xi = \phi(x)$, then we transform (7.2.1) into its another

standard form
$$u_{\xi\xi} = H(u_\xi, u_\eta, u, \xi, \eta).$$

For the elliptic case where $\Delta < 0$, the equation (7.2.11) has two complex conjugate solutions:
$$\phi(x,y) \pm i\psi(x,y) \equiv \text{const},$$
where $\phi(x,y)$ and $\psi(x,y)$ are real functions. So if set
$$\begin{cases} \Phi = \phi(x,y) + i\psi(x,y), \\ \Psi = \phi(x,y) - i\psi(x,y), \end{cases}$$
as new variables, then the elliptic PDE is transformed into the form
$$u_{\Phi\Psi} = H(u_\Phi, u_\Psi, u, \Phi, \Psi).$$
Then by setting
$$\begin{cases} \xi = \dfrac{\Phi + \Psi}{2} = \phi(x,y), \\ \eta = \dfrac{\Phi - \Psi}{2i} = \psi(x,y), \end{cases}$$
we have
$$u_{\xi\xi} + u_{\eta\eta} = H(u_\xi, u_\eta, u, \xi, \eta),$$
which returns the equation to the real plane. This is the elliptic standard form.

We now present several examples to illustrate how to simplify the linear second order PDE to its corresponding standard form.

Example 7.2.1 Determine the type of the PDE
$$x^2 u_{xx} + 2xy u_{xy} + y^2 u_{yy} = 0, \quad (x^2 + y^2 \neq 0),$$
and transform it into its standard form.

Solution The discriminant is
$$\Delta = B^2 - 4AC = 4x^2 y^2 - 4x^2 y^2 = 0.$$
Hence it is a parabolic PDE.

When $x = 0$ or $y = 0$, it is already in the standard form. When $xy \neq 0$, the characteristic equations is
$$x^2 dy^2 - 2xy dx dy + y^2 dx^2 = 0.$$
It has only one solution:
$$\frac{dy}{dx} = \frac{y}{x}.$$
By Calculus, the solution is
$$\frac{y}{x} = \text{const}.$$
So we may take the new variables as
$$\begin{cases} \xi = \dfrac{y}{x}, \\ \eta = y. \end{cases}$$
The Jacobian of the transformation is

Chapter 7 Classification and Simplification for Linear Second Order PDEs

$$J = \begin{vmatrix} \xi_x & \xi_y \\ \eta_x & \eta_y \end{vmatrix} = \begin{vmatrix} -\dfrac{y}{x^2} & \dfrac{1}{x} \\ 0 & 1 \end{vmatrix} = -\dfrac{y}{x^2} \neq 0.$$

So the variable transformation is nondegenerate. Direct calculations show that with such variable transformation, the PDE is transformed to

$$\eta^2 u_{\eta\eta} = 0,$$

or equivalently,

$$u_{\eta\eta} = 0,$$

the parabolic standard form.

Example 7.2.2 Determine the type of the PDE

$$y^2 u_{xx} - x^2 u_{yy} = 0, \quad (x^2 + y^2 \neq 0),$$

and transform it into its standard form.

Solution The discriminant is

$$\Delta = B^2 - 4AC = 4x^2 y^2.$$

When $x = 0$ or $y = 0$, $\Delta = 0$, and it is of parabolic type. Otherwise, $\Delta > 0$, and it is of hyperbolic type.

When $x = 0$ or $y = 0$, it is already in the parabolic standard form. When $xy \neq 0$, the characteristic equations is

$$y^2 \, dy^2 - x^2 \, dx^2 = 0.$$

It has two different real solutions:

$$\frac{dy}{dx} = \pm \frac{x}{y}.$$

By Calculus, the solutions are

$$\frac{1}{2} y^2 \pm \frac{1}{2} x^2 = \text{const}.$$

So by taking new variables as

$$\begin{cases} \xi = \dfrac{1}{2} y^2 - \dfrac{1}{2} x^2, \\ \eta = \dfrac{1}{2} y^2 + \dfrac{1}{2} x^2, \end{cases}$$

with corresponding Jacobian

$$J = \begin{vmatrix} \xi_x & \xi_y \\ \eta_x & \eta_y \end{vmatrix} = \begin{vmatrix} -x & y \\ x & y \end{vmatrix} = -2xy \neq 0,$$

the original PDE is transformed into

$$u_{\xi\eta} = \frac{\eta}{2(\xi^2 - \eta^2)} u_\xi - \frac{\xi}{2(\xi^2 - \eta^2)} u_\eta,$$

the hyperbolic standard form.

Example 7.2.3 Determine the type of the PDE

$$u_{xx} + x^2 u_{yy} = 0,$$

and transform it into its standard form.

Solution The discriminant is

$$\Delta = B^2 - 4AC = -4x^2.$$

So when $x=0$, it is of parabolic type, and otherwise, it is of elliptic type.

When $x=0$, it is already in the parabolic standard form. When $x \neq 0$, the characteristic equations is

$$dy^2 + x^2 dx^2 = 0.$$

It has two complex conjugate solutions:

$$\frac{dy}{dx} = \pm ix.$$

Obviously, the solutions are

$$2y \pm ix^2 = \text{const}.$$

So by taking new variables as

$$\begin{cases} \xi = 2y, \\ \eta = x^2, \end{cases}$$

with corresponding Jacobian

$$J = \begin{vmatrix} \xi_x & \xi_y \\ \eta_x & \eta_y \end{vmatrix} = \begin{vmatrix} 0 & 2 \\ 2x & 0 \end{vmatrix} = -4x \neq 0,$$

the original PDE is transformed into

$$u_{\xi\xi} + u_{\eta\eta} = -\frac{1}{2\eta} u_\eta,$$

the elliptic standard form.

Sometimes when the simplified standard form is easy to solve, we may then find the general solution of the original PDE.

Example 7.2.4 Find the general solution of

$$u_{xx} - 2\sin x \, u_{xy} - \cos^2 x \, u_{yy} - \cos x \, u_y = 0.$$

Solution The discriminant is

$$\Delta = B^2 - 4AC = 4\sin^2 x + 4\cos^2 x = 4 > 0.$$

So it is a hyperbolic PDE. The characteristic equations is

$$dy^2 + 2\sin x \, dx dy - \cos^2 x \, dx^2 = 0.$$

It has two different real solutions:

$$\frac{dy}{dx} = -\sin x \pm 1.$$

The solutions are

$$x \pm (y - \cos x) = \text{const}.$$

Let

$$\begin{cases} \xi = x - y + \cos x, \\ \eta = x + y - \cos x, \end{cases}$$

then the Jacobian is

$$J = \begin{vmatrix} \xi_x & \xi_y \\ \eta_x & \eta_y \end{vmatrix} = \begin{vmatrix} 1 + \sin x & -1 \\ 1 + \sin x & 1 \end{vmatrix} = -2 \neq 0,$$

and the original PDE is transformed into

$$u_{\xi\eta}=0,$$
whose general solution can be easily found as
$$u=f(\xi)+g(\eta).$$
So the general solution of the original PDE is
$$u(x, y)=f(x-y+\cos x)+g(x+y-\cos x),$$
where $f(\cdot)$ and $g(\cdot)$ are arbitrary smooth functions.

Example 7.2.5 Find the general solution of
$$x^2 u_{xx}-2xy u_{xy}+y^2 u_{yy}+x u_x+y u_y=0 \quad (y\neq 0).$$
Solution The discriminant is
$$\Delta=B^2-4AC=4x^2 y^2-4x^2 y^2=0.$$
So the PDE is of parabolic type. The characteristic equation is
$$x^2 \mathrm{d}y^2+2xy\,\mathrm{d}x\,\mathrm{d}y+y^2 \mathrm{d}x^2=0.$$
It has only one real solution
$$\frac{\mathrm{d}y}{\mathrm{d}x}=-\frac{y}{x}.$$
By Calculus, the solution is
$$xy=\text{const.}$$
So we may let
$$\begin{cases} \xi=xy,\\ \eta=y. \end{cases}$$
The Jacobian of the transformation is
$$J=\begin{vmatrix} \xi_x & \xi_y \\ \eta_x & \eta_y \end{vmatrix}=\begin{vmatrix} y & x \\ 0 & 1 \end{vmatrix}=y\neq 0,$$
so the variable transformation is nondegenerate, and the original PDE is transformed into
$$u_{\eta\eta}=-\frac{1}{\eta}u_\eta.$$
To solve this equation, rewrite it as
$$\eta u_{\eta\eta}+u_\eta=0,$$
or equivalently $(\eta u_\eta)_\eta=0$. So as a function of ξ and η,
$$\eta u_\eta=f(\xi),$$
where $f(\cdot)$ is an arbitrary smooth function. Then from $u_\eta=\frac{1}{\eta}f(\xi)$, we know that
$$u(\xi, \eta)=f(\xi)\ln|\eta|+g(\xi),$$
where $g(\cdot)$ is an arbitrary smooth function. Hence the general solution of the original PDE is
$$u(x,y)=f(xy)\ln|y|+g(xy).$$
Example 7.2.6 Find the general solution of
$$4u_{xx}+5u_{xy}+u_{yy}+u_x+u_y=2.$$
Solution The discriminant is
$$\Delta=B^2-4AC=25-16=9>0.$$

So this PDE is of hyperbolic type. The characteristic equation is
$$4dy^2 - 5dxdy + dx^2 = 0.$$
It has two different real solutions
$$\frac{dy}{dx} = 1 \quad \text{or} \quad \frac{1}{4},$$
that is,
$$y - x = \text{const}, \quad \text{or} \quad y - \frac{x}{4} = \text{const}.$$
So we may let
$$\begin{cases} \xi = y - x, \\ \eta = y - \dfrac{x}{4}. \end{cases}$$
The Jacobian of the transformation is
$$J = \begin{vmatrix} \xi_x & \xi_y \\ \eta_x & \eta_y \end{vmatrix} = \begin{vmatrix} -1 & 1 \\ -\dfrac{1}{4} & 1 \end{vmatrix} = -\frac{3}{4} \neq 0.$$
With this variable transformation, the PDE is transformed into
$$u_{\xi\eta} = \frac{1}{3} u_\eta - \frac{8}{9}.$$
Let $v = u_\eta$, then the equation becomes
$$v_\xi = \frac{1}{3} v - \frac{8}{9}.$$
Premultiplying $e^{-\frac{1}{3}\xi}$ on both sides of the above equation leads to
$$(e^{-\frac{1}{3}\xi} v)_\xi = -\frac{8}{9} e^{-\frac{1}{3}\xi},$$
then we have
$$e^{-\frac{1}{3}\xi} v = \frac{8}{3} e^{-\frac{1}{3}\xi} + f_1(\eta),$$
or equivalently,
$$u_\eta(\xi, \eta) = v(\xi, \eta) = \frac{8}{3} + e^{\frac{1}{3}\xi} f_1(\eta).$$
From the above equation, we can solve
$$u(\xi, \eta) = \frac{8}{3} \eta + e^{\frac{1}{3}\xi} f(\eta) + g(\xi).$$
where $f(\cdot)$ and $g(\cdot)$ are arbitrary smooth functions. Hence the general solution of the original PDE is
$$u(x, y) = \frac{8}{3} \left(y - \frac{x}{4}\right) + e^{\frac{1}{3}(y-x)} f\left(y - \frac{x}{4}\right) + g(y - x).$$

Exercises 7.2

1. Transform the PDEs in Exercises 7.1 into the standard form.

2. Find the general solution of the following PDEs:
   (a) $4u_{xx} - 7u_{xy} - 2u_{yy} - 243(2x+y)(x-4y)^2 = 0$;
   (b) $16u_{xx} - 24u_{xy} + 9u_{yy} + 36u_x - 27u_y - 9 = 0$;
   (c) $u_{xx} - 6u_{xy} + 9u_{yy} = xy^2$;
   (d) $y^5 u_{xx} - yu_{yy} + 2u_y = 0$, $(y>0)$.

Chapter 8 Integral Method on Characteristics

8.1 D'Alembert formula for one dimensional infinite string oscillation

Consider the Cauchy problem of the one dimensional infinite string oscillation equation:
$$\begin{cases} u_{tt}-a^2 u_{xx}=0, & -\infty<x<+\infty,\ t>0, \\ u(x,0)=\phi(x), & -\infty<x<+\infty, \\ u_t(x,0)=\psi(x), & -\infty<x<+\infty. \end{cases}$$
where $a>0$ is a constant, $\phi(x)$ and $\psi(x)$ are known functions. The discriminant of the PDE is
$$\Delta=B^2-4AC=4a^2>0.$$
Hence it is a hyperbolic PDE, and the characteristic equations is
$$\mathrm{d}x^2-a^2\,\mathrm{d}t^2=0,$$
which has two different real solutions
$$\frac{\mathrm{d}x}{\mathrm{d}t}=\pm a.$$
The solutions are
$$x-at=\text{const},\quad\text{or}\quad x+at=\text{const}.$$
So by letting
$$\begin{cases} \xi=x-at, \\ \eta=x+at, \end{cases}$$
the PDE is transformed to
$$u_{\xi\eta}=0,$$
whose solution is
$$u(\xi,\eta)=f(\xi)+g(\eta),$$
where $f(\cdot)$ and $g(\cdot)$ are arbitrary smooth functions. So the general solution of the PDE is
$$u(x,t)=f(x-at)+g(x+at).$$
Substituting the initial conditions into the general solution gives
$$\begin{cases} u(x,0)=f(x)+g(x)=\phi(x), \\ u_t(x,0)=-af'(x)+ag'(x)=\psi(x). \end{cases}$$

The second equation is an ordinary differential equation, whose solution is
$$-f(x)+g(x)=\frac{1}{a}\int_{x_0}^{x}\psi(z)\,dz+c,$$
where x_0 is any fixed point, and c is an arbitrary constant. Then we have
$$f(x)=\frac{1}{2}\phi(x)-\frac{1}{2a}\int_{x_0}^{x}\psi(z)\,dz-\frac{c}{2},$$
$$g(x)=\frac{1}{2}\phi(x)+\frac{1}{2a}\int_{x_0}^{x}\psi(z)\,dz+\frac{c}{2}.$$

Hence the solution to the Cauchy problem is
$$u(x,t)=f(x-at)+g(x+at)$$
$$=\frac{1}{2}\phi(x-at)-\frac{1}{2a}\int_{x_0}^{x-at}\psi(z)\,dz+\frac{1}{2}\phi(x+at)+\frac{1}{2a}\int_{x_0}^{x+at}\psi(z)\,dz$$
$$=\frac{1}{2}(\phi(x-at)+\phi(x+at))+\frac{1}{2a}\int_{x-at}^{x+at}\psi(z)\,dz.$$

This is known as the **D'Alembert formula**. The terms $\phi(x-at)$ and $\phi(x+at)$ are called the right and left travelling wave, respectively. Fig. 8.1.1 is an illustration of $\phi(x-at)$ at different time t. As time t increases, the wave is travelling right.

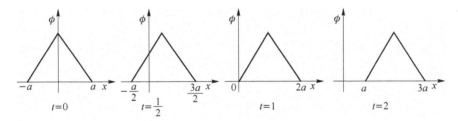

Fig. 8.1.1

Example 8.1.1 Solve the following Cauchy problem:
$$\begin{cases} u_{tt}-a^2 u_{xx}=0, & -\infty<x<+\infty,\ t>0, \\ u(x,0)=2\sin x, & -\infty<x<+\infty, \\ u_t(x,0)=2\cos x, & -\infty<x<+\infty. \end{cases}$$

Solution By D'Alembert formula, we have
$$u(x,t)=\frac{1}{2}(2\sin(x-at)+2\sin(x+at))+\frac{1}{2a}\int_{x-at}^{x+at}2\cos z\,dz$$
$$=\sin(x-at)+\sin(x+at)+\frac{1}{a}\sin z\Big|_{x-at}^{x+at}$$
$$=\left(1+\frac{1}{a}\right)\sin(x+at)+\left(1-\frac{1}{a}\right)\sin(x-at).$$

Example 8.1.2 Solve the following Cauchy problem:
$$\begin{cases} u_{tt}-u_{xx}=0, & -\infty<x<+\infty,\ t>0, \\ u(x,0)=0, & -\infty<x<+\infty, \\ u_t(x,0)=xe^{-x^2}, & -\infty<x<+\infty. \end{cases}$$

Solution By D'Alembert formula, we have

$$u(x,t) = 0 + \frac{1}{2}\int_{x-t}^{x+t} z e^{-z^2} dz$$

$$= -\frac{1}{4}e^{-z^2}\big|_{x-t}^{x+t} = \frac{1}{4}(e^{-(x-t)^2} - e^{-(x+t)^2}).$$

Exercises 8.1

1. Solve the following Cauchy problem by D'Alembert formula:
$$\begin{cases} u_{tt} - 4u_{xx} = 0, & -\infty < x < +\infty, \ t > 0, \\ u(x,0) = \sin x, & -\infty < x < +\infty, \\ u_t(x,0) = x\sin x, & -\infty < x < +\infty. \end{cases}$$

2. Let $u(x, t)$ be the solution of the Cauchy problem
$$\begin{cases} u_{tt} - 9u_{xx} = 0, & -\infty < x < +\infty, \ t > 0, \\ u(x,0) = f(x) = \begin{cases} 1, & |x| \leqslant 2, \\ 0, & |x| > 2, \end{cases} \\ u_t(x,0) = g(x) = \begin{cases} 1, & |x| \leqslant 2, \\ 0, & |x| > 2. \end{cases} \end{cases}$$

(a) Find $u\left(0, \dfrac{1}{6}\right)$.

(b) Discuss the large time behavior of the solution, that is, for fixed ξ, find $\lim\limits_{t\to\infty} u(\xi, t)$.

8.2 Small oscillations of semi-infinite string with rigidly fixed or free ends, method of prolongation

In the previous section, we discuss the problem of small oscillation of the one dimensional infinite string, that is, $-\infty < x < +\infty$. In this section, we will consider the problem of small oscillation of the semi-infinite string, that is, $0 < x < +\infty$.

The small oscillation of a semi-infinite unloaded string with rigidly fixed end $x = 0$ is described as
$$\begin{cases} u_{tt} - a^2 u_{xx} = 0, & 0 < x < +\infty, \ t > 0, \\ u(x,0) = \phi(x), & 0 \leqslant x < +\infty, \\ u_t(x,0) = \psi(x), & 0 \leqslant x < +\infty, \\ u(0,t) = 0, & t \geqslant 0. \end{cases}$$

If we ignore the last boundary condition, the D'Alembert formula in the previous section shows that the solution can be expressed as
$$u(x,t) = \frac{1}{2}(\phi(x-at) + \phi(x+at)) + \frac{1}{2a}\int_{x-at}^{x+at} \psi(z)dz.$$

However, now the functions $\phi(x)$ and $\psi(x)$ are only defined for $x \geqslant 0$. We need to continuously prolong these two functions into the left half real axis $x < 0$, such that the last

boundary condition $u(0,t)=0$ is satisfied. Setting $x=0$ in the above equation, we have
$$u(0,t)=\frac{1}{2}(\phi(-at)+\phi(at))+\frac{1}{2a}\int_{-at}^{at}\psi(z)\mathrm{d}z.$$
So if we prolong the functions $\phi(x)$ and $\psi(x)$ as odd functions, that is,
$$\phi(-x)=-\phi(x), \quad \psi(-x)=-\psi(x),$$
then the boundary condition $u(0,t)=0$ can be satisfied. So in all, the solution is
$$u(x,t)=\frac{1}{2}(\phi(x-at)+\phi(x+at))+\frac{1}{2a}\int_{x-at}^{x+at}\psi(z)\mathrm{d}z,$$
where $\phi(x)$ and $\psi(x)$ are prolonged as odd functions.

Example 8.2.1 Solve the following problem:
$$\begin{cases} u_{tt}-a^2 u_{xx}=0, & 0<x<+\infty, t>0, \\ u(x,0)=\sin^2 x, & 0\leqslant x<+\infty, \\ u_t(x,0)=0, & 0\leqslant x<+\infty, \\ u(0,t)=0, & t\geqslant 0. \end{cases}$$

Solution We need to prolong the original $\phi(x)=\sin^2 x$ and $\psi(x)=0$ as odd functions, which results in
$$\phi(x)=\begin{cases} \sin^2 x, & x\geqslant 0, \\ -\sin^2 x, & x<0, \end{cases}$$
$$\psi(x)=0, \quad -\infty<x<+\infty.$$
Hence the solution is
$$u(x,t)=\frac{1}{2}(\phi(x-at)+\phi(x+at))$$
$$=\begin{cases} \frac{1}{2}(-\sin^2(x-at)+\sin^2(x+at)), & 0<x\leqslant at, \\ \frac{1}{2}(\sin^2(x-at)+\sin^2(x+at)), & x>at. \end{cases}$$

Now we consider the small oscillation of a semi-infinite unloaded string with free end $x=0$, which is described as
$$\begin{cases} u_{tt}-a^2 u_{xx}=0, & 0<x<+\infty, t>0, \\ u(x,0)=\phi(x), & 0\leqslant x<+\infty, \\ u_t(x,0)=\psi(x), & 0\leqslant x<+\infty, \\ u_x(0,t)=0, & t\geqslant 0. \end{cases}$$
Similarly if we ignore the last boundary condition, the D'Alembert formula shows that the solution can be expressed as
$$u(x,t)=\frac{1}{2}(\phi(x-at)+\phi(x+at))+\frac{1}{2a}\int_{x-at}^{x+at}\psi(z)\mathrm{d}z.$$
Again we need to continuously prolong the functions $\phi(x)$ and $\psi(x)$ into the left half real axis $x<0$, such that the last boundary condition $u_x(0,t)=0$ is satisfied.

Differentiating $u(x,t)$ in the above equation with respect to x and setting $x=0$ give
$$u_x(0,t)=\frac{1}{2}(\phi'(-at)+\phi'(at))+\frac{1}{2a}(\psi(at)-\psi(-at)).$$

So if we prolong the functions $\phi(x)$ and $\psi(x)$ as even functions, that is,
$$\phi(-x)=\phi(x), \quad \psi(-x)=\psi(x),$$
then the boundary condition $u_x(0,t)=0$ is satisfied. In all, the solution is
$$u(x,t)=\frac{1}{2}(\phi(x-at)+\phi(x+at))+\frac{1}{2a}\int_{x-at}^{x+at}\psi(z)\mathrm{d}z,$$
where $\phi(x)$ and $\psi(x)$ are prolonged as even functions.

Example 8.2.2 Solve the following problem:
$$\begin{cases} u_{tt}=u_{xx}, & 0<x<+\infty,\ t>0, \\ u(x,0)=\mathrm{e}^{-x^2}, & 0\leqslant x<+\infty, \\ u_t(x,0)=0, & 0\leqslant x<+\infty, \\ u_x(0,t)=0, & t\geqslant 0. \end{cases}$$

Solution These $\phi(x)=\mathrm{e}^{-x^2}$ and $\psi(x)=0$ are even functions in the whole real axis. So the solution is
$$u(x,t)=\frac{1}{2}(\phi(x-t)+\phi(x+t))=\frac{1}{2}(\mathrm{e}^{-(x-t)^2}+\mathrm{e}^{-(x+t)^2}).$$

Exercises 8.2

Solve the following problems:

(a) $\begin{cases} u_{tt}-u_{xx}=0, & 0<x<+\infty,\ t>0, \\ u(x,0)=x^3+x^6, & 0\leqslant x<+\infty, \\ u_t(x,0)=\sin^2 x, & 0\leqslant x<+\infty, \\ u(0,t)=0, & t\geqslant 0. \end{cases}$

(b) $\begin{cases} u_{tt}-u_{xx}=0, & 0<x<+\infty,\ t>0, \\ u(x,0)=x^3+x^6, & 0\leqslant x<+\infty, \\ u_t(x,0)=\sin^2 x, & 0\leqslant x<+\infty, \\ u_x(0,t)=0, & t\geqslant 0. \end{cases}$

8.3 Integral method on characteristics for other second order PDEs, some examples

The integral method on characteristics can be extended to more general linear second order PDEs. In this section, we will illustrate by some examples.

Example 8.3.1 Solve the following problem:
$$\begin{cases} u_{xx}+2u_{xy}-3u_{yy}=0, \\ u(x,0)=3x^2, \\ u_y(x,0)=0. \end{cases}$$

Solution The discriminant is

$$\Delta = B^2 - 4AC = 4 + 12 > 0.$$

Hence it is a hyperbolic PDE. The characteristic equation is
$$dy^2 - 2dxdy - 3dx^2 = 0,$$
whose solutions are
$$\frac{dy}{dx} = -1 \quad \text{or} \quad \frac{dy}{dx} = 3,$$
that is,
$$x + y = \text{const} \quad \text{or} \quad 3x - y = \text{const}.$$

So we may take the new variables as
$$\begin{cases} \xi = x + y, \\ \eta = 3x - y, \end{cases}$$
then the PDE is transformed to $u_{\xi\eta} = 0$. Hence the general solution is
$$u = f(\xi) + g(\eta) = f(x+y) + g(3x-y).$$
Substituting the conditions into the general solution leads to
$$\begin{cases} u(x,0) = f(x) + g(3x) = 3x^2, \\ u_y(x,0) = f'(x) - g'(3x) = 0. \end{cases}$$
From the second equation we have
$$f(x) - \frac{1}{3} g(3x) = c,$$
where c is an arbitrary constant. Then we can find out
$$f(x) = \frac{3}{4}(x^2 + c) \quad \text{and} \quad g(3x) = \frac{3}{4}(3x^2 - c).$$
The second equation implies that $g(x) = \frac{1}{4}x^2 - \frac{3}{4}c$. Hence the solution to the problem is
$$u(x,y) = f(x+y) + g(3x-y) = 3x^2 + y^2.$$

Example 8.3.2 Solve the following problem:
$$\begin{cases} u_{tt} - u_{xx} = 0, \quad -\infty < x < +\infty, \ t > |x|, \\ u(x,t)|_{l_1} = \psi(x), \\ u(x,t)|_{l_2} = \phi(x), \end{cases}$$
where $\psi(0) = \phi(0)$ and l_1, l_2 are lines $x = t$, $x = -t$, respectively.

Solution The discriminant is $\Delta = B^2 - 4AC = 4 > 0$, so the PDE is of hyperbolic type. The characteristic equation is
$$dx^2 - dt^2 = 0,$$
whose solutions are $\frac{dx}{dt} = \pm 1$, or $x \pm t = \text{const}$. So by setting
$$\begin{cases} \xi = x - t, \\ \eta = x + t, \end{cases}$$
the PDE is transformed to $u_{\xi\eta} = 0$. The general solution is
$$u = f(\xi) + g(\eta) = f(x-t) + g(x+t).$$
Substituting the boundary conditions into the general solution shows that

$$\begin{cases} u(x,t)|_{l_1}=u(x,t)|_{x=t}=f(0)+g(2x)=\psi(x), \\ u(x,t)|_{l_2}=u(x,t)|_{x=-t}=f(2x)+g(0)=\phi(x). \end{cases}$$

Then we can find out

$$\begin{cases} f(x)=\phi\left(\dfrac{x}{2}\right)-g(0), \\ g(x)=\psi\left(\dfrac{x}{2}\right)-f(0). \end{cases}$$

Setting $x=0$ in the two equations shows that $f(0)+g(0)=\phi(0)=\psi(0)$. Hence the solution to the problem is

$$u(x,y)=f(x-t)+g(x+t)=\phi\left(\dfrac{x-t}{2}\right)+\psi\left(\dfrac{x+t}{2}\right)-\psi(0).$$

Example 8.3.3 Solve the following problem:

$$\begin{cases} u_{xx}+yu_{yy}+\dfrac{1}{2}u_y=0, & -\infty<x<+\infty,\ y<0, \\ u(x,0)=h(x), \\ |u_y(x,0)|<+\infty. \end{cases}$$

Solution Noting $y<0$, the discriminant is $\Delta=B^2-4AC=-4y>0$. So the PDE is of hyperbolic type. The characteristic equation is

$$dy^2+ydx^2=0.$$

It has two different solutions:

$$\dfrac{dy}{dx}=\pm\sqrt{-y},$$

that is,

$$x\pm 2\sqrt{-y}=\text{const.}$$

Taking the new variables as

$$\begin{cases} \xi=x-2\sqrt{-y}, \\ \eta=x+2\sqrt{-y}, \end{cases}$$

then the PDE is transformed to $u_{\xi\eta}=0$. Hence the general solution is

$$u=f(\xi)+g(\eta)=f(x-2\sqrt{-y})+g(x+2\sqrt{-y}).$$

Substituting the boundary conditions into the general solution leads to

$$u(x,0)=f(x)+g(x)=h(x),$$

$$|u_y(x,0)|=\left|\dfrac{f'(x-2\sqrt{-y})-g'(x+2\sqrt{-y})}{\sqrt{-y}}\right|_{y=0}<+\infty.$$

The second inequality shows that it must hold

$$f'(x)-g'(x)=0,$$

or

$$f(x)-g(x)=c.$$

So we have

$$f(x)=\dfrac{1}{2}h(x)+\dfrac{c}{2},\quad g(x)=\dfrac{1}{2}h(x)-\dfrac{c}{2}.$$

Hence the solution to the problem is
$$u(x,y)=f(x-2\sqrt{-y})+g(x+2\sqrt{-y})$$
$$=\frac{1}{2}[h(x-2\sqrt{-y})+h(x+2\sqrt{-y})].$$

Exercises 8.3

Solve the following problems:

(a) $\begin{cases} u_{xx}+6u_{xy}-16u_{yy}=0, \\ u(-x,\ 2x)=x, \\ u(x,\ 0)=\sin 2x. \end{cases}$

(b) $\begin{cases} u_{xx}+4u_{xy}+u_x=0, \\ u(x,\ 8x)=0, \\ u_x(x,\ 8x)=4e^{-2x}. \end{cases}$

Chapter 9　The Method of Separation of Variables on Finite Region

There are in fact only a limited number of methods available to solve partial differential equations analytically without introducing approximate or numerical techniques. For a few equations, such as wave equation, a general solution can be written which involves arbitrary functions. These arbitrary functions of the partial differential equation are equivalent to the arbitrary constant which arises in the solution of ordinary differential equation. To obtain the solution of an ordinary differential equation, the boundary conditions are used to fix the arbitrary constants. In the partial differential case, the boundary conditions lead to a functional equation for the arbitrary function. The solutions of such functional equations can be as difficult as the original partial differential equation, although some progress can be made along these lines and this is pursued in Chapter 8.

A powerful method is the method of ***separation of variables***. Here the solution of an equation, say $u(x,t)$, is expressed in the form
$$u(x,t) = X(x)T(t),$$
in which the solution can be separated into a product of a function purely of x and another purely of t. This form is substituted into the differential equation and results in ordinary differential equations for the single variable functions $X(x)$ and $T(t)$. A set of such solutions is obtained and these can be summed because of linearity to give a "general solution". The boundary conditions are applied to this solution and these restrict the summed functions to a subset, and yield the coefficients of the series. The latter process is effectively one of expressing the solution as a series of orthogonal functions. The process is the expansion of the solution as a generalized Fourier series with close links to Fourier expansion in signal analysis. There are considerable theoretical grounds for expecting such a method to be successful. Primarily, the theory of Fourier series and orthogonal functions indicate that quite general classes of functions can be approximated by convergent series of orthogonal functions, and hence one might expect the series of separated functions to be effective. Clearly if no such split exists for a given problem the approach will break down.

To describe the separation of variables method, each of the standard problems will be taken in turn, and fairly general problem will be solved by the method in each case. The wave equation is the first to be considered.

Chapter 9 The Method of Separation of Variables on Finite Region

9.1 Separation of variables for (1+1)-dimensional homogeneous equations

The partial differential equation that function $u(x,t)$ satisfies is called the (1+1)-dimensional differential equation, which denotes the partial differential equation with one dimensional space variable and one dimensional time variable. We will give the method of separation of variables to solve this problem in what follows.

9.1.1 Separation of variables for wave equation on finite region

The problem of an elastic string stretched to length l and then fixed at its endpoints constitutes a simple problem on which to consider the method of separation of variables. The wave equation is solved to find the deflection of the string at time t. From the physical derivation in chapter 6, the vibrations of an elastic string are governed by the wave equation in the form

$$\frac{\partial^2 u}{\partial t^2} = a^2 \frac{\partial^2 u}{\partial x^2},$$

where $u(x,t)$ is the deflection of the string at a distance x along the string at a time t. The string is fixed at its endpoints $x=0$ and $x=l$, yielding the boundary conditions $u(0,t) = u(l,t) = 0$ for $t \geqslant 0$. For the initial conditions, let the deflection at $t=0$ be described by the function $f(x)$, so that $u(x,0) = f(x)$, and let the velocity at $t=0$ be similarly described by the function $g(x)$. The latter condition yields $\left.\frac{\partial u}{\partial t}\right|_{t=0} = g(x)$, and both of these conditions are valid for $0 \leqslant x \leqslant l$.

Thus, the defining problem we considered is concluded as

$$\begin{cases} \dfrac{\partial^2 u}{\partial t^2} = a^2 \dfrac{\partial^2 u}{\partial x^2}, & 0 < x < l,\ t > 0, & (9.1.1) \\ u(0,t) = u(l,t) = 0, & t \geqslant 0, & (9.1.2) \\ u(x,0) = f(x),\ \left.\dfrac{\partial u}{\partial t}\right|_{t=0} = g(x), & 0 \leqslant x \leqslant l. & (9.1.3) \end{cases}$$

According to the method of separation of variables, we now assume that the solution may be expressed in the form

$$u(x,t) = X(x) T(t), \qquad (9.1.4)$$

where $X(x)$ and $T(t)$ are functions of only x and t, respectively. Substituting (9.1.4) into (9.1.1), and since $\frac{\partial^2 u}{\partial x^2} = X''(x) T(t)$, $\frac{\partial^2 u}{\partial t^2} = X(x) T''(t)$, we have

$$X(x) T''(t) = a^2 X''(x) T(t).$$

Dividing its both sides by $a^2 X(x) T(t)$, we transform the partial differential equation into

$$\frac{T''(t)}{a^2 T(t)} = \frac{X''(x)}{X(x)}.$$

It is clear that the left-hand side of this equation is a function of t and the right-hand side is independent of t. Hence the equation can only have a solution of the form (9.1.4) if both sides have the same constant value. Let this constant value be $-\lambda$ to give

$$\frac{T''(t)}{a^2 T(t)} = \frac{X''(x)}{X(x)} = -\lambda. \tag{9.1.5}$$

Then solving the solution of (9.1.1) is reduced to solving the ordinary differential equations

$$T''(t) + \lambda a^2 T(t) = 0, \tag{9.1.6}$$

and

$$X''(x) + \lambda X(x) = 0. \tag{9.1.7}$$

The solution $u(x,t)$ must now be forced to satisfy the boundary conditions, $u(0,t) = X(0)T(t) = 0$ and $u(l,t) = X(l)T(t) = 0$. Therefore either $T(t) \equiv 0$ for all t giving $u(x,t) \equiv 0$ for all t, which is a trivial solution, or $X(0) = X(l) = 0$ for all t.

Hence for non-trivial solutions, consider

$$\begin{cases} X''(x) + \lambda X(x) = 0, \\ X(0) = X(l) = 0. \end{cases} \tag{9.1.8}$$

Since λ is an arbitrary constant, we need determine its value first to derive the non-trivial solution. This problem is called the **eigenvalue problem** and λ is called **eigenvalue**, and the corresponding solution of $X(x)$, is called **eigenfunction**.

We will discuss three cases, namely $\lambda > 0$, $\lambda = 0$ and $\lambda < 0$, for which each gives a different form of solution. As this is the first example of this type, the three cases will be pursued in detail, but with experience the valid choice for the given boundary conditions is usually written without explicitly considering each avenue in detail. Hence the cases are as follows:

(1) $\lambda = 0$ gives $X''(x) = 0$ with general solution $X(x) = Ax + B$. The boundary condition $X(0) = 0$ then gives $B = 0$ and the condition $X(l) = 0$ gives $0 = Al + B$, so forcing $A = 0$. Hence this case generates the trivial solution $X = 0$.

(2) For the case $\lambda < 0$, let $\lambda = -\beta^2$ to give $X''(x) - \beta^2 X(x) = 0$. The general solution is

$$X(x) = Ae^{\beta x} + Be^{-\beta x}.$$

The boundary conditions $X(0) = 0$ and $X(l) = 0$ then give

$$0 = A + B \quad \text{and} \quad 0 = Ae^{\beta l} + Be^{-\beta l},$$

which yield the trivial solution with $A = B = 0$.

(3) The third case with $\lambda > 0$, which can be written as $\lambda = p^2$, gives $X''(x) + p^2 X(x) = 0$. Then the general solution is

$$X(x) = A\cos px + B\sin px$$

and the above boundary conditions then give

$$A = 0 \quad \text{and} \quad 0 = B\sin pl.$$

Hence either $B = 0$ or $\sin pl = 0$. $B = 0$ gives the trivial solution once again, and $\sin pl = 0$ gives $pl = n\pi$, $n = 1, 2, 3, \cdots$ or $p = \dfrac{n\pi}{l}$, $n = 1, 2, 3, \cdots$.

Therefore, we solved the eigenvalue problem (9.1.8) and obtained the eigenvalues and corresponding eigenfunctions as follows:

$$\lambda_n = \frac{n^2 \pi^2}{l^2}, \quad (n=1, 2, \cdots), \tag{9.1.9}$$

$$X_n(x) = B_n \sin \frac{n\pi}{l} x, \quad (n=1, 2, \cdots). \tag{9.1.10}$$

Substituting the values of λ_n into (9.1.6), we have

$$T_n''(t) + \frac{a^2 n^2 \pi^2}{l^2} T_n(t) = 0.$$

Its general solution is

$$T_n(t) = a_n' \cos \frac{n\pi}{l} at + b_n' \sin \frac{n\pi}{l} at. \tag{9.1.11}$$

Then

$$u_n(x,t) = X_n(x) T_n(t)$$
$$= \left(a_n \cos \frac{n\pi}{l} at + b_n \sin \frac{n\pi}{l} at \right) \sin \frac{n\pi}{l} x, \quad n=1, 2, \cdots \tag{9.1.12}$$

are solutions of the equation, where $a_n = a_n' B_n$ and $b_n = b_n' B_n$.

According to the superposition principle, a linear combination of u_n is also a solution of (9.1.1), that is

$$u(x,t) = \sum_{n=1}^{\infty} u_n(x,t) = \sum_{n=1}^{\infty} (a_n \cos \frac{n\pi}{l} at + b_n \sin \frac{n\pi}{l} at) \sin \frac{n\pi}{l} x. \tag{9.1.13}$$

It is obvious that $u(x,t)$ satisfies the homogeneous conditions. In the following, we will determine the value of the coefficients a_n and b_n.

From (9.1.3), we have

$$u\big|_{t=0} = u(x,0) = \sum_{n=1}^{\infty} a_n \sin \frac{n\pi}{l} x = f(x),$$

$$u_t\big|_{t=0} = \sum_{n=1}^{\infty} b_n \frac{an\pi}{l} \sin \frac{n\pi}{l} x = g(x).$$

The coefficients a_n and b_n may now be determined by Fourier series technique and are given by the Fourier coefficients

$$a_n = \frac{2}{l} \int_0^l f(x) \sin \frac{n\pi x}{l} dx. \tag{9.1.14}$$

and

$$b_n = \frac{2}{n\pi a} \int_0^l g(x) \sin \frac{n\pi x}{l} dx, \quad n = 1, 2, 3, \cdots. \tag{9.1.15}$$

Substituting these coefficients into (9.1.13) gives

$$u(x,t) = \sum_{n=1}^{\infty} \left[\left(\frac{2}{l} \int_0^l f(x) \sin \frac{n\pi}{l} x dx \right) \cos \frac{an\pi}{l} t + \left(\frac{2}{an\pi} \int_0^l g(x) \sin \frac{n\pi}{l} x dx \right) \sin \frac{an\pi}{l} t \right] \sin \frac{n\pi}{l} x, \tag{9.1.16}$$

which is the solution of the wave equation subject to the given conditions.

Before leaving this solution to consider some examples, it is useful at this stage to link this work to the more general solution which was obtained in Chapter 8. To this end, the

solution may be rewritten as
$$u(x,t) = \sum_{n=1}^{\infty} \left(a_n \cos \frac{an\pi t}{l} + b_n \sin \frac{an\pi t}{l} \right) \sin \frac{n\pi x}{l},$$
which yields
$$u(x,t) = \sum_{n=1}^{\infty} \begin{bmatrix} \frac{a_n}{2} \left(\sin \frac{n\pi}{l}(x+at) + \sin \frac{n\pi}{l}(x-at) \right) + \\ \frac{b_n}{2} \left(\cos \frac{n\pi}{l}(x-at) - \cos \frac{n\pi}{l}(x+at) \right) \end{bmatrix}$$
$$= \sum_{n=1}^{\infty} \begin{bmatrix} \left(\frac{a_n}{2} \sin \frac{n\pi}{l}(x+at) - \frac{b_n}{2} \cos \frac{n\pi}{l}(x+at) \right) + \\ \left(\frac{b_n}{2} \cos \frac{n\pi}{l}(x-at) + \frac{a_n}{2} \sin \frac{n\pi}{l}(x-at) \right) \end{bmatrix}$$
$$= f(x+at) + g(x-at),$$
which is the form of the general solution of (9.1.1) precisely used in Chapter 8.

9.1.2 Separation of variables for heat equation on finite region

The heat equation is the second of the canonical forms derived in the previous chapter. A typical problem would be to consider the case of heat flux of a bar of finite length l. Suppose the ends $x=0$ and $x=l$ are perfectly insulated. The flux of heat through the endpoints is proportional to the values of $\frac{\partial u}{\partial x}$ at each endpoint. Let the initial temperature distribution be
$$u(x,0) = f(x) = \beta x, \quad 0 < x < l,$$
making the full problem to solve
$$\frac{\partial u}{\partial t} = c^2 \frac{\partial^2 u}{\partial x^2} \tag{9.1.17}$$
subject to
$$\frac{\partial u}{\partial x}(0,t) = \frac{\partial u}{\partial x}(l,t) = 0 \tag{9.1.18}$$
and
$$u(x,0) = f(x) = \beta x, \quad 0 < x < l. \tag{9.1.19}$$

In the spirit of the separation of variables method, put $u(x,t) = X(x)T(t)$ where $X(x)$ and $T(t)$ are functions of x and t respectively, then the heat equation gives
$$X(x)T'(t) = c^2 T(t) X''(x) \quad \text{or} \quad \frac{T'(t)}{c^2 T(t)} = \frac{X''(x)}{X(x)} = -\lambda, \tag{9.1.20}$$
as each side is a function of a different independent variable and must therefore be a constant. Hence
$$T'(t) + c^2 \lambda T(t) = 0 \quad \text{and} \quad X''(x) + \lambda X(x) = 0. \tag{9.1.21}$$

Since $u(x,t) = X(x)T(t)$, the boundary conditions give $X'(0)T(t) = X'(l)T(t) = 0$ for all t which implies $X'(0) = X'(l) = 0$ for all t, otherwise $T(t) \equiv 0$ gives the trivial solution.

Hence $X''(x)+\lambda X(x)=0$ subject to $X'(0)=X'(l)=0$ is constructed to the eigenvalue problem, that is

$$\begin{cases} X''(x)+\lambda X(x)=0, \\ X'(0)=X'(l)=0. \end{cases} \tag{9.1.22}$$

Then again three cases arise,

(1) $\lambda=0$ gives $X''(x)=0$ with general solution $X(x)=Ax+B$. The boundary condition $X'(0)=0$ then gives $A=0$ and the condition $X'(l)=0$ also gives $A=0$, so this case generates the nontrivial solution $X(x)=B$.

(2) For the case $\lambda<0$, let $\lambda=-\beta^2$ to give $X''(x)-\beta^2 X(x)=0$. The general solution is

$$X(x)=Ae^{\beta x}+Be^{-\beta x}.$$

The boundary conditions $X'(0)=0$ and $X'(l)=0$ then give

$$\beta(A-B)=0,$$
$$\beta(Ae^{\beta l}-Be^{-\beta l})=0,$$

which yield the trivial solution $A=B=0$.

(3) The third case with $\lambda>0$, which can be written as $\lambda=p^2$, gives $X''(x)+p^2 X(x)=0$. Then the general solution is

$$X(x)=A\cos px+B\sin px$$

and the above boundary conditions then give

$$B=0 \quad \text{and} \quad 0=A\sin pl.$$

Hence either $A=0$ or $\sin pl=0$. $A=0$ gives the trivial solution once again, and $\sin pl=0$ gives $pl=n\pi$, $n=1, 2, 3, \cdots$ or $p=\dfrac{n\pi}{l}$, $n=1, 2, 3, \cdots$.

Therefore, we solved the eigenvalue problem (9.1.22) and obtained the eigenvalues and corresponding eigenfunctions as follows:

$$\lambda=0 \quad \text{and} \quad \lambda_n=\frac{n^2\pi^2}{l^2}, \quad (n=1, 2, \cdots), \tag{9.1.23}$$

$$X_0=B \quad \text{and} \quad X_n=A_n\cos\frac{n\pi x}{l}, \quad (n=1, 2, \cdots). \tag{9.1.24}$$

If $\lambda=0$, then $T'(t)=0$ implies that $T_0(t)=C_0$, and if $\lambda=\left(\dfrac{n\pi}{l}\right)^2$, then

$$T'(t)+\left(\frac{nc\pi}{l}\right)^2 T(t)=0,$$

which solves to give

$$T_n(t)=C_n e^{-\left(\frac{nc\pi}{l}\right)^2 t}, \quad n=1, 2, 3, \cdots.$$

The various solutions are then

$$u_0(x,t)=X_0(x)T_0(t)=BC_0=D_0, \tag{9.1.25}$$

and

$$u_n(x,t)=T_n(t)X_n(x)=D_n\cos\frac{n\pi x}{l}e^{-\left(\frac{nc\pi}{l}\right)^2 t} \tag{9.1.26}$$

with $D_n=A_n C_n$, $n=1,2,3,\cdots$.

As in the solution of the wave equation, the initial conditions need to be satisfied by the full separated solution which is

$$u(x, t) = \sum_{n=0}^{\infty} u_n(x, t) = D_0 + \sum_{n=1}^{\infty} D_n \cos \frac{n\pi x}{l} e^{-(\frac{n\pi}{l})^2 t}. \tag{9.1.27}$$

Hence setting $t=0$, gives

$$u(x, 0) = D_0 + \sum_{n=1}^{\infty} D_n \cos \frac{n\pi x}{l} = f(x). \tag{9.1.28}$$

The Fourier method gives

$$D_0 = \frac{1}{l} \int_0^l f(x) \, dx \tag{9.1.29}$$

and

$$D_n = \frac{2}{l} \int_0^l f(x) \cos \frac{n\pi x}{l} dx. \tag{9.1.30}$$

And for the particular case $f(x) = \beta x$, we have

$$D_0 = \frac{1}{l} \int_0^l \beta x \, dx = \frac{\beta}{l} \frac{x^2}{2} \Big|_0^l = \frac{\beta l}{2} \tag{9.1.31}$$

and

$$D_n = \frac{2}{l} \int_0^l \beta x \cos \frac{n\pi x}{l} dx = \frac{2\beta l}{n^2 \pi^2} [(-1)^n - 1] \tag{9.1.32}$$

with

$$D_n = \begin{cases} -\dfrac{4l\beta}{n^2 \pi^2} & n=1, 3, 5, \cdots, \\ 0 & n=2, 4, 6, \cdots, \end{cases} \tag{9.1.33}$$

which can be written as

$$D_{2m-1} = -\frac{4l\beta}{(2m-1)^2 \pi^2}, \quad m=1, 2, 3, \cdots. \tag{9.1.34}$$

Hence

$$u(x,t) = \frac{\beta l}{2} - \sum_{m=1}^{\infty} \frac{4l\beta}{(2m-1)^2 \pi^2} \cos \frac{(2m-1)\pi x}{l} \exp\left(-\frac{(2m-1)^2 \pi^2 c^2}{l^2} t\right). \tag{9.1.35}$$

Note that as $t \to \infty$, $u(x,t) \to \frac{\beta l}{2}$ which is the mean value of $f(x) = \beta x$ on $[0, l]$.

Exercises 9.1

1. Use the method of separation of the variables to obtain a solution of the wave equation

$$\frac{\partial^2 u}{\partial x^2} = \frac{1}{c^2} \frac{\partial^2 u}{\partial t^2}$$

on the interval $x \in [0, L]$ subject to the conditions that

$$u(0,t) = 0, \quad u(L,t) = 0,$$

and

$$u(x,0) = \sin \frac{\pi x}{L} + \sin \frac{2\pi x}{L}, \quad u_t(x,0) = 0.$$

2. Use the method of separation of variables to solve
$$\frac{\partial^2 u}{\partial t^2}=a^2\frac{\partial^2 u}{\partial x^2}$$
on the interval $x\in[0,2]$ satisfying the boundary conditions
$$\frac{\partial u}{\partial x}(0,t)=\frac{\partial u}{\partial x}(2,t)=0$$
for all t and the initial conditions
$$\frac{\partial u}{\partial t}=0, \quad (t=0, \ 0<x<2)$$
and
$$u(x,0)=\begin{cases}kx, & 0\leqslant x\leqslant 1,\\ k(2-x), & 1\leqslant x\leqslant 2.\end{cases}$$

3. Solve the following mixed problems in finite region for wave equations:

(a) $\begin{cases}\dfrac{\partial^2 u}{\partial t^2}=c^2\dfrac{\partial^2 u}{\partial x^2}, & 0<x<1, \ t>0,\\ u(x,0)=x(1-x), & 0\leqslant x\leqslant 1,\\ \dfrac{\partial u}{\partial t}(x,0)=0, & 0\leqslant x\leqslant 1,\\ u(0,t)=0, \ u(1,t)=0, & t\geqslant 0.\end{cases}$

(b) $\begin{cases}\dfrac{\partial^2 u}{\partial t^2}=c^2\dfrac{\partial^2 u}{\partial x^2}, & 0<x<\pi, \ t>0,\\ u(x,0)=0, & 0\leqslant x\leqslant \pi,\\ \dfrac{\partial u}{\partial t}(x,0)=8\sin^2 x, & 0\leqslant x\leqslant \pi,\\ u(0,t)=0, \ u(\pi,t)=0, & t\geqslant 0.\end{cases}$

(c) $\begin{cases}\dfrac{\partial^2 u}{\partial t^2}=c^2\dfrac{\partial^2 u}{\partial x^2}, & 0<x<\pi, \ t>0,\\ u(x,0)=\sin x, & 0\leqslant x\leqslant \pi,\\ \dfrac{\partial u}{\partial t}(x,0)=x^2, & 0\leqslant x\leqslant \pi,\\ u(0,t)=0, \ u(\pi,t)=0, & t\geqslant 0.\end{cases}$

(d) $\begin{cases}\dfrac{\partial^2 u}{\partial t^2}=c^2\dfrac{\partial^2 u}{\partial x^2}, & 0<x<\pi, \ t>0,\\ u(x,0)=x^3, & 0\leqslant x\leqslant \pi,\\ \dfrac{\partial u}{\partial t}(x,0)=0, & 0\leqslant x\leqslant \pi,\\ u(0,t)=0, \ u_x(\pi,t)=0, & t\geqslant 0.\end{cases}$

4. Use separation of the variables to solve the heat equation
$$\frac{\partial u}{\partial t}=a^2\frac{\partial^2 u}{\partial x^2}$$
in the region $0<x<\pi$ and $t>0$ satisfying the boundary conditions
$$u(x,t)=0 \quad \text{when} \quad x=0 \quad \text{and} \quad x=\pi,$$
$$u(x,t)\to 0 \quad \text{as} \quad t\to\infty,$$

$$u(x,t)=x+1 \quad \text{when} \quad t=0 \quad \text{for} \quad 0<x<\pi.$$

5. Solve the following mixed problem in finite region for heat equations:

(a) $\begin{cases} \dfrac{\partial u}{\partial t}=4\dfrac{\partial^2 u}{\partial x^2}, & 0<x<1, t>0, \\ u(x,0)=x^2(1-x), & 0\leqslant x\leqslant 1, \\ u(0,t)=0, u(1,t)=0, & t\geqslant 0. \end{cases}$

(b) $\begin{cases} \dfrac{\partial u}{\partial t}=k\dfrac{\partial^2 u}{\partial x^2}, & 0<x<l, t>0, \\ u(x,0)=x(l-x), & 0\leqslant x\leqslant l, \\ u_x(0,t)=0, u_x(l,t)=0, & t\geqslant 0. \end{cases}$

(c) $\begin{cases} \dfrac{\partial u}{\partial t}=k\dfrac{\partial^2 u}{\partial x^2}, & 0<x<\pi, t>0, \\ u(x,0)=\sin^2 x, & 0\leqslant x\leqslant \pi, \\ u(0,t)=0, u(\pi,t)=0, & t\geqslant 0. \end{cases}$

9.2 Separation of variables for 2-dimensional Laplace equations

The physical derivation of Laplace's equation in Chapter 6 resulted in the equation:

$$\nabla^2 u = \frac{\partial^2 u}{\partial x^2} + \frac{\partial^2 u}{\partial y^2} = 0.$$

For this equation, the boundary conditions are given on a closed curve. This is analogous to the boundary value problems of ordinary differential equations, which means that the solution has to be found simultaneously over the whole region, as opposed to the step-by-step methods which are feasible in the time direction in the heat equation and in both variables in the wave equation. In this section, we will consider two cases, that is, Laplace equation with rectangular boundary and circular one.

9.2.1 Laplace equation with rectangular boundary

Example 9.2.1 Solve the Laplace equation:

$$\nabla^2 u = \frac{\partial^2 u}{\partial x^2} + \frac{\partial^2 u}{\partial y^2} = 0 \tag{9.2.1}$$

in the rectangular domain $\begin{cases} 0\leqslant x\leqslant a, \\ 0\leqslant y\leqslant b, \end{cases}$ with the boundary conditions

$$u\big|_{x=0}=0, \quad u\big|_{x=a}=Ay, \tag{9.2.2}$$
$$u\big|_{y=0}=0, \quad u\big|_{y=b}=0. \tag{9.2.3}$$

Solution To solve this equation, the separation of variables approach suggests the trial solution $u(x,y)=X(x)Y(y)$ giving $X''(x)Y(y)+X(x)Y''(y)=0$ which results in the differential equations

$$\frac{X''(x)}{X(x)}+\frac{Y''(y)}{Y(y)}=0 \quad \text{or} \quad \frac{X''(x)}{X(x)}=-\frac{Y''(y)}{Y(y)}=\lambda \tag{9.2.4}$$

where λ is arbitrary, which gives the problems
$$X''(x)-\lambda X(x)=0 \quad \text{and} \quad Y''(y)+\lambda Y(y)=0 \tag{9.2.5}$$
with zero boundary conditions
$$\left.\begin{array}{l}u(0,y)=0 \Rightarrow X(0)Y(y)=0 \Rightarrow X(0)=0,\\ u(x,0)=0 \Rightarrow X(x)Y(0)=0 \Rightarrow Y(0)=0,\\ u(x,b)=0 \Rightarrow X(x)Y(b)=0 \Rightarrow Y(b)=0\end{array}\right\} \tag{9.2.6}$$
for a non-trivial solution. Consider first the two boundary conditions associated with the function $Y(y)$, then once more there are three cases which may arise. The non-trivial solution arises when $\lambda>0$ or $\lambda=\beta^2$, and gives the equation $Y''(y)+\beta^2 Y(y)=0$ with solution
$$Y(y)=A\cos(\beta y)+B\sin(\beta y). \tag{9.2.7}$$
Now the boundary condition $Y(0)=0$ gives $A=0$, and the condition $Y(b)=0$ gives $B\sin\beta b=0$. For non-trivial solutions $\sin\beta b=0$ which implies that $\beta b=n\pi$, $n=1,2,3,\cdots$, or $\beta=\dfrac{n\pi}{b}$, $n=1,2,3,\cdots$.

Hence if
$$\lambda=\left(\frac{n\pi}{b}\right)^2, \quad n=1,2,3,\cdots, \tag{9.2.8}$$
then
$$Y_n(y)=B_n\sin\frac{n\pi y}{b}. \tag{9.2.9}$$

Now return to the equation for $X(x)$ with $\lambda=\left(\dfrac{n\pi}{b}\right)^2$, that is $X''(x)-\left(\dfrac{n\pi}{b}\right)^2 X(x)=0$. Its solution is
$$X_n(x)=C_n e^{\frac{n\pi x}{b}}+D_n e^{\frac{-n\pi x}{b}}. \tag{9.2.10}$$
The boundary condition here is $X(0)=0$ or $C_n+D_n=0$, which gives
$$X_n(x)=C_n(e^{\frac{n\pi x}{b}}-e^{\frac{-n\pi x}{b}})=2C_n\sinh\frac{n\pi x}{b}. \tag{9.2.11}$$
The full solution is then
$$u_n(x,y)=X_n(x)Y_n(y)=E_n\sinh\frac{n\pi x}{b}\sin\frac{n\pi y}{b} \tag{9.2.12}$$
where
$$E_n=2C_n B_n.$$
To satisfy $u(a,y)=f(y)$, the solutions are summed over n to give
$$u(x,y)=\sum_{n=1}^{\infty}E_n\sinh\frac{n\pi x}{b}\sin\frac{n\pi y}{b}. \tag{9.2.13}$$
The remaining boundary condition becomes
$$u(a,y)=\sum_{n=1}^{\infty}E_n\sinh\frac{n\pi a}{b}\sin\frac{n\pi y}{b}=f(y). \tag{9.2.14}$$
By the usual Fourier approach, we have
$$E_n\sinh\frac{n\pi a}{b}=\frac{2}{b}\int_0^b f(y)\sin\frac{n\pi y}{b}dy, \tag{9.2.15}$$
which yields the complete solution

$$u(x,y) = \sum_{n=1}^{\infty} \frac{2}{b\sinh\frac{n\pi a}{b}} \left\{ \int_0^b f(y) \sin\frac{n\pi y}{b} dy \right\} \sinh\frac{n\pi x}{b} \sin\frac{n\pi y}{b}. \qquad (9.2.16)$$

For a specific function $f(y)$, the integral in (9.2.16) can be evaluated to yield the solution as an infinite series.

A second example will now be considered with boundary conditions involving partial derivatives of u, and an extra term in the equation.

Example 9.2.2 Consider the problem of solving

$$\frac{\partial^2 u}{\partial x^2} + \frac{\partial^2 u}{\partial y^2} + u = 0 \qquad (9.2.17)$$

in $0 < x < 1$, $0 < y < 1$ subject to the boundary condition $u(0,y) = 0$, $\frac{\partial u}{\partial x}(1,y) = 0$ and $u(x,0) = 0$ and $u(x,1) = x$. Again in the spirit of separation of the variables, let $u(x,y) = X(x)Y(y)$ where $X(x)$ and $Y(y)$ are functions of x and y respectively, to give

$$\frac{X''(x)}{X(x)} + \frac{Y''(y)}{Y(y)} + 1 = 0. \qquad (9.2.18)$$

Now $\frac{Y''(y)}{Y(y)}$ is a function of y alone and the rest of the expression is a function of x only, hence

$$\frac{Y''(y)}{Y(y)} = \lambda \quad \text{and} \quad \frac{X''(x)}{X(x)} + 1 = -\lambda \qquad (9.2.19)$$

yield the two equations

$$X''(x) + (\lambda+1)X(x) = 0 \quad \text{and} \quad Y'''(y) = \lambda Y(y). \qquad (9.2.20)$$

The zero boundary conditions give

$$\begin{cases} u(0,y) = 0 \Rightarrow X(0)Y(y) = 0 \Rightarrow X(0) = 0, \\ \frac{\partial u}{\partial x}(1,y) = 0 \Rightarrow X'(1)Y(y) = 0 \Rightarrow X'(1) = 0, \\ u(x,0) = 0 \Rightarrow X(x)Y(0) = 0 \Rightarrow Y(0) = 0, \end{cases} \qquad (9.2.21)$$

for the non-trivial solution, leaving

$$X''(x) + (\lambda+1)X(x) = 0, \quad X(0) = X'(1) = 0 \qquad (9.2.22)$$

whose solution depends on the sign of $\lambda+1$. The non-trivial solution arises when $\lambda+1 > 0$, hence setting $\lambda+1 = \beta^2$ gives $X''(x) + \beta^2 X(x) = 0$ with solution

$$X(x) = A\cos\beta x + B\sin\beta x. \qquad (9.2.23)$$

The boundary condition $X(0) = 0$ then gives $A = 0$, and $X'(1) = 0$ gives $\beta B\cos\beta = 0$. But $\beta > 0$ and hence $\cos\beta = 0$ gives $\beta = (2n+1)\frac{\pi}{2}$, for $n = 0, 1, 2, 3, \cdots$. Hence the non-trivial solution arises when

$$\lambda = (2n+1)^2 \frac{\pi^2}{4} - 1 \qquad (9.2.24)$$

and has the form

$$X_n(x) = B_n \sin(2n+1)\frac{\pi x}{2}. \qquad (9.2.25)$$

The final step is to solve the equation for $Y(y)$ which assumes the form

$$Y''(y) = \left((2n+1)^2 \frac{\pi^2}{4} - 1\right) Y(y) = \frac{1}{4}[(2n+1)^2 \pi^2 - 4] Y(y) = \alpha_n^2 Y(y) \quad (9.2.26)$$

which is solved to give

$$Y_n(y) = C_n e^{\alpha_n y} + D_n e^{-\alpha_n y} \quad (9.2.27)$$

where

$$\alpha_n = \frac{1}{2}[(2n+1)^2 \pi^2 - 4]^{\frac{1}{2}}. \quad (9.2.28)$$

The boundary condition $Y(0) = 0$ then gives $0 = C_n + D_n$, and hence

$$Y_n(y) = C_n(e^{\alpha_n y} - e^{-\alpha_n y}) = 2C_n \sinh(\alpha_n y). \quad (9.2.29)$$

The general separated solution is then

$$u(x, y) = \sum_{n=0}^{\infty} E_n \sin(2n+1) \frac{\pi x}{2} \sinh(\alpha_n y) \quad (9.2.30)$$

and hence the final boundary condition requires

$$u(x, 1) = \sum_{n=0}^{\infty} E_n \sin(2n+1) \frac{\pi x}{2} \sinh \alpha_n = x. \quad (9.2.31)$$

Hence the usual Fourier expansion has coefficients given by

$$E_n \sinh \alpha_n = 2\int_0^1 x \sin(2n+1) \frac{\pi x}{2} dx$$

$$= \frac{8}{(2n+1)^2 \pi^2} \sin(2n+1) \frac{\pi}{2}, \quad (9.2.32)$$

which leaves

$$E_n = \frac{1}{\sinh \alpha_n} \frac{8}{(2n+1)^2 \pi^2}(-1)^n. \quad (9.2.33)$$

Hence the complete solution is

$$u(x, y) = \sum_{n=0}^{\infty} \frac{8}{(2n+1)^2 \pi^2} (-1)^n \frac{1}{\sinh \frac{1}{2} \sqrt{(2n+1)^2 \pi^2 - 4}}$$

$$\sin(2n+1) \frac{\pi x}{2} \sinh \frac{1}{2} \sqrt{(2n+1)^2 \pi^2 - 4} \, y. \quad (9.2.34)$$

9.2.2 Laplace equation with circular boundary

Sometimes it is simpler to consider some problems in polar coordinate, so it is important to solve some defining problem in polar coordinate.

Example 9.2.3 Suppose that there is a thin circular disk with radius R and it is insolate at its upper and lower sides. If the distribution of the temperature of its boundary is given, please determine the distribution of the temperature in the disk.

Solution In the stationary state, the temperature in the disk has no dependence on time, and its distribution satisfies Laplace equation

$$\nabla^2 u = 0. \quad (9.2.35)$$

Since the boundary is a circle, whose equation in polar coordinate is $\rho = \rho_0$, the boundary condition in polar coordinate is

$$u|_{\rho=\rho_0}=f(\theta).\tag{9.2.36}$$

To solve this problem, we introduce polar coordinate and transform Laplace equation (9.2.35) into

$$\nabla^2 u=\frac{\partial^2 u}{\partial \rho^2}+\frac{1}{\rho}\frac{\partial u}{\partial \rho}+\frac{1}{\rho^2}\frac{\partial^2 u}{\partial \theta^2}=0.\tag{9.2.37}$$

On the other hand, the range of the dependent variables (ρ,θ) are $[0,\rho_0]$ and $[0,2\pi]$, respectively, while the temperature in the center is finite, and (ρ,θ) and $(\rho,\theta+2\pi)$ represent the same point with the same temperature, that is

$$|u(0,\theta)|<+\infty,\tag{9.2.38}$$
$$u(\rho,\theta)=u(\rho,\theta+2\pi).\tag{9.2.39}$$

Now the defining problem we want to solve is

$$\begin{cases}\nabla^2 u=\dfrac{\partial^2 u}{\partial \rho^2}+\dfrac{1}{\rho}\dfrac{\partial u}{\partial \rho}+\dfrac{1}{\rho^2}\dfrac{\partial^2 u}{\partial \theta^2}=0,\\ |u(0,\theta)|<+\infty,\\ u(\rho,\theta)=u(\rho,\theta+2\pi),\\ u|_{\rho=\rho_0}=f(\theta).\end{cases}$$

Let $u(\rho,\theta)=P(\rho)\Theta(\theta)$, substituting it into (9.2.37), we have

$$P''(\rho)\Theta(\theta)+\frac{1}{\rho}P'(\rho)\Theta(\theta)+\frac{1}{\rho^2}P(\rho)\Theta''(\theta)=0.$$

Dividing it by $\dfrac{P(\rho)\Theta(\theta)}{\rho^2}$, we obtain

$$\frac{\rho^2 P''(\rho)}{P(\rho)}+\frac{\rho P'(\rho)}{P(\rho)}+\frac{\Theta''(\theta)}{\Theta(\theta)}=0.$$

That is

$$\frac{\rho^2 P''(\rho)+\rho P'(\rho)}{P(\rho)}=-\frac{\Theta''(\theta)}{\Theta(\theta)}=\lambda.$$

And then we have two ordinary differential equations

$$\rho^2 P''(\rho)+\rho P'(\rho)-\lambda P(\rho)=0,\tag{9.2.40}$$
$$\Theta''(\theta)+\lambda\Theta(\theta)=0.\tag{9.2.41}$$

The conditions (9.2.38) and (9.2.39) become

$$\begin{cases}|P(\rho)|_{\rho=0}|<+\infty,\\ \Theta(\theta)=\Theta(\theta+2\pi).\end{cases}$$

Then we have the following two problems:

$$\begin{cases}\rho^2 P''(\rho)+\rho P'(\rho)-\lambda P(\rho)=0,\\ |P(\rho)|_{\rho=0}|<+\infty,\end{cases}\tag{9.2.42}$$

and

$$\begin{cases}\Theta''(\theta)+\lambda\Theta(\theta)=0,\\ \Theta(\theta)=\Theta(\theta+2\pi).\end{cases}\tag{9.2.43}$$

Because the conditions in (9.2.43) satisfy the superposition principle, (9.2.43) is an eigenvalue problem.

Now we discuss the value of the parameter λ.

(1) If $\lambda<0$, then $\Theta(\theta)=Ae^{\sqrt{-\lambda}\theta}+Be^{-\sqrt{-\lambda}\theta}$.

From the initial condition, we have
$$Ae^{\sqrt{-\lambda}(\theta+2\pi)}+Be^{-\sqrt{-\lambda}(\theta+2\pi)}=Ae^{\sqrt{-\lambda}\theta}+Be^{-\sqrt{-\lambda}\theta}$$
which means $A=B=0$.

(2) If $\lambda=0$, then $\Theta(\theta)=A\theta+B$.

From the initial condition, we have
$$A(\theta+2\pi)+B=A\theta+B$$
which means $A=0$, and $\Theta(\theta)=B$.

(3) If $\lambda>0$, then $\Theta(\theta)=A\cos\sqrt{\lambda}\theta+B\sin\sqrt{\lambda}\theta$,

and now $\Theta(\theta)=\Theta(\theta+2\pi)$ implies
$$A\cos\sqrt{\lambda}(\theta+2\pi)+B\sin\sqrt{\lambda}(\theta+2\pi)=A\cos\sqrt{\lambda}(\theta)+B\sin\sqrt{\lambda}(\theta),$$
which is true if $\sqrt{\lambda}=n$, $(n=1, 2, \cdots)$.

Combining the above analysis, we have
$$\lambda=n^2, \quad (n=0, 1, 2, \cdots) \quad \text{and} \quad \Theta_n(\theta)=A_n\cos n\theta+B_n\sin n\theta. \tag{9.2.44}$$

Then the other problem (9.2.42) becomes
$$\rho^2 P''(\rho)+\rho P'(\rho)-n^2 P(\rho)=0, \quad (n=1, 2, \cdots) \tag{9.2.45}$$
and in the special case of $\lambda=0$,
$$\frac{d}{d\rho}(\rho P'(\rho))=0. \tag{9.2.46}$$

To solve the general solution of (9.2.45), we need to seek $P(\rho)$ in the form
$$P(\rho)=A\rho^m$$
to get $m(m-1)\rho^m+m\rho^m-n^2\rho^m=0$, which reduces to $m^2=n^2$ or $m=\pm n$.

Thus the solution is
$$P_n(\rho)=C_n\rho^{-n}+D_n\rho^n, \quad (n=1, 2, \cdots) \tag{9.2.47}$$
where C_n and D_n are arbitrary constants.

For the special case, equation (9.2.46) gives the solution
$$P_0(\rho)=C_0\ln\rho+D_0. \tag{9.2.48}$$

From the condition in (9.2.42) $|P(0)|<+\infty$, we have $D_n=0$ $(n=0, 1, 2, \cdots)$, then
$$P_n(\rho)=C_n\rho^n, \quad (n=0, 1, 2, \cdots).$$

Combining and superpositing, we have
$$u(\rho,\theta) = \frac{A_0}{2}+\sum_{n=1}^{\infty}\rho^n(A_n\cos n\theta+B_n\sin n\theta). \tag{9.2.49}$$

From the boundary condition (9.2.36), we have
$$\frac{A_0}{2}+\sum_{n=1}^{\infty}\rho_0^n(A_n\cos n\theta+B_n\sin n\theta)=f(\theta). \tag{9.2.50}$$

Hence
$$A_0 = \frac{1}{\pi}\int_0^{2\pi} f(\theta)d\theta,$$
$$A_n = \frac{1}{\rho_0^n\pi}\int_0^{2\pi} f(\theta)\cos n\theta d\theta, \tag{9.2.51}$$
$$B_n = \frac{1}{\rho_0^n\pi}\int_0^{2\pi} f(\theta)\sin n\theta d\theta.$$

Example 9.2.4 Solve the following defining problem:
$$\begin{cases} \dfrac{\partial^2 u}{\partial \rho^2} + \dfrac{1}{\rho}\dfrac{\partial u}{\partial \rho} + \dfrac{1}{\rho^2}\dfrac{\partial^2 u}{\partial \theta^2} = 0, & \rho < R,\ 0 \leqslant \theta \leqslant 2\pi, \\ u\big|_{\rho=R} = A\cos\theta, & (A \text{ is a constant}). \end{cases}$$

Solution According to the above example, we have
$$u(\rho,\theta) = \dfrac{A_0}{2} + \sum_{n=1}^{\infty}\rho^n(A_n\cos n\theta + B_n\sin n\theta).$$

From the boundary condition $u(R,\theta) = A\cos\theta$, we have
$$A\cos\theta = \dfrac{A_0}{2} + \sum_{n=1}^{\infty}R^n(A_n\cos n\theta + B_n\sin n\theta)$$

which means $B_n = 0, (n=1,2,\cdots)$.

Then $\dfrac{A_0}{2} + RA_1\cos\theta + R^2 A_2\cos 2\theta + \cdots = A\cos\theta$ gives
$$A_0 = 0, \quad RA_1 = 1, \quad A_2 = A_3 = \cdots = A_n = 0 \quad (n=2,3,\cdots),$$

which implies $A_1 = \dfrac{1}{R}$. Hence the solution of this problem is
$$u(\rho,\theta) = \dfrac{A}{R}\rho\cos\theta.$$

Exercises 9.2

1. Solve the following boundary value problems for Laplace equation:

(a) $\begin{cases} \dfrac{\partial^2 u}{\partial x^2} + \dfrac{\partial^2 u}{\partial y^2} = 0, & 0<x<a,\ 0<y<b, \\ u(x,0) = u(x,b) = f(x), & 0 \leqslant x \leqslant a, \\ u(0,y) = 0, u(a,y) = 0, & 0 \leqslant y \leqslant b. \end{cases}$

(b) $\begin{cases} \dfrac{\partial^2 u}{\partial x^2} + \dfrac{\partial^2 u}{\partial y^2} = 0, & 0<x<1,\ 0<y<1, \\ u_y(x,0) = 0, u_y(x,1) = 0, & 0 \leqslant x \leqslant 1, \\ u(0,y) = 0, u(1,y) = 1-y, & 0 \leqslant y \leqslant 1. \end{cases}$

(c) $\begin{cases} \dfrac{\partial^2 u}{\partial x^2} + \dfrac{\partial^2 u}{\partial y^2} = 0, & 0<x<1,\ 0<y<1, \\ u(x,0) = 100, u(x,1) = 200, & 0 \leqslant x \leqslant 1, \\ u(0,y) = 0, u(1,y) = 0, & 0 \leqslant y \leqslant 1. \end{cases}$

2. Solve the following defining problem in the annular domain:
$$\begin{cases} \nabla^2 u = 0, & 0<\rho<a,\ 0<\theta<2\pi, \\ u\big|_{\theta=0} = u\big|_{\theta=2\pi} = 0, & 0 \leqslant \rho \leqslant a, \\ u\big|_{\rho=a} = f(\theta), & 0 \leqslant \theta \leqslant 2\pi. \end{cases}$$

9.3 Nonhomogeneous equations and nonhomogeneous boundary conditions

In the above sections, we discussed the method of separation of variables in terms of wave equation, heat equation and Laplace equation. In these problems we were solving a homogeneous linear differential equation $L(u)=0$ (either the wave or the heat equation) for a function $u(x,t)$ on the region $a<x<b$, $t>0$. We imposed some homogeneous linear boundary conditions on $u(x,t)$ at $x=a$ and $x=b$, and some linear initial conditions on $u(x,t)$ at $t=0$.

Let us write the boundary conditions as $B(u)=0$ and the initial conditions as $I(u)=h(x)$, with the understanding that each of these equations may stand for several equations grouped together. (For example, for the vibrating string problem, "$B(u)=0$" stands for "$u(0,t)=0$ and $u(l,t)=0$", and "$I(u)=h(x)$" stands for "$u(x,0)=f(x)$ and $u_t(x,0)=g(x)$".) Thus the boundary value problem has the form.

$$L(u)=0, \quad B(u)=0, \quad I(u)=h(x). \tag{9.3.1}$$

The technique for solving (9.3.1) is to use the method of separation of variables to produce an infinite family of functions $u(x,t)=\sum C_n \Phi_n(t) \Psi_n(x)$ that satisfy $L(u)=0$ and $B(u)=0$, and then to choose the constants C_n appropriately to obtain $I(u)=h(x)$.

In the above sections, the boundary conditions were such as to lead to Fourier sine or cosine series. In this part, we shall consider nonhomogeneous equations and nonhomogeneous boundary conditions:

$$L(u)=F(x,t), \quad B(u)=g(t), \quad I(u)=h(x). \tag{9.3.2}$$

There are several techniques for reducing such problems to more manageable ones. We will now discuss these techniques on the general level and show its validity with specific examples.

Technique 1 Use the superposition principle to deal with nonhomogeneous terms once at a time.

In problem (9.3.2), there are three nonhomogeneous terms $F(x,t)$, $g(t)$ and $h(x)$. Suppose that we can solve the three problems obtained by replacing all but one of these functions by zero:

$$L(u)=0, \quad B(u)=0, \quad I(u)=h(x), \tag{9.3.3}$$

$$L(u)=0, \quad B(u)=g(t), \quad I(u)=0, \tag{9.3.4}$$

$$L(u)=F(x,t), \quad B(u)=0, \quad I(u)=0. \tag{9.3.5}$$

If $u_1(x,t)$, $u_2(x,t)$ and $u_3(x,t)$ are the solutions to (9.3.3), (9.3.4) and (9.3.5), then $u(x,t)=u_1(x,t)+u_2(x,t)+u_3(x,t)$ will be the solution of (9.3.2). In particular, (9.3.3) is just (9.3.1), which we already know how to deal with, so it suffices to solve (9.3.4) and (9.3.5).

For example, if we are working on the interval $a<x<b$, the boundary condition

$B(u)=g(t)$ generally stands for two conditions $B_a(u)=g_a(t)$ and $B_b(u)=g_b(t)$. If we can solve the problems by replacing $g_a(t)$ and $g_b(t)$ with zero, we can solve the original problem by adding the solutions to the two simpler problems.

First of all, we consider the nonhomogeneous equation with homogeneous boundary conditions.

Technique 2 The Fourier expansions used to solve $L(u)=0$ with homogeneous boundary conditions $B(u)=0$ can also be used to solve the nonhomogeneous equation $L(u)=F(x,t)$ with the same boundary conditions.

Consider the following problem:
$$L(u)=F(x,t), \quad B(u)=0, \quad I(u)=h(x). \tag{9.3.6}$$

It can be divided into two problems with one nonhomogeneous term
$$L(u)=F(x,t), \quad B(u)=0, \quad I(u)=0,$$
$$L(u)=0, \quad B(u)=0, \quad I(u)=h(x).$$

Suppose the homogeneous equation $L(u)=0$ with homogeneous boundary conditions $B(u)=0$ can be handled by separation of variables, leading to solutions $u(x,t)=\sum C_n \Phi_n(t) \Psi_n(x)$, where the $\Phi_n(x)$ are the eigenfunctions for the eigenvalue problem. Then the same sort of eigenfunction expansion can be used to produce solutions of the nonhomogeneous equation $L(u)=F(x,t)$ subject to the same boundary conditions $B(u)=0$. Namely, for each t we expand the function $F(x,t)$ in terms of the eigenfunction $\Phi_n(x)$, that is, $F(x,t)=\sum f_n(t)\Phi_n(x)$, and we try to find a solution u in the form
$$u(x,t)=\sum \omega_n(t)\Phi_n(x),$$
where the functions $\omega_n(t)$ are to be determined. If we plug these series into the differential equation $L(u)=F(x,t)$, the result will be a sequence of ordinary differential equations for the unknown functions $\omega_n(t)$ in terms of the known functions $\Phi_n(x)$. These equations can be solved subject to whatever initial conditions at $t=0$ one may require. The resulting function $u(x,t)$ then satisfies the differential equation $L(u)=F(x,t)$ and the desired initial conditions; it satisfies the boundary conditions $B(u)=0$ because they are built into the eigenfunction $\Phi_n(x)$.

Example 9.3.1 Forced vibrating string problem is as follows:
$$\begin{cases} \dfrac{\partial^2 u}{\partial t^2}=a^2 \dfrac{\partial^2 u}{\partial x^2}+F(x,t), & 0<x<l,\ t>0, \tag{9.3.7} \\ u\big|_{x=0}=u\big|_{x=l}=0, & t\geqslant 0, \tag{9.3.8} \\ u\big|_{t=0}=f(x),\ \dfrac{\partial u}{\partial t}\bigg|_{t=0}=g(x), & 0\leqslant x\leqslant l. \tag{9.3.9} \end{cases}$$

In order to solve this nonhomogeneous equation with homogeneous boundary conditions, firstly, we consider the homogeneous wave equation with the same boundary conditions, that is

$$\begin{cases} \dfrac{\partial^2 u}{\partial t^2} = a^2 \dfrac{\partial^2 u}{\partial x^2}, & 0<x<l, \ t>0, \\ u|_{x=0} = u|_{x=l} = 0, & t \geqslant 0, \\ u|_{t=0} = f(x), \ \dfrac{\partial u}{\partial t}\bigg|_{t=0} = g(x), & 0 \leqslant x \leqslant l. \end{cases}$$

For this problem, we take the method of separation of variables and suppose that $u(x,t) = X(x)T(t)$, and then we can obtain the eigenvalue problem

$$\begin{cases} X''(x) + \lambda X(x) = 0, \\ X(0) = X(l) = 0. \end{cases}$$

By solving this problem, we get the eigenvalues $\lambda_n = \dfrac{n^2 \pi^2}{l^2}$, $(n=1, 2, \cdots)$ and the eigenfunctions $X_n(x) = \sin \dfrac{n\pi}{l} x$, $(n=1, 2, \cdots)$.

For the nonhomogeneous problem, we begin by expanding everything in sight in a Fourier sine series, that is

$$u(x,t) = \sum_{n=1}^{\infty} b_n(t) \sin \frac{n\pi}{l} x, \qquad (9.3.10)$$

and

$$F(x,t) = \sum_{n=1}^{\infty} \beta_n(t) \sin \frac{n\pi}{l} x. \qquad (9.3.11)$$

Here, the coefficients $\beta_n(t)$ are computed from the known function $F(x,t)$ in the usual way, and the coefficients $b_n(t)$ are to be determined. Substituting these series into equation (9.3.7), we obtain

$$\sum_{n=1}^{\infty} b''_n(t) \sin \frac{n\pi}{l} x = \sum_{n=1}^{\infty} a^2 \frac{-n^2 \pi^2}{l^2} b_n(t) \sin \frac{n\pi}{l} x + \sum_{n=1}^{\infty} \beta_n(t) \sin \frac{n\pi}{l} x.$$

That is

$$b''_n(t) + a^2 \frac{n^2 \pi^2}{l^2} b_n(t) = \beta_n(t), \quad (n=1,2,\cdots). \qquad (9.3.12)$$

These equations can be solved by standard techniques such as variation of parameters or Laplace transformation. The solution with initial conditions $b_n(0) = b'_n(0) = 0$ is

$$b_n(t) = \frac{l}{n\pi a} \int_0^t \sin \frac{n\pi a(t-s)}{l} \beta_n(s) \, ds. \qquad (9.3.13)$$

This formula leads to the solution of the nonhomogeneous equation with initial condition $f(x) = g(x) = 0$, that is

$$\begin{cases} \dfrac{\partial^2 u}{\partial t^2} = a^2 \dfrac{\partial^2 u}{\partial x^2} + F(x, t), & 0<x<l, \ t>0, \\ u(0,t) = u(l,t) = 0, & t \geqslant 0, \\ u|_{t=0} = 0, \ \dfrac{\partial u}{\partial t}\bigg|_{t=0} = 0, & 0 \leqslant x \leqslant l. \end{cases}$$

But for the problem (9.3.7)~(9.3.9), by the superposition principle (Technique 1), one is reduced to solve the homogeneous wave equation with initial conditions (9.3.9), and this we have already done.

Summary For the nonhomogeneous equation with homogeneous boundary conditions

$$\begin{cases} \dfrac{\partial^2 u}{\partial t^2}=a^2\dfrac{\partial^2 u}{\partial x^2}+F(x,t), & 0<x<l,\ t>0,\\ u|_{x=0}=u|_{x=l}=0, & t\geq 0,\\ u|_{t=0}=f(x),\ \dfrac{\partial u}{\partial t}\bigg|_{t=0}=g(x), & 0\leq x\leq l, \end{cases}$$

we can separate this problem into two problems, that is

(I) $\begin{cases} \dfrac{\partial^2 w}{\partial t^2}=a^2\dfrac{\partial^2 w}{\partial x^2}, & 0<x<l,\ t>0,\\ w|_{x=0}=w|_{x=l}=0, & t\geq 0,\\ w|_{t=0}=f(x),\ \dfrac{\partial w}{\partial t}\bigg|_{t=0}=g(x), & 0\leq x\leq l, \end{cases}$

and

(II) $\begin{cases} \dfrac{\partial^2 v}{\partial t^2}=a^2\dfrac{\partial^2 v}{\partial x^2}+F(x,t), & 0<x<l,\ t>0,\\ v|_{x=0}=v|_{x=l}=0, & t\geq 0,\\ v|_{t=0}=0,\ \dfrac{\partial v}{\partial t}\bigg|_{t=0}=0, & 0\leq x\leq l. \end{cases}$

Then the solution of the original problem has the following relationship with the solutions of the above problems

$$u(x,t)=w(x,t)+v(x,t).$$

Appendix The method of variation of parameters

This method is to solve the nonhomogeneous ordinary differential equation

$$y''+p(x)y'+q(x)y=r(x). \qquad (9.3.14)$$

Its corresponding homogeneous equation is

$$y''+p(x)y'+q(x)y=0, \qquad (9.3.15)$$

and its general solution is $y(x)=c_1 y_1(x)+c_2 y_2(x)$, where c_1 and c_1 are constants.

Suppose that $y_*(x)=c_1(x)y_1(x)+c_2(x)y_2(x)$ is a solution of (9.3.14), then

$$y'_*(x)=c'_1(x)y_1(x)+c'_2(x)y_2(x)+c_1(x)y'_1(x)+c_2(x)y'_2(x).$$

Set

$$c'_1(x)y_1(x)+c'_2(x)y_2(x)=0, \qquad (*)$$

and then

$$y'_*(x)=c_1(x)y'_1(x)+c_2(x)y'_2(x),$$

and

$$y''_*(x)=c'_1(x)y'_1(x)+c'_2(x)y'_2(x)+c_1(x)y''_1(x)+c_2(x)y''_2(x).$$

Substituting the above derivatives into (9.3.14), we have

$$c'_1(x)y'_1(x)+c'_2(x)y'_2(x)+c_1(x)y''_1(x)+c_2(x)y''_2(x)+$$
$$p(x)[c_1(x)y'_1(x)+c_2(x)y'_2(x)]+$$
$$q(x)[c_1(x)y_1(x)+c_2(x)y_2(x)]=r(x).$$

That is

$$c'_1(x)y'_1(x)+c'_2(x)y'_2(x)=r(x). \qquad (**)$$

Solving (*) and (* *), we can obtain $c_1(x)$ and $c_2(x)$, and then the solution of (9.3.14) is derived.

For example 9.3.1, we obtain the following nonhomogeneous ordinary differential equation
$$b_n''(t)+a^2\frac{n^2\pi^2}{l^2}b_n(t)=\beta_n(t).$$
Its corresponding homogeneous equation is
$$b_n''(t)+a^2\frac{n^2\pi^2}{l^2}b_n(t)=0,$$
and the general solution of this equation is $b_n(t)=A\cos\frac{n\pi a}{l}t+B\sin\frac{n\pi a}{l}t$.

Then suppose the solution of the nonhomogeneous equation is
$$b_n(t)=A(t)\cos\frac{n\pi a}{l}t+B(t)\sin\frac{n\pi a}{l}t,$$
and then
$$b_n'(t)=-A(t)\frac{n\pi a}{l}\sin\frac{n\pi a}{l}t+B(t)\frac{n\pi a}{l}\cos\frac{n\pi a}{l}t.$$
Here we set
$$A'(t)\cos\frac{n\pi a}{l}t+B'(t)\sin\frac{n\pi a}{l}t=0, \qquad (*)$$
then
$$b_n''(t)=-A'(t)\frac{n\pi a}{l}\sin\frac{n\pi a}{l}t-A(t)\frac{n^2\pi^2 a^2}{l^2}\cos\frac{n\pi a}{l}t+$$
$$B'(t)\frac{n\pi a}{l}\cos\frac{n\pi a}{l}t-B(t)\frac{n^2\pi^2 a^2}{l^2}\sin\frac{n\pi a}{l}t.$$
Since
$$b_n''(t)+\frac{a^2\pi^2 n^2}{l^2}b_n(t)=\beta_n(t),$$
then we have
$$-A'(t)\frac{n\pi a}{l}\sin\frac{n\pi a}{l}t+B'(t)\frac{n\pi a}{l}\cos\frac{n\pi a}{l}t=\beta_n(t). \qquad (**)$$
Solving $(*)$ and $(**)$, we have
$$B'(t)=\frac{l}{an\pi}\beta_n(t)\cos\frac{an\pi}{l}t,$$
$$A'(t)=\frac{-l}{an\pi}\beta_n(t)\sin\frac{an\pi}{l}t.$$
Hence
$$A(t)=\frac{l}{an\pi}\int_0^t\beta_n(s)\left(-\sin\frac{an\pi}{l}s\right)ds,$$
$$B(t)=\frac{l}{an\pi}\int_0^t\beta_n(s)\cos\frac{an\pi}{l}s\,ds.$$
And then
$$b_n(t)=A(t)\cos\frac{an\pi}{l}t+B(t)\sin\frac{an\pi}{l}t$$
$$=\frac{l}{an\pi}\int_0^t\beta_n(s)\left(\cos\frac{an\pi}{l}s\sin\frac{an\pi}{l}t-\cos\frac{an\pi}{l}t\sin\frac{an\pi}{l}s\right)ds$$
$$=\frac{l}{an\pi}\int_0^t\beta_n(s)\sin\frac{an\pi(t-s)}{l}ds.$$

Example 9.3.2 Solve the following problem in the annular domain $a \leqslant \sqrt{x^2+y^2} \leqslant b$,

$$\begin{cases} \dfrac{\partial^2 u}{\partial x^2}+\dfrac{\partial^2 u}{\partial y^2}=12(x^2-y^2), & a<\sqrt{x^2+y^2}<b, \\ u\big|_{\sqrt{x^2+y^2}=a}=0, \quad \dfrac{\partial u}{\partial n}\bigg|_{\sqrt{x^2+y^2}=b}=0. \end{cases}$$

Solution It is simpler for the annular domain to take polar coordinate. Let $x=\rho\cos\theta$, $y=\rho\sin\theta$, then the above problem can be transformed into

$$\begin{cases} \dfrac{\partial^2 u}{\partial \rho^2}+\dfrac{1}{\rho}\dfrac{\partial u}{\partial \rho}+\dfrac{1}{\rho^2}\dfrac{\partial^2 u}{\partial \theta^2}=12\rho^2\cos 2\theta, & a<\rho<b, \end{cases} \tag{9.3.16}$$

$$\begin{cases} u\big|_{\rho=a}=0, \quad \dfrac{\partial u}{\partial \rho}\bigg|_{\rho=b}=0. \end{cases} \tag{9.3.17}$$

This is a Laplace equation in polar coordinates. According to the results we obtained in Example 9.2.3, the eigenvalues of homogeneous Laplace equation is $\lambda=m^2$, $(m=0,1,2,\cdots)$ and corresponding eigenfunctions are

$$\Theta_m(\theta)=A_m\cos m\theta+B_m\sin m\theta, \quad (m=0,1,2,\cdots).$$

Suppose that the solution of (9.3.16) is

$$u(\rho,\theta)=\sum_{m=0}^{\infty}[A_m(\rho)\cos m\theta+B_m(\rho)\sin m\theta]. \tag{9.3.18}$$

The nonhomogeneous term in (9.3.16) is

$$f(\rho,\theta)=12\rho^2\cos 2\theta, \tag{9.3.19}$$

which has the Fourier series form.

Substituting (9.3.18) and (9.3.19) into (9.3.16), we have

$$\sum_{m=0}^{\infty}[A''_m(\rho)\cos m\theta+B''_m(\rho)\sin m\theta]+\frac{1}{\rho}\sum_{m=0}^{\infty}[A'_m(\rho)\cos m\theta+B'_m(\rho)\sin m\theta]+$$

$$\frac{1}{\rho^2}\sum_{m=0}^{\infty}[-m^2 A_m(\rho)\cos m\theta-m^2 B_m(\rho)\sin m\theta]=12\rho^2\cos 2\theta,$$

that is

$$\sum_{m=0}^{\infty}\left[A''_m(\rho)+\frac{1}{\rho}A'_m(\rho)-\frac{m^2}{\rho^2}A_m(\rho)\right]\cos m\theta+$$

$$\sum_{m=0}^{\infty}\left[B''_m(\rho)+\frac{1}{\rho}B'_m(\rho)-\frac{m^2}{\rho^2}B_m(\rho)\right]\sin m\theta=12\rho^2\cos 2\theta.$$

Comparing the coefficients of two sides, we obtain

$$B''_m(\rho)+\frac{1}{\rho}B'_m(\rho)-\frac{m^2}{\rho^2}B_m(\rho)=0, \quad (m=0,1,\cdots), \tag{9.3.20}$$

$$A''_m(\rho)+\frac{1}{\rho}A'_m(\rho)-\frac{m^2}{\rho^2}A_m(\rho)=0, \quad (m\neq 2), \tag{9.3.21}$$

$$A''_2(\rho)+\frac{1}{\rho}A'_2(\rho)-\frac{4}{\rho^2}A_2(\rho)=12\rho^2. \tag{9.3.22}$$

(9.3.20) and (9.3.21) are homogeneous Euler's equation, and the general solutions of them are

$$A_0(\rho)=C_0+D_0\ln\rho,$$
$$B_0(\rho)=C_0'+D_0'\ln\rho,$$
$$A_m(\rho)=C_m\rho^m+D_m\rho^{-m}, \quad (m\neq 2),$$
$$B_m(\rho)=C_m'\rho^m+D_m'\rho^{-m}.$$

(9.3.22) is a nonhomogeneous Euler's equation and the general solution of it is
$$A_2(\rho)=C_2\rho^2+D_2\rho^{-2}+A_2^*(\rho),$$
where $A^*(\rho)$ is a special solution of (9.3.22).

Suppose that $A_2^*(\rho)=\alpha\rho^4$ is a solution of (9.3.22), then substituting it into (9.3.22), we can obtain $\alpha=1$, then
$$A_2(\rho)=C_2\rho^2+D_2\rho^{-2}+\rho^4.$$

The boundary conditions (9.3.17) can be transformed into
$$A_m(a)=A_m'(b)=0,$$
$$B_m(a)=B_m'(b)=0. \tag{9.3.23}$$

From (9.3.23), we can get
$$A_m(\rho)\equiv 0, \quad (m\neq 2),$$
$$B_m(\rho)\equiv 0.$$

The following task is to determine C_2 and D_2.

Since $A_2(a)=A_2'(b)=0$, then
$$C_2a^2+D_2a^{-2}+a^4=0,$$
$$2C_2b-2D_2b^{-3}+b^4=0,$$
that is
$$\begin{cases} C_2=-\dfrac{a^6+2b^6}{a^4+b^4}, \\ D_2=-\dfrac{a^4b^4(a^2-2b^2)}{a^4+b^4}. \end{cases}$$

And then $A_2(\rho)=-\dfrac{a^6+2b^6}{a^4+b^4}\rho^2-\dfrac{a^4b^4(a^2-2b^2)}{a^4+b^4}\rho^{-2}+\rho^4.$

Hence the solution of this problem is
$$u(\rho,\theta)=-\frac{1}{a^4+b^4}[(a^6+2b^6)\rho^2+a^4b^4(a^2-2b^2)\rho^{-2}-(a^4+b^4)\rho^4]\cos 2\theta.$$

Technique 3 To solve a nonhomogeneous problem with time independent data, reduce to the homogeneous case by finding a steady-state solution.

Consider the following problem:
$$L(u)=F(x,t), \quad B(u)=g(t), \quad I(u)=h(x). \tag{9.3.24}$$
The superposition principle can be used to trade off nonhomogeneous boundary conditions for nonhomogeneous equations. Namely, suppose we wish to solve
$$L(u)=0, \quad B(u)=g(t), \quad I(u)=h(x). \tag{9.3.25}$$
Let $w(x,t)$ be any smooth function that satisfies the boundary conditions $B(w)=g(t)$, and the initial conditions $I(w)=0$; such functions are relatively easy to construct because no differential equation needs to be solved. Then u satisfies (9.3.24) if and only if $v=u-w$ satisfies

$$L(v)=F(x,t), \quad B(v)=0, \quad I(v)=h(x), \qquad (9.3.26)$$

which we have already discussed.

In what follows, we solve the following problem

$$\begin{cases} \dfrac{\partial^2 u}{\partial t^2}=a^2\dfrac{\partial^2 u}{\partial x^2}+f(x,t), \\ u\big|_{x=0}=\mu_1(t), \ u\big|_{x=l}=\mu_2(t), \\ u\big|_{t=0}=\varphi(x), \ \dfrac{\partial u}{\partial t}\bigg|_{t=0}=\psi(x). \end{cases}$$

We need transform the boundary conditions into homogeneous one, and so we can set

$$u(x,t)=v(x,t)+w(x,t),$$

in which $w(x,t)$ is chosen such that $v\big|_{x=0}=0$, $v\big|_{x=l}=0$, and then $w(x,t)$ must satisfy $w\big|_{x=0}=\mu_1(t)$, $w\big|_{x=l}=\mu_2(t)$.

In order to reach this object, we choose $w(x,t)$ such that

$$w(x,t)=A(t)x+B(t),$$

then

$$A(t)=\frac{1}{l}[\mu_2(t)-\mu_1(t)],$$

$$B(t)=\mu_1(t).$$

It is obvious that

$$w(x,t)=\mu_1(t)+\frac{\mu_2(t)-\mu_1(t)}{l}x.$$

Hence

$$u(x,t)=v(x,t)+\left[\mu_1(t)+\frac{\mu_2(t)-\mu_1(t)}{l}x\right],$$

and $v(x,t)$ satisfies the following equation:

$$\begin{cases} \dfrac{\partial^2 v}{\partial t^2}=a^2\dfrac{\partial^2 v}{\partial x^2}+f_1(x,t), \\ v\big|_{x=0}=v\big|_{x=l}=0, \\ v\big|_{t=0}=\varphi_1(x), \ \dfrac{\partial v}{\partial t}\bigg|_{t=0}=\psi_1(x), \end{cases}$$

where

$$\begin{cases} f_1(x,t)=f(x,t)-\left[\mu_1''(t)+\dfrac{\mu_2''(t)-\mu_1''(t)}{l}x\right], \\ \varphi_1(x)=\varphi(x)-\left[\mu_1(0)+\dfrac{\mu_2(0)-\mu_1(0)}{l}x\right], \\ \psi_1(x)=\psi(x)-\left[\mu_1'(0)+\dfrac{\mu_2'(0)-\mu_1'(0)}{l}x\right]. \end{cases}$$

Note that if the nonhomogeneous terms are independent on t, that is, $f(x,t)=f(x)$, $\mu_1(t)=\mu_1$ (constant), $\mu_2(t)=\mu_2$ (constant), then $w(x,t)$ is chosen as a function independent on t, that is, $w(x,t)=w(x)$, such that the partial differential equation and the boundary conditions about $v(x,t)$ are homogeneous. We can explain this method by the following example.

Chapter 9 The Method of Separation of Variables on Finite Region

Example 9.3.3 Solve the following problem

$$\begin{cases} \dfrac{\partial^2 u}{\partial t^2} = a^2 \dfrac{\partial^2 u}{\partial x^2} + A, & 0<x<l,\ t>0, \quad (9.3.27) \\ u\big|_{x=0}=0,\ u\big|_{x=l}=B, & t\geqslant 0, \quad (9.3.28) \\ u\big|_{t=0}=\dfrac{\partial u}{\partial t}\bigg|_{t=0}=0, & 0\leqslant x\leqslant l. \quad (9.3.29) \end{cases}$$

Solution This is a nonhomogeneous equation with nonhomogeneous boundary conditions. The first thing we have to do is to transform the nonhomogeneous boundary conditions into homogeneous ones. Since the nonhomogeneous terms are all independent on t, we can transform these terms into homogeneous ones once at a time.

Let

$$u(x,t) = v(x,t) + w(x), \quad (9.3.30)$$

and substituting it into (9.3.27), we have

$$\dfrac{\partial^2 v}{\partial t^2} = a^2 \left[\dfrac{\partial^2 v}{\partial x^2} + w''(x)\right] + A. \quad (9.3.31)$$

We can choose $w(x)$ such that

$$\begin{cases} a^2 w''(x) + A = 0, & 0<x<l, \\ w\big|_{x=0}=0,\ w\big|_{x=l}=B, & t>0. \end{cases} \quad (9.3.32)$$

(9.3.32) is an ordinary differential equation with some boundary conditions, its solution can be easily solved and

$$w(x) = -\dfrac{A}{2a^2}x^2 + \left(\dfrac{Al}{2a^2} + \dfrac{B}{l}\right)x. \quad (9.3.33)$$

With this $w(x)$, the problem $v(x,t)$ satisfied is

$$\begin{cases} \dfrac{\partial^2 v}{\partial t^2} = a^2 \dfrac{\partial^2 v}{\partial x^2}, & 0<x<l,\ t>0, \\ v\big|_{x=0}=0,\ v\big|_{x=l}=0, & t\geqslant 0, \\ v\big|_{t=0}=-w(x),\ \dfrac{\partial v}{\partial t}\bigg|_{t=0}=0, & 0\leqslant x\leqslant l. \end{cases} \quad (9.3.34)$$

We can solve the problem (9.3.34) by the method of separation of variables, that is, let $v(x,t) = X(x)T(t)$ and we can obtain

$$\begin{cases} X''(x) + \lambda X(x) = 0, \\ X(0) = X(l) = 0 \end{cases} \quad (9.3.35)$$

and

$$T''(t) + \lambda a^2 T(t) = 0. \quad (9.3.36)$$

Then according to the principle analysis, the equation in (9.3.34) with homogeneous boundary conditions has the following solution

$$v(x,t) = \sum_{n=1}^{\infty}\left(C_n \cos\dfrac{n\pi a}{l}t + D_n \sin\dfrac{n\pi a}{l}t\right)\sin\dfrac{n\pi}{l}x. \quad (9.3.37)$$

From $\dfrac{\partial v}{\partial t}\bigg|_{t=0} = 0$, we obtain $D_n = 0$.

From $v\big|_{t=0} = -w(x)$, we obtain

· 189 ·

that is

$$-w(x) = \sum_{n=1}^{\infty} C_n \sin \frac{n\pi}{l} x,$$

$$\frac{A}{2a^2} x^2 - \left(\frac{Al}{2a^2} + \frac{B}{l}\right) x = \sum_{n=1}^{\infty} C_n \sin \frac{n\pi}{l} x,$$

where

$$\begin{aligned} C_n &= \frac{2}{l} \int_0^l \left[\frac{A}{2a^2} x^2 - \left(\frac{Al}{2a^2} + \frac{B}{l}\right) x\right] \sin \frac{n\pi}{l} x \, dx \\ &= \frac{A}{a^2 l} \int_0^l x^2 \sin \frac{n\pi}{l} x \, dx - \left(\frac{A}{a^2} + \frac{2B}{l^2}\right) \int_0^l x \sin \frac{n\pi}{l} x \, dx \\ &= -\frac{2Al^2}{a^2 n^3 \pi^3} + \frac{2}{n\pi}\left(\frac{Al^2}{a^2 n^2 \pi^2} + B\right) \cos n\pi. \end{aligned} \tag{9.3.38}$$

Finally, the solution of the original problem is

$$u(x, t) = -\frac{A}{2a^2} x^2 + \left(\frac{Al}{2a^2} + \frac{B}{l}\right) x + \sum_{n=1}^{\infty} C_n \cos \frac{n\pi a}{l} t \sin \frac{n\pi}{l} x, \tag{9.3.39}$$

where C_n is given in (9.3.38).

Example 9.3.4 Solve the following problem

$$\begin{cases} \dfrac{\partial u}{\partial t} = K \dfrac{\partial^2 u}{\partial x^2} + R, & 0 < x < l, \ t > 0, & (9.3.40) \\ u\big|_{x=0} = 0, \ u\big|_{x=l} = A, & t \geqslant 0, & (9.3.41) \\ u\big|_{t=0} = f(x), & 0 \leqslant x \leqslant l. & (9.3.42) \end{cases}$$

Solution Let $u(x,t) = v(x,t) + w(x)$, and substituting it into (9.3.40), we obtain

$$\frac{\partial v}{\partial t} = K\left(\frac{\partial^2 v}{\partial x^2} + w''(x)\right) + R. \tag{9.3.43}$$

We can choose $w(x)$ such that

$$\begin{cases} Kw''(x) + R = 0, & 0 < x < l, \\ w\big|_{x=0} = 0, \ w\big|_{x=l} = A. \end{cases} \tag{9.3.44}$$

The solution of (9.3.44) is

$$w(x) = -\frac{R}{2K} x^2 + \left(\frac{Rl}{2K} + \frac{A}{l}\right) x. \tag{9.3.45}$$

Then the problem of $v(x,t)$ is

$$\begin{cases} \dfrac{\partial v}{\partial t} = K \dfrac{\partial^2 v}{\partial x^2}, & 0 < x < l, \ t > 0, \\ v\big|_{x=0} = 0, \ v\big|_{x=l} = 0, & t \geqslant 0, \\ v_{t=0} = f(x) - w(x), & 0 \leqslant x \leqslant l. \end{cases} \tag{9.3.46}$$

We can solve it by separation of variables. Let $v(x,t) = X(x)T(t)$, then the corresponding eigenvalue problem for (9.3.46) is

$$\begin{cases} X''(x) + \lambda X(x) = 0, \\ X(0) = X(l) = 0. \end{cases} \tag{9.3.47}$$

Solving it, we have the eigenvalues as $\lambda_n = \dfrac{n^2 \pi^2}{l^2}$, $(n = 1, 2, \cdots)$ and the eigenfunctions as

$$X_n(x) = A_n \sin \frac{n\pi}{l} x.$$

Then another equation becomes
$$T'(t)+\frac{n^2\pi^2 K}{l^2}T(t)=0. \tag{9.3.48}$$

The general solution of (9.3.48) is
$$T_n(t)=C'_n e^{-\frac{n^2\pi^2 K}{l^2}t}. \tag{9.3.49}$$

Then
$$v(x,t)=\sum_{n=1}^{\infty}X_n(x)T_n(t)=\sum_{n=1}^{\infty}C_n e^{-\frac{n^2\pi^2 K}{l^2}t}\sin\frac{n\pi}{l}x. \tag{9.3.50}$$

With the initial condition $v|_{t=0}=f(x)-w(x)$, we have
$$\sum_{n=1}^{\infty}C_n\sin\frac{n\pi}{l}x=f(x)+\frac{R}{2K}x^2-\left(\frac{Rl}{2K}+\frac{A}{l}\right)x,$$

which means
$$\begin{aligned}C_n &= \frac{2}{l}\int_0^l\left[f(x)+\frac{R}{2K}x^2-\left(\frac{Rl}{2K}+\frac{A}{l}\right)x\right]\sin\frac{n\pi}{l}x\,dx \\ &= \frac{2}{l}\int_0^l f(x)\sin\frac{n\pi}{l}x\,dx+\frac{2}{n\pi}\left(\frac{Rl^2}{Kn^2\pi^2}+A\right)\cos n\pi-\frac{2Rl^2}{Kn^3\pi^3}.\end{aligned} \tag{9.3.51}$$

Finally, the solution of the original problem is
$$u(x,t)=-\frac{R}{2K}x^2+\left(\frac{Rl}{2K}+\frac{A}{l}\right)x+\sum_{n=1}^{\infty}C_n e^{-\frac{Kn^2\pi^2}{l^2}t}\sin\frac{n\pi}{l}x, \tag{9.3.52}$$

with C_n being giving by (9.3.51).

Now we give a summary about separation of variables for all possible cases.

1. Homogeneous equation with homogeneous boundary conditions
$$L(u)=0,\quad B(u)=0,\quad I(u)=h(x).$$

(1) Separation of variables

Let $u(x,t)=X(x)T(t)$, we can obtain two ordinary differential equations.

(2) One of the ordinary differential equation with the homogeneous boundary conditions constitute the eigenvalue problem. Solving this problem, one can obtain the eigenvalues and the eigenfunctions.

(3) Solve the remaining one ordinary differential equation.

(4) Combining and superpositing $u(x,t)=\sum_{n=0}^{\infty}X_n(x)T_n(t)$, one can determine the constant coefficients in the series solution by the initial conditions.

2. Nonhomogeneous equation with homogeneous boundary conditions
$$L(u)=F(x,t),\quad B(u)=0,\quad I(u)=h(x).$$

Divide this problem into two parts:

(1) $L(v)=0$, $B(v)=0$, $I(v)=h(x)$. (Same as 1)

(2) $L(w)=F(x,t)$, $B(w)=0$, $I(w)=0$.

From 1, we can obtain eigenvalues and eigenfunctions.

From 2, we expand $w(x,t)$ and the nonhomogeneous term $F(x,t)$ as series about the eigenfunctions, and then solve 2. (Generally by variation of parameters.)

Finally, $u(x,t)=w(x,t)+v(x,t)$.

3. Nonhomogeneous equation with nonhomogeneous boundary conditions

$$L(u)=F(x,t), \quad B(u)=g(t), \quad I(u)=h(x).$$

Construct a transformation $w(x,t)$ such that $B(w)=g(t)$.

Then let $u(x,t)=w(x,t)+v(x,t)$, where $v(x,t)$ satisfies the following problem

$$L(v)=F(x,t)-L(w), \quad B(v)=0, \quad I(v)=h(x)-I(w),$$

which is the problem we discussed in 2.

If $F(x,t)=F(x)$, $g(t)=C$, then we can choose $w(x,t)=w(x)$, such that $L(w)=F(x)$ and $B(w)=0$, then $v(x,t)$ satisfies $L(v)=0$, $B(v)=0$ and $I(v)=h(x)-w(x)$.

Exercises 9.3

Solve the following nonhomogeneous problems:

(a) $\begin{cases} \dfrac{\partial u}{\partial t}=\dfrac{1}{a^2}\dfrac{\partial^2 u}{\partial x^2}+Ae^{-ax}, & 0<x<l, \ t>0, \\ u|_{x=0}=0, \ u|_{x=l}=0, & t\geqslant 0, \\ u|_{t=0}=T, \ (T \text{ is a constant}), & 0\leqslant x\leqslant l. \end{cases}$

(b) $\begin{cases} \dfrac{\partial u}{\partial t}=a^2\dfrac{\partial^2 u}{\partial x^2}+A, & 0<x<l, \ t>0, \\ u|_{x=0}=u|_{x=l}=0, & t\geqslant 0, \\ u|_{t=0}=0, & 0\leqslant x\leqslant l. \end{cases}$

9.4 Sturm-Liouville eigenvalue problem

The key step in the method of separation of variables we presented in the above sections is solving the eigenvalue problem. The eigenvalue problem is consisting of a homogeneous ordinary differential equation with one parameter and some boundary conditions. We can obtain the eigenvalues and eigenfunctions by discussing the different values of the parameter in the eigenvalue problem to satisfy the corresponding boundary conditions. In this section, we focus on this eigenvalue problem and called it as **Sturm-Liouville eigenvalue problem**.

Consider the general homogeneous ordinary differential equation with parameter λ,

$$c_1(x)\frac{d^2 u}{dx^2}+c_2(x)\frac{du}{dx}+[c_3(x)+\lambda]u=0, \tag{9.4.1}$$

where $c_1(x)\neq 0$.

Set $p(x)=e^{\int_0^x \frac{c_2(\eta)}{c_1(\eta)}d\eta}$, $q(x)=\dfrac{c_3(x)}{c_1(x)}p(x)$, $s(x)=\dfrac{1}{c_1(x)}p(x)$. From (9.4.1), we have

$$\frac{d}{dx}\left(p(x)\frac{du}{dx}\right)+[q(x)+\lambda s(x)]u=0. \tag{9.4.2}$$

The equation (9.4.2) is called the **Sturm-Liouville normal form** for (9.4.1), in which $p(x)$ is called kernel function and $s(x)$ is called the weight function.

Applying the self-adjoint operator

$$L = \frac{d}{dx}\left(p(x)\frac{d}{dx}\right) + q(x), \qquad (9.4.3)$$

(9.4.2) can be written as follows:

$$L(u) + \lambda s(x) u = 0. \qquad (9.4.4)$$

In the above equation, the parameter λ is independent of x, $p(x) \in c^{(1)}[a,b]$, $q(x), s(x) \in c[a,b]$.

If $p(x)$ and $s(x)$ are positive on $[a,b]$ and $-\infty < a < b < +\infty$, then the Sturm-Liouville equation is called **regular** on $[a,b]$. Otherwise, namely $p(a)s(a) = 0$ or $p(b)s(b) = 0$ or $a = -\infty$ or $b = +\infty$, then the Sturm-Liouville equation is called **singular**. The conditions at end points on the Sturm-Liouville equation are given as follows.

(1) General boundary conditions ($p(a) \neq 0$)

$$\partial_1 u(a) + \beta_1 u'(a) = 0, \qquad (9.4.5)$$

$$\partial_2 u(b) + \beta_2 u'(b) = 0, \qquad (9.4.6)$$

where $|\partial_i| + |\beta_i| > 0, \quad (i=1, 2)$.

(2) Periodic boundary condition ($p(x) \neq 0$ and $p(a) \neq p(b)$)

$$u(a) = u(b),$$
$$u'(a) = u'(b). \qquad (9.4.7)$$

(3) Natural boundary condition ($p(a) = 0$ or $p(b) = 0$)

$$|u(a)| < +\infty,$$
$$|u(b)| < +\infty. \qquad (9.4.8)$$

(9.4.2) and (9.4.5), (9.4.6) as well as (9.4.2) and (9.4.7) or (9.4.8) are called the **Sturm-Liouville problem**. The value of parameter λ, for which the corresponding Sturm-Liouville problem has non-trivial solutions, is called **eigenvalue** and corresponding non-trivial solutions are called **eigenfunctions** of this problem.

Example 9.4.1 Find the eigenvalues and eigenfunctions of the following Sturm-Liouville problem:

$$\begin{cases} u''(x) + \lambda u(x) = 0, & 0 < x < \pi, \qquad (9.4.9) \\ u(0) = 0, & (9.4.10) \\ u'(\pi) = 0. & (9.4.11) \end{cases}$$

Solution For this problem, it is obvious that $p(x) = 1$, $q(x) = 0$, $s(x) = 1$.

Then we discuss the value of the parameter λ.

(i) If $\lambda = 0$, then $u(x) = c_1 x + c_2$. From (9.4.10) and (9.4.11), we have $c_2 = c_1 = 0$, then $u(x) \equiv 0$.

(ii) If $\lambda < 0$, then $u(x) = c_1 e^{\sqrt{-\lambda} x} + c_2 e^{\sqrt{-\lambda} x}$. From (9.4.10) and (9.4.11), we have

$$\begin{cases} c_1 + c_2 = 0, \\ \sqrt{-\lambda}(c_1 e^{\sqrt{-\lambda}\pi} - c_2 e^{\sqrt{-\lambda}\pi}) = 0, \end{cases}$$

which means $c_1 = c_2 = 0$, and then $u(x) \equiv 0$.

(iii) If $\lambda > 0$, then $u(x) = c_1 \cos\sqrt{\lambda} x + c_2 \sin\sqrt{\lambda} x$. From (9.4.10) and (9.4.11), we have

$$\begin{cases} c_1 = 0, \\ \sqrt{\lambda} c_2 \cos\sqrt{\lambda}\pi = 0, \end{cases}$$

then if $c_2 \neq 0$, we have $\cos\sqrt{\lambda}\pi = 0$, which means $\lambda_n = \left(n - \dfrac{1}{2}\right)^2$, $(n=1, 2, \cdots)$.

Hence, for this problem, the eigenvalues are $\lambda_n = \left(n - \dfrac{1}{2}\right)^2$, $(n=1, 2, \cdots)$, and $u_n(x) = \sin\left(n - \dfrac{1}{2}\right)x = \sin\dfrac{2n-1}{2}x$ $(n=1, 2, \cdots)$ are eigenfunctions.

Example 9.4.2 Find the eigenvalues and eigenfunctions of the following Sturm-Liouville problem on Cauchy Euler equation

$$\begin{cases} x^2 \dfrac{d^2 u}{dx^2} + x \dfrac{du}{dx} + \lambda u = 0, & 1 < x < e, \\ u(1) = 0, \\ u(e) = 0. \end{cases} \quad (9.4.12) \\ (9.4.13) \\ (9.4.14)$$

Solution The equation (9.4.12) can be written as follows:

$$\dfrac{d}{dx}\left(x \dfrac{du}{dx}\right) + \dfrac{1}{x}\lambda u = 0, \quad (9.4.15)$$

where $p(x) = x$, $q(x) = 0$, $s(x) = \dfrac{1}{x}$. Now we discuss the value of the parameter λ.

(i) If $\lambda = 0$, then $\dfrac{d}{dx}\left(x \dfrac{du}{dx}\right) = 0$, which means $x \dfrac{du}{dx} = c_1$, and then $u(x) = c_1 \ln x + c_2$. By (9.4.13) and (9.4.14), we have $c_2 = c_1 = 0$, with which we conclude $u(x) \equiv 0$.

(ii) If $\lambda < 0$, let $u(x) = x^\beta$, then by (9.4.12) we have

$$(\beta^2 + \lambda)x^\beta = 0,$$

that is $\beta = \pm\sqrt{-\lambda}$, and the general solution of (9.4.12) is

$$u(x) = c_1 x^{\sqrt{-\lambda}} + c_2 x^{-\sqrt{-\lambda}}.$$

By (9.4.13) and (9.4.14), we have

$$\begin{cases} c_1 + c_2 = 0, \\ c_1 e^{\sqrt{-\lambda}} + c_2 e^{-\sqrt{-\lambda}} = 0, \end{cases}$$

which means $c_1 = c_2 = 0$, with which we conclude $u(x) \equiv 0$.

(iii) If $\lambda > 0$, then by (ii), we have $\bar{u}(x) = c_1 x^{i\sqrt{\lambda}} + c_2 x^{-i\sqrt{\lambda}}$.

Since $x^{\pm i\sqrt{\lambda}} = e^{\ln x^{\pm i\sqrt{\lambda}}} = e^{\pm i\sqrt{\lambda} \ln x} = \cos(\sqrt{\lambda} \ln x) \pm i \sin(\sqrt{\lambda} \ln x)$, then

$$u(x) = c_1 \cos(\sqrt{\lambda}\ln x) + c_2 (\sin\sqrt{\lambda} \ln x).$$

By (9.4.13), $u(1) = 0$, which means $c_1 = 0$.

And by (9.4.14), $u(e) = 0$, which means $c_2 \sin\sqrt{\lambda} = 0$.

If $c_2 \neq 0$, then $\sin\sqrt{\lambda} = 0$, and we derive the eigenvalues $\lambda_n = (n\pi)^2$, $(n=1, 2, \cdots)$, and the eigenfunctions are $\sin(n\pi\ln x)$.

Example 9.4.3 Solve the periodic Sturm-Liouville problem

$$\begin{cases} u'' + \lambda u = 0, & -\pi < x < \pi, \\ u(-\pi) = u(\pi), \\ u'(-\pi) = u'(\pi). \end{cases} \quad (9.4.16) \\ (9.4.17) \\ (9.4.18)$$

Solution In the following we discuss the value of the parameter λ.

(i) If $\lambda<0$, then $u(x)=c_1 e^{\sqrt{-\lambda}x}+c_2 e^{-\sqrt{-\lambda}x}$.

By (9.4.17), we have
$$c_1 e^{-\sqrt{-\lambda}\pi}+c_2 e^{\sqrt{-\lambda}\pi}=c_1 e^{\sqrt{-\lambda}\pi}+c_2 e^{-\sqrt{-\lambda}\pi}$$
which means $c_1=c_2$, and then $u(x)=c_1(e^{\sqrt{-\lambda}x}+e^{-\sqrt{\lambda}x})$.

By (9.4.18), we have $c_1=0$, and it is concluded that $u(x)\equiv 0$.

(ii) If $\lambda=0$, then $u(x)=c_1 x+c_2$.

By (9.4.17), we have $c_1=0$.

And by (9.4.18), we have $c_2\neq 0$ and it is an arbitrary constant.

Hence $\lambda_0=0$ is an eigenvalue, $u_0(x)=1$ is an eigenfunction.

(iii) If $\lambda>0$, then $u(x)=c_1\cos\sqrt{\lambda}x+c_2\sin\sqrt{\lambda}x$.

By (9.4.17) and (9.4.18), we have
$$c_1\cos\sqrt{\lambda}\pi-c_2\sin\sqrt{\lambda}\pi=c_1\cos\sqrt{\lambda}\pi+c_2\sin\sqrt{\lambda}\pi,$$
and $\sqrt{\lambda}(c_1\sin\sqrt{\lambda}\pi+c_2\cos\sqrt{\lambda}\pi)=\sqrt{\lambda}(-c_1\sin\sqrt{\lambda}\pi+c_2\cos\sqrt{\lambda}\pi)$, with which it can be concluded that $2c_2\sin\sqrt{\lambda}\pi=0$, $2c_1\sin\sqrt{\lambda}\pi=0$.

If $c_2\neq 0$, $\sin\sqrt{\lambda}\pi=0$, then $\lambda_n=n^2$, $(n=1, 2, \cdots)$.

If $c_1\neq 0$, $\sin\sqrt{\lambda}\pi=0$, then $\lambda_n=n^2$, $(n=1, 2, \cdots)$.

Namely, only if $\lambda_n=n^2$, $(n=1, 2, \cdots)$, then $c_1\neq 0$ and $c_2\neq 0$ and they are arbitrary constants. Therefore $\lambda_n=n^2$, $(n=1, 2, \cdots)$, are eigenvalues, and $u_n^{(1)}(x)=\cos nx$, and $u_n^{(2)}(x)=\sin nx$, $(n=1, 2, \cdots)$ are eigenfunctions.

In words, the eigenvalues are $\lambda_n=n^2$, $(n=1, 2, \cdots)$, and the eigenfunctions are
$$\{\cos nx\}_{n=0}^{\infty}\cup\{\sin nx\}_{n=1}^{\infty} \text{ or } \{1\}\cup\{\cos nx\}_{n=1}^{\infty}\cup\{\sin nx\}_{n=1}^{\infty}.$$

For $\lambda_0=0$, there is only one corresponding eigenfunctions $u_0(x)=1$, and then $\lambda_0=0$ is called a simple eigenvalue; for $\lambda_n=n^2$, $(n=1, 2, \cdots)$, there are two linear independent eigenfunctions $\cos nx$ and $\sin nx$, then $\lambda_n=n^2$, $(n=1, 2, \cdots)$, are called two repeated eigenvalues.

Now, we consider the basic properties on the system of eigenfunctions for the Sturm-Liouville problem.

Theorem 9.4.1 Assume $p(x)$, $q(x)$ and $s(x)\in C[a,b]$ and the eigenfunctions $u_j(x)$ and $u_k(x)$ which correspond to eigenvalues λ_j and $\lambda_k (\lambda_j\neq\lambda_k)$, respectively, and then $u_j(x)$ and $u_k(x)$ are orthogonal with weight function $s(x)$ on $[a,b]$, namely,
$$\int_a^b s(x)u_j(x)u_k(x)\mathrm{d}x = 0.$$

Proof Since $u_j(x)$, $u_k(x)$ are the solutions of the Sturm-Liouville problem corresponding to $\lambda_j(x)$ and $\lambda_k(x)$ respectively, we have

$$\frac{\mathrm{d}}{\mathrm{d}x}(p(x)u_j')+(q(x)+\lambda_j s(x))u_j=0, \quad (9.4.19)$$

$$\frac{\mathrm{d}}{\mathrm{d}x}(p(x)u_k')+(q(x)+\lambda_k s(x))u_k=0. \quad (9.4.20)$$

The result of multiplying (9.4.19) by u_k subtracts the result of multiplying (9.4.20) by

u_j, namely, $(9.4.19) \times u_k - (9.4.20) \times u_j$, gives

$$(\lambda_k - \lambda_j) s(x) u_k u_j = u_k \frac{d}{dx}(p(x) u_j') - u_j \frac{d}{dx}(p(x) u_k')$$

$$= \frac{d}{dx}[(p(x) u_j') u_k - u_j (p(x) u_k')].$$

Integrating two sides of above equality on $[a, b]$, we get

$$(\lambda_k - \lambda_j) \int_a^b s(x) u_j u_k dx = [(p(x) u_j') u_k - (p(x) u_k') u_j]\Big|_a^b$$
$$= p(b)[u_j'(b) u_k(b) - u_j(b) u_k'(b)] - \qquad (9.4.21)$$
$$p(a)[u_j'(a) u_k(a) - u_j(a) u_k'(a)].$$

By the boundary conditions of the Sturm-Liouville problem, we have

$$\alpha_1 u_j(a) + \beta_1 u_j'(a) = 0, \qquad (9.4.22)$$
$$\alpha_1 u_k(a) + \beta_1 u_k'(a) = 0, \qquad (9.4.23)$$
$$\alpha_2 u_j(b) + \beta_2 u_j'(b) = 0, \qquad (9.4.24)$$
$$\alpha_2 u_k(b) + \beta_2 u_k'(b) = 0. \qquad (9.4.25)$$

(i) If $\alpha_1 \neq 0$, then $(9.4.22) \times u_k'(a) - (9.4.23) \times u_j'(a)$, we get

$$\alpha_1 [u_j(a) u_k'(a) - u_k(a) u_j'(a)] = 0 \Rightarrow u_j(a) u_k'(a) - u_k(a) u_j'(a) = 0.$$

(ii) If $\beta_1 \neq 0$, then $(9.4.22) \times u_k(a) - (9.4.23) \times u_j(a)$, we get

$$\beta_1 [u_k(a) u_j'(a) - u_j(a) u_k'(a)] = 0 \Rightarrow u_k(a) u_j'(a) - u_j(a) u_k'(a) = 0.$$

It follows that

$$p(a)[u_j'(a) u_k(a) - u_j(a) u_k'(a)] = 0.$$

In the same way, we can obtain

$$p(b)[u_j'(b) u_k(b) - u_j(b) u_k'(b)] = 0.$$

Therefore, from (9.4.21) and the last two equalities, we have

$$(\lambda_k - \lambda_j) \int_a^b s(x) u_j(x) u_k(x) dx = 0.$$

Since $\lambda_k - \lambda_j \neq 0$, we obtain

$$\int_a^b s(x) u_j(x) u_k(x) dx = 0.$$

This completes the proof of Theorem 9.4.1.

Corollary 9.4.1 The system of the eigenfunctions on periodic Sturm-Liouville problem is orthogonal with the weight function $s(x)$ on $[a, b]$. Namely

$$\int_a^b s(x) u_j(x) u_k(x) dx = 0,$$

where $u_k(x)$ and $u_j(x)$ are arbitrary eigenfuntions corresponding the eigenvalue λ_j and $\lambda_k (\lambda_j \neq \lambda_k)$ respectively.

Proof By the discussion on Theorem 9.4.1, we know that for $\lambda_j \neq \lambda_k$,

$$(\lambda_k - \lambda_j) \int_a^b s(x) u_j(x) u_k(x) dx = [p(x)(u_k u_j' - u_j u_k')]\Big|_a^b.$$

Applying the periodic boundary conditions:

$$p(a) = p(b),$$
$$u_k(a) = u_k(b) u_k'(a) = u_k'(b),$$

$$u_j(a) = u_j(b) u_j'(a) = u_j'(b).$$

We see that
$$p(b)[u_j'(b)u_k(b) - u_j(b)u_k'(b)] - p(a)[u_j'(a)u_k(a) - u_j(a)u_k'(a)] = 0,$$
and then by $\lambda_j - \lambda_k \neq 0$, we get
$$\int_a^b s(x) u_j(x) u_k(x) \, dx = 0.$$

Theorem 9.4.2 The eigenvalues $\{\lambda_n\}_{n=1}^{\infty}$ of the regular Sturm-Liouville problem are real.

Proof Suppose a complex eigenvalue $\bar{\lambda}_j = \alpha + i\beta$ corresponding eigenfunction is $u_j(x)$. Since the coefficients of the equation are real, we can see that the conjugate complex number $\bar{\lambda}_j = \alpha - i\beta$ is also an eigenvalue and the corresponding conjugate function $\bar{u}_j(x)$ is also an eigenfunction.

If $\beta \neq 0$, then $\lambda_j \neq \bar{\lambda}_j$, and then by Theorem 9.4.1, we know that u_j and $\bar{u}_j$ are orthogonal with the weight function $s(x)$ on $[a,b]$. Namely,
$$(\lambda_k - \lambda_j) \int_a^b s(x) u_j(x) u_k(x) \, dx = 0.$$
Then
$$2\beta \int_a^b s(x) |u_j(x)|^2 \, dx = 0.$$
Since $s(x) > 0$, $u_j(x) \neq 0$, we have $\int_a^b s(x) |u_j(x)|^2 \, dx > 0$ and $\beta = 0$. This is in contradiction to the above assumption of $\beta \neq 0$.

Theorem 9.4.3 For the regular Sturm-Liouville problem, there is a real infinite sequence of eigenvalue $\{\lambda_n\}_{n=1}^{\infty}$, which satisfies that $\lambda_1 < \lambda_2 < \cdots < \lambda_n < \cdots$, and $\lim_{n \to \infty} \lambda_n = +\infty$, and corresponding eigenfunctions $\{u_n(x)\}_{n=1}^{\infty}$ construct one completely orthogonal system with the weight $s(x)$ on $[a,b]$. Namely,
$$\int_a^b s(x) u_k(x) u_j(x) \, dx = \begin{cases} 0, & \text{as } k \neq j, \lambda_k \neq \lambda_j, \\ M_k, & \text{as } k = j, M_k > 0. \end{cases}$$
Furthermore, any function $f(x)$ which is piecewise smooth on $[a,b]$ and satisfies the end point conditions of Sturm-Liouville problem can be expanded in an absolutely and uniformly convergent series by the eigenfunction system $\{u_n(x)\}_{n=1}^{\infty}$ on $[a,b]$. Namely,
$$f(x) = \sum_{n=1}^{\infty} f_n u_n(x),$$
where
$$f_n = \frac{\int_a^b s(x) f(x) u_n(x) \, dx}{\int_a^b s(x) u_n^2(x) \, dx}, \quad (n = 1, 2, \cdots).$$

Exercises 9.4

Find the eigenvalues and eigenfunctions of the following Sturm-Liouville problems:

(a) $\begin{cases} u''+\lambda u=0, \ 0<x<\pi, \\ u(0)=0, \ u(\pi)=0. \end{cases}$

(b) $\begin{cases} u''+\lambda u=0, \ 0<x<1, \\ u(0)=0, \ u'(1)=0. \end{cases}$

(c) $\begin{cases} u''+\lambda u=0, \ 0<x<\pi, \\ u'(0)=0, \ u'(\pi)=0. \end{cases}$

(d) $\begin{cases} u''+\lambda u=0, \ 0<x<\pi, \\ u(0)=u(\pi), \ u'(0)=u'(\pi). \end{cases}$

Chapter 10 Special Functions

In last chapter, we solved some defining problems by the method of separation of variables. We only considered some lower dimensional problems in Chapter 9. If we consider $(2+1)$-dimensional heat equation or 3-dimensional Laplace equation, we can obtain some special ordinary differential equations by separation of variables. In this chapter, we will introduce the Bessel equation and Legendre equation by some examples, and discuss some properties and applications for the solutions of these equations.

10.1 Bessel function

In this section, we will introduce the Bessel equation from an example and derive the Bessel function by solving the Bessel equation.

10.1.1 Introduction to the Bessel equation

We will introduce the Bessel equation by the following example.

Example 10.1.1 Suppose that there is a thin circular disk with radius R and it is insolate at its ends. If the temperature of its boundary keeps 0 ℃ and its initial temperature is given, how to determine the distribution of the temperature in the disk?

To solve this problem is equivalent to solve the following defining problem:

$$\begin{cases} \dfrac{\partial u}{\partial t} = a^2 \left(\dfrac{\partial^2 u}{\partial x^2} + \dfrac{\partial^2 u}{\partial y^2} \right), & x^2 + y^2 < R^2,\ t > 0, & (10.1.1) \\ u(x,y,0) = \varphi(x,y), & x^2 + y^2 \leqslant R^2, & (10.1.2) \\ u\big|_{x^2+y^2=R^2} = 0, & t \geqslant 0. & (10.1.3) \end{cases}$$

We can solve this problem by the method of separation of variables. Let

$$u(x,y,t) = V(x,y)T(t),$$

and substituting it into (10.1.1), we have

$$VT' = a^2 \left(\dfrac{\partial^2 V}{\partial x^2} + \dfrac{\partial^2 V}{\partial y^2} \right) T$$

or

$$\frac{T'}{a^2 T} = \frac{\frac{\partial^2 V}{\partial x^2} + \frac{\partial^2 V}{\partial y^2}}{V} = -\lambda, \quad (\lambda > 0).$$

Then we obtain the following equations about $V(x,y)$ and $T(t)$, that is,

$$T'(t) + a^2 \lambda T(t) = 0, \tag{10.1.4}$$

$$\frac{\partial^2 V}{\partial x^2} + \frac{\partial^2 V}{\partial y^2} + \lambda V = 0. \tag{10.1.5}$$

From (10.1.4), we can find $T(t) = A e^{-a^2 \lambda t}$.

And equation (10.1.5) is called the **Helmboltz equation**. In order to solve the nontrivial solution of this equation with initial condition

$$V\big|_{x^2+y^2=R^2} = 0, \tag{10.1.6}$$

we can introduce polar coordinates $x = r\cos\theta$, $y = r\sin\theta$, and rewrite the equation (10.1.5) and the condition (10.1.6) into

$$\begin{cases} \dfrac{\partial^2 V}{\partial r^2} + \dfrac{1}{r}\dfrac{\partial V}{\partial r} + \dfrac{1}{r^2}\dfrac{\partial^2 V}{\partial \theta^2} + \lambda V = 0, & r < R,\ 0 \leqslant \theta \leqslant 2\pi, \\ V\big|_{r=R} = 0, & 0 \leqslant \theta \leqslant 2\pi. \end{cases} \tag{10.1.7}$$
$$\tag{10.1.8}$$

Let $V(r,\theta) = P(r)\Theta(\theta)$, and substituting it into (10.1.7) and applying the separation of variables again, we have

$$\frac{r^2 P''(r)}{P(r)} + \frac{r P'(r)}{P(r)} + \lambda r^2 = -\frac{\Theta''(\theta)}{\Theta(\theta)} = \mu.$$

And it is equivalent to

$$\begin{cases} r^2 P''(r) + r P'(r) + (\lambda r^2 - \mu) P(r) = 0, & \tag{10.1.9} \\ \Theta''(\theta) + \mu \Theta(\theta) = 0. & \tag{10.1.10} \end{cases}$$

Since $u(x,y,t)$ is a single-valued function, $V(x,y)$ is also a single-valued function. Thus, $\Theta(\theta)$ is a periodic function with period 2π. According to the periodic boundary condition $\Theta(\theta + 2\pi) = \Theta(\theta)$, we can derive the value of μ with

$$\mu = 0,\ 1^2,\ 2^2,\ \cdots,\ n^2,\ \cdots.$$

In correspondence with $\mu = n^2$,

$$\Theta_0(\theta) = \frac{A_0}{2},$$

$$\Theta_n(\theta) = A_n \cos n\theta + B_n \sin n\theta, \quad (n = 1, 2, \cdots).$$

Substituting $\mu = n^2$ into (10.1.9), we have

$$r^2 P''(r) + r P'(r) + (\lambda r^2 - n^2) P(r) = 0. \tag{10.1.11}$$

Taking the transformation $\rho = \sqrt{\lambda} r$ and denoting $F(\rho) = P\left(\dfrac{1}{\sqrt{\lambda}}\rho\right)$, we obtain

$$\rho^2 F''(\rho) + \rho F'(\rho) + (\rho^2 - n^2) F(\rho) = 0. \tag{10.1.12}$$

This equation is called the **n-th order Bessel equation**.

From condition (10.1.8) and noticing that the temperature at the center is always finite, we can have

$$\begin{cases} P(R) = 0, \\ |P(0)| < +\infty. \end{cases}$$

Thus, we must derive the eigenvalues and eigenfunctions of the Bessel equation (10.1.11) with the above conditions in order to solve the example 10.1.1. In the next subsection, we will give the solution of equation (10.1.11), and then discuss this eigenvalue problem.

10.1.2 The solution of the Bessel equation

In order to solve the Bessel equation, we introduce the following theorem firstly.

Theorem 10.1.1 (Fuchs theorem) Suppose that

$$y''(x) + \frac{P(x)}{x-x_0}y'(x) + \frac{Q(x)}{(x-x_0)^2}y(x) = 0,$$

where $P(x)$ and $Q(x)$ are analytic for $|x-x_0| < R$. Then x_0 is called the **regular singular point** of the equation, and the power solution of this equation in $|x-x_0| < R$ is $y(x) = (x-x_0)^s \sum_{n=0}^{\infty} a_n (x-x_0)^n$, where $a_0 \neq 0$.

Remarks

1. In the series solution, s and a_n are to be determined.

2. For the standard Bessel equation $x^2 y''(x) + x y'(x) + (x^2 - v^2) y(x) = 0$, which can be translated into

$$y''(x) + \frac{1}{x}y'(x) + \frac{x^2 - v^2}{x^2}y(x) = 0$$

where $P(x) \equiv 1$, $Q(x) = x^2 - v^2$, $x = 0$ is its regular singular point. So we expect to find solution of the form $y(x) = x^s \sum_{k=0}^{\infty} a_k x^k$, where $a_0 \neq 0$, and we need to determine s and a_k.

Now we consider the standard Bessel equation

$$x^2 y''(x) + x y'(x) + (x^2 - v^2) y(x) = 0. \qquad (10.1.13)$$

Suppose that the equation (10.1.13) has a series solution $y(x) = x^s \sum_{k=0}^{\infty} a_k x^k$ with $a_0 \neq 0$.

Substituting it and its derivatives into (10.1.13), we have

$$x^2 \sum_{k=0}^{\infty} a_k (k+s)(k+s-1) x^{k+s-2} + x \sum_{k=0}^{\infty} a_k (k+s) x^{k+s-1} + (x^2 - v^2) \sum_{k=0}^{\infty} a_k x^{k+s} = 0,$$

(10.1.14)

that is,

$$\sum_{k=0}^{\infty} a_k [(k+s)(k+s-1) x^{k+s} + (k+s) x^{k+s} + x^{k+s+2} - v^2 x^{k+s}] = 0. \qquad (10.1.15)$$

We separate out the term x^{k+s+2} and relabel the index of summation, then

$$\sum_{k=0}^{\infty} a_k x^{k+s+2} = \sum_{k=2}^{\infty} a_{k-2} x^{k+s} = a_0 x^{2+s} + a_0 x^{2+s} + a_1 x^{3+s} + \cdots. \qquad (10.1.16)$$

Thus, the equation (10.1.15) becomes

$$\sum_{k=0}^{\infty} [(k+s)^2 - v^2] a_k x^{k+s} + \sum_{k=2}^{\infty} a_{k-2} x^{k+s} = 0. \qquad (10.1.17)$$

Now, a power series can evanish identically only when all of its coefficients are zero, so we can obtain the following sequence of equations:

(1) For $k=0$, $(s^2-v^2)a_0=0$,
(2) For $k=1$, $[(1+s)^2-v^2]a_1=0$,
(3) For $k\geq 2$, $[(k+s)^2-v^2]a_k+a_{k-2}=0$.

Since $a_0\neq 0$, (1) gives $s=\pm v$. Firstly we take $s=v$, then (2) becomes $(2v+1)a_1=0$, that is $v=-\frac{1}{2}$ or $a_1=0$.

So we must have $a_1=0$ except when $v=-\frac{1}{2}$. Even when $v=-\frac{1}{2}$, it is consistent to take $a_1=0$. Then (3) says that

$$a_k=\frac{-a_{k-2}}{(k+v)^2-v^2}=-\frac{a_{k-2}}{k(2v+k)}, \quad k=2, 3, \cdots \qquad (10.1.18)$$

From this recurrence formula, we can solve for all the even-numbered coefficients in terms of a_0:

$$a_2=-\frac{a_0}{2(2v+2)},$$

$$a_4=-\frac{a_2}{4(2v+4)}=-\frac{a_0}{2\cdot 4\cdot(2v+2)(2v+4)}, \cdots,$$

$$a_{2k}=\frac{(-1)^k a_0}{2\cdot 4\cdot\cdots\cdot(2k)(2v+2)(2v+4)\cdots(2v+2k)}=\frac{(-1)^k a_0}{2^{2k}k!(v+1)\cdots(v+k)}.$$

$$\qquad (10.1.19)$$

In the same way, we obtain all of the odd-numbered coefficients in terms of a_1, that is $a_{2k-1}=0$, $(k=1, 2, \cdots)$. Then we have the solution

$$y(x)=a_0\sum_{k=0}^{\infty}\frac{(-1)^k x^{2k+v}}{2^{2k}k!(v+1)\cdots(v+k)}. \qquad (10.1.20)$$

It remains to pick up the constant a_0, and the standard choice is

$$a_0=\frac{1}{2^v\Gamma(v+1)}. \qquad (10.1.21)$$

Then the solution defined by (10.1.20) becomes

$$y(x)=\sum_{k=0}^{\infty}\frac{(-1)^k x^{2k+v}}{2^{2k}k!((v+1)\cdots(v+k))}\frac{1}{2^v\Gamma(v+1)}$$

$$=\sum_{k=0}^{\infty}\frac{(-1)^k x^{2k+v}\Gamma(v+1)}{2^{2k}k!\Gamma(v+k+1)}\frac{1}{2^v\Gamma(v+1)} \qquad (10.1.22)$$

$$=\sum_{k=0}^{\infty}\frac{(-1)^k}{k!\Gamma(v+k+1)}\left(\frac{x}{2}\right)^{2k+v}.$$

Let $J_v(x)$ denote the above solution, and it is called **the first kind of v order Bessel function**. Thus

$$J_v(x)=\sum_{k=0}^{\infty}\frac{(-1)^k}{k!\Gamma(v+k+1)}\left(\frac{x}{2}\right)^{2k+v}. \qquad (10.1.23)$$

We arrived at $J_v(x)$ by taking $s=v$ in the recurrence formula. If we choose $s=-v$, and by the same way, we can obtain the similar relation with

$$(-2v+1)a_1=0.$$

From this, one can get $a_1 = 0$, $\left(v \neq \dfrac{1}{2}\right)$.

By recurrence formular $a_k = -\dfrac{a_{k-2}}{k(k-2v)}$, $k \geq 2$, we can obtain the non-regular solution

$$J_{-v}(x) = \sum_{k=0}^{\infty} \dfrac{(-1)^k}{k!\,\Gamma(k-v+1)} \left(\dfrac{x}{2}\right)^{2k-v}. \tag{10.1.24}$$

This solution is called **the first kind of $-v$ order Bessel function**.

A simple application of the ratio test can show that these two series solutions (10.1.23) and (10.1.24) are absolutely convergent for all $x \neq 0$ (and also for $x = 0$ when Re $v > 0$ or $v = 0$).

If v is not an integer, $J_v(x)$ and $J_{-v}(x)$ are linear independent. Then the general solution of the Bessel equation is

$$y(x) = C_1 J_v(x) + C_2 J_{-v}(x). \tag{10.1.25}$$

If v is an integer, that is, $v = n$, then

$$J_n(x) = \sum_{k=0}^{\infty} \dfrac{(-1)^k}{k!\,\Gamma(k+n+1)} \left(\dfrac{x}{2}\right)^{2k+n}.$$

Since $\Gamma(j+1) = j!$, $\Gamma(k+n+1) = (k+n)!$, we have

$$J_n(x) = \sum_{k=0}^{\infty} \dfrac{(-1)^k}{k!\,(k+n)!} \left(\dfrac{x}{2}\right)^{2k+n}, \quad (n = 0, 1, 2, \cdots). \tag{10.1.26}$$

For $v = -n$, $J_{-n}(x) = \sum_{k=0}^{\infty} \dfrac{(-1)^k}{k!\,\Gamma(k-n+1)} \left(\dfrac{x}{2}\right)^{2k-n}$.

Since $\dfrac{1}{\Gamma(z)} = 0$ when $z = 0, -1, -2, \cdots$, then $\dfrac{1}{\Gamma(k-n+1)} = 0$ when $k = 0, 1, 2, \cdots, n-1$. Thus

$$J_{-n}(x) = \sum_{k=n}^{\infty} \dfrac{(-1)^k}{k!\,(k-n)!} \left(\dfrac{x}{2}\right)^{2k-n}.$$

If we take the transformation $k = j + n$, then

$$J_{-n}(x) = \sum_{j=0}^{\infty} \dfrac{(-1)^j (-1)^n}{j!\,(j+n)!} \left(\dfrac{x}{2}\right)^{2j+n} \tag{10.1.27}$$

$$= (-1)^n J_n(x).$$

It is obvious that $J_n(x)$ and $J_{-n}(x)$ are linear dependent. Thus, it is necessary to find the second solution which is independent to $J_n(x)$. Generally, we define the **Bessel function of second kind $Y_v(x)$** by

$$Y_v(x) = \dfrac{(\cos v\pi) J_v(x) - J_{-v}(x)}{\sin v\pi}. \tag{10.1.28}$$

$Y_v(x)$ is a combination of $J_v(x)$ and $J_{-v}(x)$, so it satisfies Bessel equation. Also $J_v(x)$ and $Y_v(x)$ are linearly independent. But if $v = n$, $Y_n(x)$ turns into the intermediate $\dfrac{0}{0}$, we define

$$Y_n(x) = \lim_{v \to n} Y_v(x),$$

and $Y_n(x)$ exists and is finite for all $x \neq 0$ and is a solution of Bessel equation.

In summary, whenever the order of the Bessel equation is an integer or not, the general solution of the Bessel equation is

$$y(x) = C_1 J_v(x) + C_2 Y_v(x), \tag{10.1.29}$$

where C_1 and C_2 are two constants. Especially, if the order of the Bessel equation is an integer, its general solution can be written as

$$y(x)=C_1 J_n(x)+C_2 Y_n(x),$$

where C_1 and C_2 are two constants.

10.1.3 The recurrence formula of the Bessel function

The Bessel functions with different orders are not independent, and they have some relationship. We will establish some recurrence formula in this subsection.

(1) $\dfrac{\mathrm{d}}{\mathrm{d}x}[x^v J_v(x)]=x^v J_{v-1}(x)$ for all x and v. (10.1.30)

Proof

$$\frac{\mathrm{d}}{\mathrm{d}x}[x^v J_v(x)] = \frac{\mathrm{d}}{\mathrm{d}x}\sum_{k=0}^{\infty}\frac{(-1)^k x^{2k+2v}}{2^{2k+v}k!\,\Gamma(k+v+1)}$$

$$=\sum_{k=0}^{\infty}\frac{(-1)^k(2k+2v)x^{2k+2v-1}}{2^{2k+v}k!\,\Gamma(k+v+1)}$$

$$=x^v\sum_{k=0}^{\infty}\frac{(-1)^k \cdot 2 \cdot (k+v)x^{2k+v-1}}{2^{2k+v-1}\cdot 2 \cdot k!\,\Gamma(k+v+1)}$$

$$=x^v\sum_{k=0}^{\infty}\frac{(-1)^k x^{2k+(v-1)}}{2^{2k+(v-1)}k!\,\Gamma(k+(v-1)+1)}=x^v J_{v-1}(x).$$

(2) $\dfrac{\mathrm{d}}{\mathrm{d}x}[x^{-v} J_v(x)]=-x^{-v} J_{v+1}(x)$ for all x and v. (10.1.31)

Proof $\dfrac{\mathrm{d}}{\mathrm{d}x}[x^{-v} J_v(x)] = \dfrac{\mathrm{d}}{\mathrm{d}x}\sum_{k=0}^{\infty}\dfrac{(-1)^k x^{2k}}{2^{2k+v}k!\,\Gamma(k+v+1)}$

$$=\sum_{k=1}^{\infty}\frac{(-1)^k(2k)x^{2k-1}}{2^{2k+v}k!\,\Gamma(k+v+1)}$$

$$=\sum_{k=1}^{\infty}\frac{(-1)^k x^{2k-1}}{2^{2k+v-1}(k-1)!\,\Gamma(k+v+1)}$$

$$\xrightarrow{j=k-1}\sum_{j=0}^{\infty}\frac{(-1)^{j+1} x^{2j+1}}{2^{2j+v+1}j!\,\Gamma(j+v+2)}$$

$$=-x^{-v}\sum_{j=0}^{\infty}\frac{(-1)^j x^{2j+v+1}}{2^{2j+v+1}j!\,\Gamma(j+v+2)}=-x^{-v}J_{v+1}(x).$$

(3) $xJ'_v(x)+vJ_v(x) = xJ_{v-1}(x)$ (10.1.32)

Proof

$$\frac{\mathrm{d}}{\mathrm{d}x}[x^v J_v(x)]=vx^{v-1}J_v(x)+x^v J'_v(x)=x^v J_{v-1}(x),$$

that is, $\qquad xJ'_v(x)+vJ_v(x)=xJ_{v-1}(x).$

(4) $xJ'_v(x)-vJ_v(x)=-xJ_{v+1}(x)$ (10.1.33)

Proof $\dfrac{\mathrm{d}}{\mathrm{d}x}[x^{-v} J_v(x)]=-vx^{-v-1}J_v(x)+x^{-v}J'_v(x)=-x^{-v}J_{v+1}(x),$

that is, $\qquad xJ'_v(x)-vJ_v(x)=-xJ_{v+1}(x).$

Taking the sum and the difference of (10.1.32) and (10.1.33), we can obtain
$$J_{v-1}(x) - J_{v+1}(x) = 2J'_v(x), \tag{10.1.34}$$
$$xJ_{v-1}(x) + xJ_{v+1}(x) = 2vJ_v(x). \tag{10.1.35}$$

The above formula (10.1.30) ~ (10.1.35) are recurrence formulas for the Bessel functions, and they are also valid when $v = n$.

It is important to apply the recurrence formula in some computation about the Bessel functions. Specially, we can obtain higher order Bessel function by lower order Bessel function with (10.1.35). Thus, if we have the table of the values of the zero order and first order Bessel functions, we can compute the values of any integer order Bessel functions. As the opposite application, (10.1.30) and (10.1.31) can also be rewritten as the integration formulas

$$\int x^n J_{n-1}(x) dx = x^n J_n(x) + C. \tag{10.1.36}$$

and

$$\int \frac{J_{n+1}(x)}{x^n} dx = -\frac{J_n(x)}{x^n} + C, \tag{10.1.37}$$

Example 10.1.2 Evaluate $\int_0^x xJ_2(x) dx$.

Solution
$$\int_0^x xJ_2(x) dx = \int_0^x (-x^2)(-x^{-1} J_{1+1}(x)) dx$$
$$= -\int_0^x x^2 (x^{-1} J_1(x))' dx = -\int_0^x x^2 d(x^{-1} J_1(x))$$
$$= -xJ_1(x) \Big|_0^x + 2\int_0^x J_1(x) dx$$
$$= -xJ_1(x) - 2\int_0^x (-x^0 J_{0+1}(x)) dx$$
$$= -xJ_1(x) - 2 J_0(x) \Big|_0^x = -xJ_1(x) - 2J_0(x) + 2.$$

Example 10.1.3 Evaluate the integral $\int_0^x x^4 J_1(x) dx$.

Solution $\int_0^x x^4 J_1(x) dx = \int_0^x x^2 [x^2 J_{2-1}(x)] dx$
$$= \int_0^x x^2 d[x^2 J_2(x)] = x^4 J_2(x) \Big|_0^x - 2\int_0^x x^3 J_2(x) dx$$
$$= x^4 J_2(x) - 2x^3 J_3(x).$$

Example 10.1.4 Show that
$$\int_0^x x^n J_0(x) dx = x^n J_1(x) + (n-1)x^{n-1} J_0(x) - (n-1)^2 \int_0^x x^{n-2} J_0(x) dx.$$

Proof
$$\int_0^x x^n J_0(x) dx = \int_0^x x^{n-1} [xJ_0(x)] dx = \int_0^x x^{n-1} d[xJ_1(x)]$$
$$= x^n J_1(x) \Big|_0^x - \int_0^x x^{n-1} (n-1) J_1(x) dx = x^n J_1(x) + (n-1)\int_0^x x^{n-1} dJ_0(x)$$
$$= x^n J_1(x) + (n-1)x^{n-1} J_0(x) - (n-1)\int_0^x J_0(x) dx^{n-1}$$
$$= x^n J_1(x) + (n-1)x^{n-1} J_0(x) - (n-1)2\int_0^x J_0(x) dx.$$

As an application of these recurrence formulas, we shall show that the Bessel functions of half-integer order can be expressed in terms of familiar elementary functions.

From (10.1.23), we have

$$J_{\frac{1}{2}}(x) = \sum_{k=0}^{\infty} \frac{(-1)^k}{2^{2k+\frac{1}{2}} k! \, \Gamma(k+\frac{1}{2}+1)} \cdot x^{2k+\frac{1}{2}}$$

$$= \sum_{k=0}^{\infty} \frac{(-1)^k}{2^{\frac{1}{2}}[2^k k!][2^k \Gamma(k+\frac{3}{2})]} \cdot x^{2k+\frac{1}{2}}.$$

Since $2^k k! = 2^k (1 \cdot 2 \cdot 3 \cdot \cdots \cdot k) = 2 \cdot 4 \cdot 6 \cdot \cdots \cdot (2k)$, and

$$\Gamma\left(\frac{1}{2}\right) = \sqrt{\pi}, \quad \Gamma(z+1) = z\Gamma(z),$$

then

$$2^{k+1} \Gamma(k+\frac{3}{2}) = 2^{k+1}(k+\frac{1}{2})(k-\frac{1}{2})\cdots\frac{3}{2}\cdot\frac{1}{2}\cdot \Gamma\left(\frac{1}{2}\right)$$

$$= (2k+1)(2k-1) \cdot \cdots \cdot 3 \cdot 1 \cdot \sqrt{\pi}.$$

So we have

$$J_{\frac{1}{2}}(x) = \sum_{k=0}^{\infty} \frac{(-1)^k}{2^{-\frac{1}{2}}[2^k k!][2^{k+1} \Gamma(k+\frac{3}{2})]} \cdot x^{2k+1} \cdot x^{-\frac{1}{2}}$$

(10.1.38)

$$= \sqrt{\frac{2}{\pi x}} \sum_{k=0}^{\infty} \frac{(-1)^k}{(2k+1)!} \cdot x^{2k+1} = \sqrt{\frac{2}{\pi x}} \sin x.$$

With a similar calculation for $J_{-\frac{1}{2}}(x)$, we have

$$J_{-\frac{1}{2}}(x) = \sqrt{\frac{2}{\pi x}} \cos x.$$

(10.1.39)

By applying (10.1.35), we have

$$J_{\frac{3}{2}}(x) = \frac{1}{x} J_{\frac{1}{2}}(x) - J_{-\frac{1}{2}}(x)$$

$$= \sqrt{\frac{2}{\pi x}} \left(\frac{\sin x}{x} - \cos x\right)$$

$$= -\sqrt{\frac{2}{\pi}} x^{\frac{3}{2}} \frac{1}{x} \frac{d}{dx}\left(\frac{\sin x}{x}\right)$$

$$= -\sqrt{\frac{2}{\pi}} x^{\frac{3}{2}} \left(\frac{1}{x}\frac{d}{dx}\right)\left(\frac{\sin x}{x}\right).$$

Similarly, we can have

$$J_{-\frac{3}{2}}(x) = \sqrt{\frac{2}{\pi}} x^{\frac{3}{2}} \left(\frac{1}{x}\frac{d}{dx}\right)\left(\frac{\cos x}{x}\right).$$

Generally, we can have

$$J_{n+\frac{1}{2}}(x) = (-1)^n \sqrt{\frac{2}{\pi}} x^{n+\frac{1}{2}} \left(\frac{1}{x}\frac{d}{dx}\right)^n \left(\frac{\sin x}{x}\right);$$

$$J_{-(n+\frac{1}{2})}(x) = \sqrt{\frac{2}{\pi}} x^{n+\frac{1}{2}} \left(\frac{1}{x}\frac{d}{dx}\right)^n \left(\frac{\cos x}{x}\right).$$

For simplicity, we introduce the differential operator $\left(\dfrac{1}{x}\dfrac{d}{dx}\right)^n$, which means that we apply the operator $\dfrac{1}{x}\dfrac{d}{dx}$ continuously for n times. For example,

$$\left(\frac{1}{x}\frac{d}{dx}\right)^2\left(\frac{\sin x}{x}\right)=\frac{1}{x}\frac{d}{dx}\left[\frac{1}{x}\frac{d}{dx}\left(\frac{\sin x}{x}\right)\right].$$

10.1.4 The properties of the Bessel function

We need expand any given function into a series with respect to the Bessel function system when we solve some defining problem in applying the Bessel function. So in this subsection, we will give some properties of the Bessel function and prove that the Bessel function system is an orthogonal system.

In subsection 10.1.1, we transformed the example 10.1.1 into an eigenvalue problem

$$\begin{cases} r^2 P''(r)+rP'(r)+(\lambda r^2-n^2)P(r)=0, & 0<r<R, & (10.1.40)\\ P(r)|_{r=R}=0, & & (10.1.41)\\ |P(0)|<+\infty. & & (10.1.42) \end{cases}$$

The general solution of (10.1.40) is

$$P(r)=C_1 J_n(\sqrt{\lambda}r)+C_2 Y_n(\sqrt{\lambda}r),$$

and we have $C_2=0$ from condition (10.1.42), that is,

$$P(r)=C_1 J_n(\sqrt{\lambda}r),$$

and then from (10.1.41)

$$J_n(\sqrt{\lambda}R)=0. \qquad (10.1.43)$$

It means that we must evaluate the zeros of the Bessel function $J_n(x)$ to find the eigenvalues of the above eigenvalue problem. In what follows, we will discuss the zeros of the Bessel function and how to expand a given function into a series with respect to the Bessel function.

The figures of the Bessel functions of the first kind and the second kind are given in Fig. 10.1.1(a) and Fig. 10.1.1(b):

From the figures, it is obvious that

(1) $J_n(x)$ has finite values at $x=0$ and $Y_n(x)$ has infinite values at $x=0$;

(2) $J_n(x)$ and $Y_n(x)$ have an infinite number of positive zeros;

(3) $J_0(0)=1$, $J_n(0)=0(n\neq 0)$, $Y_n(x)\xrightarrow{x\to 0}-\infty$;

(4) If we denote the m-th zero of $J_n(x)$ as $x_m^{(n)}$, then $x_{m+1}^{(n)}-x_m^{(n)}\to \pi$ as $m\to\infty$, that is, $J_n(x)$ is almost a periodic function with period 2π. The asymptotical formula of $J_n(x)$ and $Y_n(x)$ are

$$J_n(x)\approx\sqrt{\frac{2}{\pi x}}\cos\left(x-\frac{1}{4}\pi-\frac{n}{2}\pi\right),$$

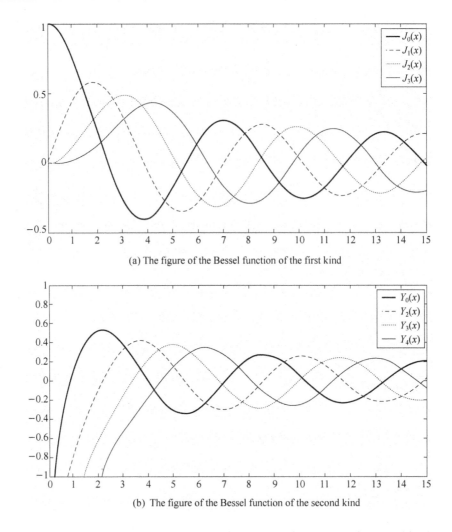

(a) The figure of the Bessel function of the first kind

(b) The figure of the Bessel function of the second kind

Fig. 10.1.1 The figures of the Bessel functions of the first and the second kind

$$Y_n(x) \approx \sqrt{\frac{2}{\pi x}} \sin\left(x - \frac{1}{4}\pi - \frac{n}{2}\pi\right).$$

In order to solve some defining problems, we need to evaluate the zeros of the Bessel functions. For convenience, some approximate values of the zeros of the Bessel function $J_0(x)$ and $J_1(x)$ are presented in the following table.

Table 10.1.1 Zeros of $J_0(x)$ and $J_1(x)$

n	$x_n^{(0)}$	$x_n^{(1)}$	n	$x_n^{(0)}$	$x_n^{(1)}$
1	2.404 8	3.831 7	6	18.071 1	19.615 9
2	5.520 1	7.015 6	7	21.211 6	22.760 1
3	8.653 7	10.173 5	8	24.352 5	25.903 7
4	11.792 5	13.323 7	9	27.493 5	29.046 8
5	14.930 9	16.470 6	10	30.634 6	32.189 7

According to the above results, the solution of (10.1.43) is
$$\sqrt{\lambda}R = x_m^{(n)}, \quad m=1, 2, \cdots,$$
that is,
$$\lambda_m^{(n)} = \left(\frac{x_m^{(n)}}{R}\right)^2, \quad m=1, 2, \cdots. \tag{10.1.44}$$

And the eigenfunctions in accordance with these eigenvalues are
$$P_m(r) = J_n\left(\frac{x_m^{(n)}}{R}r\right), \quad m=1, 2, \cdots. \tag{10.1.45}$$

Next, we discuss the orthogonality of the eigenfunction system $\left\{J_n\left(\frac{x_m^{(n)}}{R}r\right)\right\}$, $m=1, 2, \cdots$.

According to the property of the Sturm-Liouville problem, Bessel functions are orthogonal with weight function $s(r) = r$, that is
$$\int_0^R r J_n(\mu_m^{(n)} r) J_n(\mu_k^{(n)} r) dr = 0, \quad (m \neq k),$$
where $\mu_m^{(n)} = \frac{x_m^{(n)}}{R}$.

However, $\int_0^R r J_n(\mu_m^{(n)} r) J_n(\mu_m^{(n)} r) dr = [N_m^{(n)}]^2 = \frac{R^2}{2}[J_n'(\mu_m^{(n)} R)]^2 + \frac{1}{2}\left[R^2 - \left(\frac{n}{\mu_m^{(n)}}\right)^2\right][J_n(\mu_m^{(n)} R)]^2.$

If we denote $\int_0^R r J_n(\mu_m^{(n)} r) J_n(\mu_m^{(n)} r) dr = [N_m^{(n)}]^2$, then
$$[N_m^{(n)}]^2 = \frac{R^2}{2}[J_n'(\mu_m^{(n)} R)]^2 + \frac{1}{2}\left[R^2 - \left(\frac{n}{\mu_m^{(n)}}\right)^2\right][J_n(\mu_m^{(n)} R)]^2.$$

This formula is applicable for different homogeneous boundary conditions. It can be simplified according to different boundary condition.

Case 1 For first boundary condition $J_n(\mu_m^{(n)} R) = 0$,
$$[N_m^{(n)}]^2 = \frac{R^2}{2}[J_n'(\mu_m^{(n)} R)]^2$$
$$= \frac{R^2}{2}[J_{n-1}(\mu_m^{(n)} R)]^2$$
$$= \frac{R^2}{2}[J_{n+1}(\mu_m^{(n)} R)]^2.$$

Case 2 For second boundary condition $J_n'(\mu_m^{(n)} a) = 0$,
$$[N_m^{(n)}]^2 = \frac{1}{2}\left[R^2 - \left(\frac{n}{\mu_m^{(n)}}\right)^2\right][J_n(\mu_m^{(n)} R)]^2.$$

With these results, if there is a function $f(r)$ defined in $[0, R]$, and it has continuous second-order derivative, then $f(r)$ can be expanded in a Fourier-Bessel series
$$f(r) = \sum_{m=1}^{\infty} f_m J_n(\mu_m^{(n)} r)$$
where $f_m = \frac{1}{[N_m^{(n)}]^2} \int_0^R r f(r) J_n(\mu_m^{(n)} r) dr.$

Example 10.1.5 Suppose that $x_k^{(0)}$ ($k=1, 2, \cdots$) are positive roots of $J_0(x) = 0$. Please expand $f(x)=1$ $(0\leqslant x\leqslant R)$ with respect to zero order Bessel function $J_0\left(\dfrac{x_k^{(0)}}{R}x\right)$.

Solution Let $f(x)=\sum\limits_{k=1}^{\infty}C_k J_0\left(\dfrac{x_k^{(0)}}{R}x\right)$, where $x_k^{(0)}$ is the k-th zero of $J_0(x)$, then

$$C_k = \dfrac{1}{[N_k^{(0)}]^2}\int_0^R xf(x)J_0\left(\dfrac{x_k^{(0)}}{R}x\right)\mathrm{d}x = \dfrac{1}{[N_k^{(0)}]^2}\int_0^R xJ_0\left(\dfrac{x_k^{(0)}}{R}x\right)\mathrm{d}x$$

$$= \dfrac{1}{[N_k^{(0)}]^2}\left(\dfrac{R}{x_k^{(0)}}\right)^2\int_0^R \left(\dfrac{x_k^{(0)}}{R}x\right)J_0\left(\dfrac{x_k^{(0)}}{R}x\right)\mathrm{d}\left(\dfrac{x_k^{(0)}}{R}x\right).$$

Since $\dfrac{\mathrm{d}}{\mathrm{d}x}[xJ_1(x)]=xJ_0(x)$, then

$$C_k = \dfrac{1}{[N_k^{(0)}]^2}\left(\dfrac{R}{x_k^{(0)}}\right)^2\left[\dfrac{x_k^{(0)}}{R}xJ_1\left(\dfrac{x_k^{(0)}}{R}x\right)\right]\Big|_0^R$$

$$= \dfrac{1}{[N_k^{(0)}]^2}\left(\dfrac{R}{x_k^{(0)}}\right)^2 x_k^{(0)}J_1(x_k^{(0)})$$

$$= \dfrac{R^2}{x_k^{(0)}[N_k^{(0)}]^2}J_1(x_k^{(0)}).$$

Thus

$$f(x)=1=\sum_{k=1}^{\infty}\dfrac{2}{x_k^{(0)}J_1(x_k^{(0)})}J_0\left(\dfrac{x_k^{(0)}}{R}x\right).$$

10.1.5 Application of Bessel function

In this section, we consider some defining problems which are related with the Bessel function.

Example 10.1.6 Solve the defining problem

$$\begin{cases} \dfrac{\partial^2 u}{\partial t^2}=c^2\left(\dfrac{\partial^2 u}{\partial r^2}+\dfrac{1}{r}\dfrac{\partial u}{\partial r}\right), & 0<r<a,\ t>0, \\ u(a,t)=0,\ \lim\limits_{r\to 0^+}u(r,t)<\infty, & t\geqslant 0, \\ u(r,0)=1,\ \dfrac{\partial u}{\partial t}(r,0)=0, & 0\leqslant r\leqslant a. \end{cases}$$

Solution By separation of variables, let $u(r,t)=R(r)T(t)$.
Substituting it into the equation, we have

$$R(r)T''(t)=c^2\left(R''(r)T(t)+\dfrac{1}{r}R'(r)T(t)\right),$$

that is,

$$\dfrac{T''(t)}{c^2 T(t)}=\dfrac{R''(r)+\dfrac{1}{r}R'(r)}{R(r)}=-\beta^2.$$

Then we have $\begin{cases} T''(t)+c^2\beta^2 T(t)=0, \\ r^2 R''(r)+rR'(r)+\beta^2 r^2 R(r)=0. \end{cases}$

And the boundary conditions become

$$R(a)=0,\quad \lim_{r\to 0^+}R(r)<\infty.$$

Then the eigenvalue problem becomes

$$\begin{cases} r^2 R''(r) + r R'(r) + (\beta^2 r^2 - 0^2) R(r) = 0, \\ R(a) = 0, \ \lim_{r \to 0^+} R(r) < \infty, \end{cases}$$

which is the Bessel equation of order zero, and its solution is
$$R(r) = C_0 J_0(\beta r) + D_0 Y_0(\beta r).$$

From $\lim_{r \to 0^+} R(r) < \infty$, we have $D_0 = 0$. From $R(a) = 0$, we have $J_0(\beta a) = 0$, that is $\beta_m = \dfrac{x_m^{(0)}}{a}$, $(m=1, 2, \cdots)$.

So the eigenvalues are $\beta_m^2 = \dfrac{[x_m^{(0)}]^2}{a^2}$, $(m=1, 2, \cdots)$, and the eigenfunctions are $R_m(r) = C_m J_0\left(\dfrac{x_m^{(0)}}{a} r\right)$.

Then the equation for $T(t)$ becomes $T_m''(t) + c^2 \dfrac{[x_m^{(0)}]^2}{a^2} T_m(t) = 0$, $(m=1, 2, \cdots)$ and its solution is $T_m(t) = A_m \cos \dfrac{c x_m^{(0)}}{a} t + B_m \sin \dfrac{c x_m^{(0)}}{a} t$.

Hence, the solution of this problem is
$$u(r,t) = \sum_{m=1}^{\infty} \left[E_m \cos \dfrac{c x_m^{(0)}}{a} t + F_m \sin \dfrac{c x_m^{(0)}}{a} t \right] J_0\left(\dfrac{x_m^{(0)}}{a} r\right).$$

From $\dfrac{\partial u}{\partial t}(r, 0) = 0$, we have $F_m = 0$, $(m=1, 2, \cdots)$. From $u(r, 0) = 1$, we have $\sum_{m=1}^{\infty} E_m \cos\left(\dfrac{c x_m^{(0)}}{a} \cdot 0\right) J_0\left(\dfrac{x_m^{(0)}}{a} r\right) = 1$, that is

$$\sum_{m=1}^{\infty} E_m J_0\left(\dfrac{x_m^{(0)}}{a} r\right) = 1,$$

and
$$E_m = \dfrac{1}{[N_m^{(0)}]^2} \int_0^a r J_0\left(\dfrac{x_m^{(0)}}{a} r\right) dr.$$

In this problem, we have the first boundary condition, then
$$[N_m^{(0)}]^2 = \dfrac{a^2}{2} \left[J_0'\left(\dfrac{x_m^{(0)}}{a} a\right) \right]^2 = \dfrac{a^2}{2} [J_0'(x_m^{(0)})]^2 = \dfrac{a^2}{2} J_1^2(x_m^{(0)}),$$

and then
$$E_m = \dfrac{1}{\dfrac{a^2}{2} J_1^2(x_m^{(0)})} \int_0^a r J_0\left(\dfrac{x_m^{(0)}}{a} r\right) dr.$$

If we set $\dfrac{x_m^{(0)}}{a} r = t$, then $dr = \dfrac{a}{x_m^{(0)}} dt$, and

$$\begin{aligned} E_m &= \dfrac{2}{a^2 J_1^2(x_m^{(0)})} \int_0^{x_m^{(0)}} \left[\dfrac{a}{x_m^{(0)}}\right]^2 t J_0(t) dt \\ &= \dfrac{2}{[x_m^{(0)} J_1(x_m^{(0)})]^2} t J_1(t) \Big|_0^{x_m^{(0)}} \\ &= \dfrac{2}{x_m^{(0)} J_1(x_m^{(0)})}. \end{aligned}$$

Finally, it is concluded that the solution of this problem is
$$u(r,t) = \sum_{m=1}^{\infty} \frac{2}{x_m^{(0)} J_1(x_m^{(0)})} \cos \frac{cx_m^{(0)}}{a} t J_0\left(\frac{x_m^{(0)}}{a} r\right).$$

Example 10.1.7 To solve the defining problem
$$\begin{cases} \dfrac{\partial u}{\partial t} = a^2 \left(\dfrac{\partial^2 u}{\partial r^2} + \dfrac{1}{r}\dfrac{\partial u}{\partial r}\right), & 0 < r < 1, \ t > 0, \\ u(1,t) = 0, \ \lim_{r \to 0} u(r,t) < \infty, & t \geq 0, \\ u(r, 0) = 1 - r^2, & 0 \leq r \leq 1. \end{cases}$$

Solution By separation of variables, let $u(r,t) = R(r)T(t)$.
Substituting it into the equation, we have
$$R(r)T'(t) = a^2\left(R''(r)T(t) + \frac{1}{r}R'(r)T(t)\right),$$
that is,
$$\frac{T'(t)}{a^2 T(t)} = \frac{R''(r) + \dfrac{1}{r}R'(r)}{R(r)} = -\beta^2.$$

Then we have $\begin{cases} T'(t) + a^2 \beta^2 T(t) = 0, \\ r^2 R''(r) + r R'(r) + \beta^2 r^2 R(r) = 0. \end{cases}$

And the boundary conditions become
$$R(1) = 0, \quad \lim_{r \to 0} R(r) < \infty.$$

Then the eigenvalue problem becomes
$$\begin{cases} r^2 R''(r) + r R'(r) + (\beta^2 r^2 - 0^2) R(r) = 0, \\ R(1) = 0, \ \lim_{r \to 0} R(r) < \infty, \end{cases}$$

which is the Bessel equation of order zero, and its general solution is
$$R(r) = C_0 J_0(\beta r) + D_0 Y_0(\beta r).$$

From $\lim_{r \to 0} R(r) < \infty$, we have $D_0 = 0$. From $R(1) = 0$, we have $J_0(\beta) = 0$, that is $\beta_m = x_m^{(0)}$, $(m = 1, 2, \cdots)$.

So the eigenvalues are $\beta_m^2 = [x_m^{(0)}]^2$, $(m = 1, 2, \cdots)$, and the eigenfunctions are $R_m(r) = C_m J_0(x_m^{(0)} r)$.

Then the equation for $T(t)$ becomes $T_m'(t) + a^2 [x_m^{(0)}]^2 T_m(t) = 0$, $(m = 1, 2, \cdots)$ and its general solution is $T_m(t) = A_m e^{-a^2 [x_m^{(0)}]^2 t}$.

Hence, the solution of this problem is
$$u(r,t) = \sum_{m=1}^{\infty} E_m e^{-a^2 [x_m^{(0)}]^2 t} J_0(x_m^{(0)} r).$$

From $u(r,0) = 1 - r^2$, we have
$$\sum_{m=1}^{\infty} E_m J_0(x_m^{(0)} r) = 1 - r^2,$$

then $E_m = \dfrac{1}{[N_m^{(0)}]^2} \displaystyle\int_0^1 r(1 - r^2) J_0(x_m^{(0)} r) dr.$

In this problem, we have the first boundary condition, then

$$[N_m^{(0)}]^2 = \frac{1}{2}[J_0'(x_m^{(0)})]^2 = \frac{1}{2}J_1^2(x_m^{(0)}),$$

and then

$$E_m = \frac{2}{J_1^2(x_m^{(0)})}\left[\int_0^1 rJ_0(x_m^{(0)}r)\,dr - \int_0^1 r^3 J_0(x_m^{(0)}r)\,dr\right].$$

Since

$$d[(x_m^{(0)}r)J_1(x_m^{(0)}r)] = (x_m^{(0)}r)[J_0(x_m^{(0)}r)d(x_m^{(0)}r)],$$

that is

$$d\left[\frac{rJ_1(x_m^{(0)}r)}{x_m^{(0)}}\right] = rJ_0(x_m^{(0)}r)\,dr,$$

we have $\displaystyle\int_0^1 rJ_0(x_m^{(0)}r)\,dr = \frac{rJ_1(x_m^{(0)}r)}{x_m^{(0)}}\bigg|_0^1 = \frac{J_1(x_m^{(0)})}{x_m^{(0)}}.$

On the other hand,

$$\int_0^1 r^3 J_0(x_m^{(0)}r)\,dr = \int_0^1 r^2 d\left[\frac{rJ_1(x_m^{(0)}r)}{x_m^{(0)}}\right] = \frac{r^3 J_1(x_m^{(0)}r)}{x_m^{(0)}}\bigg|_0^1 - \frac{2}{x_m^{(0)}}\int_0^1 r^2 J_1(x_m^{(0)}r)\,dr$$

$$= \frac{J_1(x_m^{(0)})}{x_m^{(0)}} - \frac{2}{x_m^{(0)}} r^2 J_2(x_m^{(0)}r)\big|_0^1$$

$$= \frac{J_1(x_m^{(0)})}{x_m^{(0)}} - \frac{2J_2(x_m^{(0)})}{(x_m^{(0)})^2}.$$

Thus,

$$E_m = \frac{4J_2(x_m^{(0)})}{(x_m^{(0)})^2 J_1^2(x_m^{(0)})}.$$

Finally, it is concluded that the solution of this problem is

$$u(r,t) = \sum_{m=1}^{\infty} \frac{4J_2(x_m^{(0)})}{(x_m^{(0)})^2 J_1^2(x_m^{(0)})} e^{-a^2 (x_m^{(0)})^2 t} J_0(x_m^{(0)}r).$$

Exercises 10.1

1. Evaluate $\dfrac{d}{dx}J_0(\alpha x)$, $\dfrac{d}{dx}[xJ_1(\alpha x)]$.

2. Prove that $y = J_n(\alpha x)$ is the solution of the equation
$$x^2 y'' + xy' + (\alpha^2 x^2 - n^2)y = 0.$$

3. Evaluate the following integral:

(1) $\displaystyle\int_0^x x^{n+1} J_n(x)\,dx;$

(2) $\displaystyle\int_0^x x^4 J_1(x)\,dx;$

(3) $\displaystyle\int_0^x x^{n+1} J_n(\alpha x)\,dx.$

4. To evaluate $J_{-\frac{3}{2}}(x)$ with $J_{-\frac{1}{2}}(x) = \sqrt{\dfrac{2}{\pi x}}\cos x.$

5. Prove the following equations:

(1) $J_2(x) = J_0''(x) - \dfrac{1}{x} J_0'(x)$;

(2) $J_3(x) + 3J_0'(x) + 4J_0'''(x) = 0$.

6. Prove
$$\int x^n J_0(x)\,dx = x^n J_1(x) + (n-1)x^{n-1} J_0(x) - (n-1)^2 \int x^{n-2} J_0(x)\,dx.$$

7. Suppose that $x_k^{(1)}$, $(k=1, 2, \cdots)$ are the positive roots of $J_1(x)=0$, please expand $f(x)=x$, $(0 \leqslant x \leqslant 1)$ with respect to $J_1(x_k^{(1)} x)$.

8. Solve the following defining problem
$$\begin{cases} \dfrac{\partial^2 u}{\partial t^2} = a^2 \left(\dfrac{\partial^2 u}{\partial r^2} + \dfrac{1}{r} \dfrac{\partial u}{\partial r} \right), & 0<r<a,\ t>0, \\ u(0,t)<\infty,\ u(R,t)=0, & t \geqslant 0, \\ u(r,0)=1-\dfrac{r^2}{R^2},\ \dfrac{\partial u}{\partial t}(r,0)=0, & 0 \leqslant r \leqslant a. \end{cases}$$

10.2 Legendre polynomial

In this section, we will introduce the Legendre equation by an example of the Laplace equation in spherical coordinates. Then the solution of this equation and the properties of the solution will be discussed. The bounded solutions of the Legendre equation in the interval $[-1, 1]$ form another orthogonal function system, that is, the **Legendre polynomial**.

10.2.1 Introduction of the Legendre equation

Consider Laplace equation in spherical coordinates whose form is as follows

$$\frac{1}{r^2} \frac{\partial}{\partial r}\left(r^2 \frac{\partial u}{\partial r}\right) + \frac{1}{r^2 \sin\theta} \frac{\partial}{\partial \theta}\left(\sin\theta \frac{\partial u}{\partial \theta}\right) + \frac{1}{r^2 \sin^2\theta} \frac{\partial^2 u}{\partial \varphi^2} = 0. \tag{10.2.1}$$

By the method of separation of variables, we can set $u(r,\theta,\varphi) = R(r)\Theta(\theta)\Phi(\varphi)$.

Substituting it into (10.2.1), we have

$$\Theta(\theta)\Phi(\varphi) \frac{1}{r^2} \frac{d}{dr}\left(r^2 \frac{dR}{dr}\right) + \frac{1}{r^2 \sin\theta} R(r)\Phi(\varphi)\left(\sin\theta \frac{d\Theta}{d\theta}\right) + \frac{1}{r^2 \sin^2\theta} R(r)\Theta(\theta) \frac{d^2\Phi}{d\varphi^2} = 0.$$

Dividing it by $\dfrac{R(r)\Theta(\theta)\Phi(\varphi)}{r^2}$, we have

$$\frac{1}{R(r)} \frac{d}{dr}\left(r^2 \frac{dR}{dr}\right) + \frac{1}{\Theta(\theta)\sin\theta} \frac{d}{d\theta}\left(\sin\theta \frac{d\Theta}{d\theta}\right) + \frac{1}{\Phi(\varphi)\sin^2\theta} \frac{d^2\Phi}{d\varphi^2} = 0,$$

that is,

$$\frac{1}{R(r)} \frac{d}{dr}\left(r^2 \frac{dR}{dr}\right) = -\frac{1}{\Theta(\theta)\sin\theta} \frac{d}{d\theta}\left(\sin\theta \frac{d\Theta}{d\theta}\right) - \frac{1}{\Phi(\varphi)\sin^2\theta} \frac{d^2\Phi}{d\varphi^2} = \lambda.$$

For convenience, we denote $\lambda = v(v+1)$. Then we obtain two equations

$$\begin{cases} r^2 \dfrac{d^2 R}{dr^2} + 2r \dfrac{dR}{dr} - v(v+1)R(r) = 0, & (10.2.2) \\ \dfrac{1}{\Theta(\theta)\sin\theta}\dfrac{d}{d\theta}\left(\sin\theta \dfrac{d\Theta}{d\theta}\right) + \dfrac{1}{\Phi(\varphi)\sin^2\theta}\dfrac{d^2\Phi}{d\varphi^2} + v(v+1) = 0. & (10.2.3) \end{cases}$$

Equation (10.2.2) is an Euler equation, and its general solution is

$$R(r) = C_1 r^v + C_2 r^{-(v+1)}.$$

On the other hand, for equation (10.2.3), we can multiply it by $\sin^2\theta$, then

$$\frac{\sin\theta}{\Theta(\theta)}\frac{d}{d\theta}\left(\sin\theta \frac{d\Theta}{d\theta}\right) + v(v+1)\sin^2\theta + \frac{\Phi''(\varphi)}{\Phi(\varphi)} = 0,$$

that is,

$$\frac{\sin\theta}{\Theta(\theta)}\frac{d}{d\theta}\left(\sin\theta \frac{d\Theta}{d\theta}\right) + v(v+1)\sin^2\theta = -\frac{\Phi''(\varphi)}{\Phi(\varphi)}.$$

For this equation, its left side is only related with θ and its right side is only related with φ, so these two sides are equal when they are constants. Similar to the above section, $\Phi(\varphi)$ is a periodic function with period 2π, which means that the constant must be m^2, $(m=0, 1, 2\cdots)$.

Then we have the following two equations

$$\begin{cases} \dfrac{\sin\theta}{\Theta(\theta)}\dfrac{d}{d\theta}\left(\sin\theta \dfrac{d\Theta}{d\theta}\right) + v(v+1)\sin^2\theta = m^2, & (10.2.4) \\ \dfrac{\Phi''(\varphi)}{\Phi(\varphi)} = -m^2. & (10.2.5) \end{cases}$$

From (10.2.5), we have $\Phi_m(\varphi) = C_m\cos m\varphi + D_m\sin m\varphi$. Then the equation of $\Theta(\theta)$ becomes

$$\frac{\sin\theta}{\Theta(\theta)}\frac{d}{d\theta}\left(\sin\theta \frac{d\Theta}{d\theta}\right) + v(v+1)\sin^2\theta - m^2 = 0.$$

Evaluating the derivative in the above equation and simplifying it, we have

$$\sin^2\theta \frac{d^2\Theta}{d\theta^2} + \sin\theta\cos\theta \frac{d\Theta}{d\theta} - m^2\Theta(\theta) + v(v+1)\sin^2\theta\Theta(\theta) = 0,$$

that is,

$$\frac{d^2\Theta}{d\theta^2} + \cot\theta \frac{d\Theta}{d\theta} + \left[v(v+1) - \frac{m^2}{\sin^2\theta}\right]\Theta(\theta) = 0. \tag{10.2.6}$$

Let $x = \cos\theta$, then $\theta = \arccos x$, and if we denote $\Theta(\theta) = \Theta(\arccos x) = P(x)$, then

$$\frac{d\Theta}{d\theta} = \frac{d\Theta}{dx}\frac{dx}{d\theta} = -\sin\theta \frac{d\Theta}{dx}, \quad \frac{d^2\Theta}{d\theta^2} = (-\sin\theta)^2 \frac{d^2\Theta}{dx^2} - \cos\theta \frac{d\Theta}{dx}.$$

Now (10.2.6) can be transformed into

$$(1-x^2)\frac{d^2 P}{dx^2} - 2x \frac{dP}{dx} + \left[v(v+1) - \frac{m^2}{1-x^2}\right]P(x) = 0. \tag{10.2.7}$$

This equation is called **associated Legendre equation**, and it is dependent on v and m.

If we consider a simplified situation, that is, the boundary we considered is a sphere with axial symmetry. We can assume that $\theta = 0$ is an axis of symmetry, which implies that there is no dependence on φ and what we considered becomes

$$\frac{1}{r^2}\frac{\partial}{\partial r}\left(r \frac{\partial u}{\partial r}\right) + \frac{1}{r^2\sin\theta}\frac{\partial}{\partial \theta}\left(\sin\theta \frac{\partial u}{\partial \theta}\right) = 0.$$

By separation of variables, we have $m=0$ and (10.2.7) can be simplified into

$$(1-x^2)\frac{d^2 P}{dx^2} - 2x\frac{dP}{dx} + v(v+1)P(x) = 0. \tag{10.2.8}$$

This equation is called the **Legendre equation of order v**. In order to solve some defining problem, we need to solve the Legendre equation and give its eigenvalues and eigenfunctions. We will discuss the solution to this equation in the next subsection.

10.2.2 The solution of the Legendre equation

In order to solve the Legendre equation, we introduce the following theorem firstly.

Theorem 10.2.1 (Cauchy theorem) Suppose that $y''(x) + p(x)y'(x) + q(x)y(x) = 0$ where $p(x)$, $q(x)$ are analytic in $|x-x_0|<R$, is the equation $y(x)$ is satisfied. Then x_0 is called the **regular point** of this equation, and the power solution of this equation in $|x-x_0|<R$ is $y(x) = \sum_{n=0}^{\infty} C_n (x-x_0)^n$.

Remarks

(1) C_n is to be determined.

(2) For Legendre equation, we can transform it into

$$\frac{d^2 P}{dx^2} - \frac{2x}{(1-x^2)}\frac{dP}{dx} + \frac{v(v+1)}{(1-x^2)}P(x) = 0,$$

and it is obvious that $x=0$ is a regular point.

So we can suppose that the series solution of the Legendre equation is $P(x) = \sum_{k=0}^{\infty} a_k x^k$, then

$$\frac{dP}{dx} = \sum_{k=1}^{\infty} k a_k x^{k-1} = \sum_{k=0}^{\infty} (k+1) a_{k+1} x^k,$$

$$\frac{d^2 P}{dx^2} = \sum_{k=2}^{\infty} k(k-1) a_k x^{k-2} = \sum_{k=0}^{\infty} (k+2)(k+1) a_{k+2} x^k.$$

Thus

$$(1-x^2)P''(x) = \sum_{k=0}^{\infty} (k+2)(k+1) a_{k+2} x^k - \sum_{k=0}^{\infty} (k+2)(k+1) a_{k+2} x^{k+2}$$

$$= \sum_{k=0}^{\infty} (k+2)(k+1) a_{k+2} x^k - \sum_{k=2}^{\infty} k(k+1) a_k x^k,$$

$$-2xP'(x) = -\sum_{k=0}^{\infty} 2(k+1) a_{k+1} x^{k+1} = -\sum_{k=1}^{\infty} 2k a_k x^k,$$

$$v(v+1)P(x) = \sum_{k=0}^{\infty} v(v+1) a_k x^k.$$

Substituting them into (10.2.8), we obtain

$$(k+2)(k+1) a_{k+2} - k(k-1) a_k - 2k a_k + v(v+1) a_k = 0, \quad (k \geq 2),$$

that is,

$$a_{k+2} = \frac{k(k-1) - v(v+1)}{(k+2)(k+1)} a_k = -\frac{(v-k)(v+k+1)}{(k+2)(k+1)} a_k. \tag{10.2.9}$$

According to this recurrence formula, one can determine a_{2k} and a_{2k-1} ($k=1, 2, \cdots$) by a_0 and a_1, that is

$$a_2 = \frac{-v(v+1)}{2!} a_0,$$

$$a_4 = \frac{(2-v)(v+3)}{4 \times 3} a_2 = \frac{(2-v)(-v)(v+1)(v+3)}{4!} a_0,$$

$$\vdots$$

$$a_{2k} = \frac{(2k-2-v)(2k-4-v)\cdots(2-v)(-v)(v+1)\cdots(v+2k-1)}{(2k)!} a_0,$$

$$a_3 = \frac{(1-v)(v+2)}{3!} a_1,$$

$$a_5 = \frac{(3-v)(v+4)}{5 \times 4} a_3 = \frac{(3-v)(1-v)(v+2)(v+4)}{5!} a_1,$$

$$\vdots$$

$$a_{2k+1} = \frac{(2k-1-v)(2k-3-v)\cdots(1-v)(v+2)\cdots(v+2k)}{(2k+1)!} a_1.$$

Then the series solution of Legendre equation is

$$P(x) = a_0 \left[1 + \sum_{k=1}^{\infty} \frac{(-1)^k v(v-2)\cdots(v-2k+2)(v+1)\cdots(v+2k-1)}{(2k)!} x^{2k} \right] +$$

$$a_1 \left[x + \sum_{k=1}^{\infty} \frac{(-1)^k (v-1)(v-3)\cdots(v-2k+1)(v+2)\cdots(v+2k)}{(2k+1)!} x^{2k+1} \right]$$

$$= a_0 p_v(x) + a_1 q_v(x), \tag{10.2.10}$$

where a_0, a_1 are two arbitrary constants. $p_v(x)$, $q_v(x)$ are solutions of Legendre equation, and they are convergent for $|x| < 1$ and linear independent. But $p_v(x)$ and $q_v(x)$ are divergent at $x = \pm 1$.

We will discuss the solution for two cases.

Case 1 v is not an integer.

For this case, $p_v(x)$ and $q_v(x)$ are two infinite power series, and they are convergent for $|x| < 1$ and divergent at $x = \pm 1$.

Case 2 v is an integer, that is $v = n$.

When $v = 2k-2$, $a_{2k} = 0$, $a_{2k-2} \neq 0$, $p_v(x)$ is a polynomial of order $2k-2$;

When $v = 2k-1$, $a_{2k+1} = 0$, $a_{2k-1} \neq 0$, $q_v(x)$ is a polynomial of order $2k-1$.

In this case, the recurrence formula becomes

$$a_k = \frac{-(k+2)(k+1)}{(n-k)(k+n+1)} a_{k+2}, \quad (k \leq n-2),$$

$$a_{n-2} = -\frac{n(n-1)}{2(2n-1)} a_n, \quad a_{n-4} = -\frac{(n-2)(n-3)}{4(2n-3)} a_{n-2} = \frac{n(n-1)(n-2)(n-3)}{2 \times 4 \times (2n-1) \times (2n-3)} a_n.$$

$$\vdots$$

In order to simplify the expression, we take

$$a_n = \frac{(2n)!}{2^n (n!)^2} = \frac{1 \times 3 \times 5 \cdots (2n-1)}{n!}, \quad (n=1, 2, \cdots).$$

Then

$$a_{n-2} = -\frac{n(n-1)}{2(2n-1)} \frac{2n!}{2^n (n!)^2} = \frac{-(2n-2)!}{2^n (n-1)!(n-2)!},$$

$$a_{n-4} = \frac{(2n-4)!}{2^n \cdot 2!(n-2)!(n-4)!},$$

$$\vdots$$

Generally, when $n-2m \geq 0$, we have

$$a_{n-2m} = (-1)^m \frac{(2n-2m)!}{2^n \cdot m!(n-m)!(n-2m)!}.$$

When n is an even number, then

$$P_n(x) = \frac{2n!}{2^n (n!)^2} x^n - \frac{(2n-2)!}{2^n (n-1)!(n-2)!} x^{n-2} + \cdots$$

$$= \sum_{m=0}^{\frac{n}{2}} (-1)^m \frac{(2n-2m)!}{2^n m!(n-m)!(n-2m)!} x^{n-2m}; \quad (10.2.11)$$

and when n is an odd number, then

$$q_n(x) = \sum_{m=0}^{\frac{n-1}{2}} (-1)^m \frac{(2n-2m)!}{2^n m!(n-m)!(n-2m)!} x^{n-2m}. \quad (10.2.12)$$

Denote these polynomials as $P_n(x)$ with

$$P_n(x) = \sum_{k=0}^{N} \frac{(-1)^k (2n-2k)!}{2^n k!(n-k)!(n-2k)!} x^{n-2k} \quad (10.2.13)$$

with $N = \begin{cases} \dfrac{n}{2}, & n=2k, \ k=0, 1, \cdots, \\ \dfrac{n-1}{2}, & n=2k+1, \ k=0, 1, \cdots, \end{cases}$ then $P_n(x)$ is called **the first kind of n-th Legendre polynomial or function**.

For another infinite series of $p_n(x)$ and $q_n(x)$, we can denote them as $Q_n(x)$, which is called **the second kind of Legendre function**.

Then, the general solution of the Legendre equation for $v=n$ is

$$P(x) = C_1 P_n(x) + C_2 Q_n(x). \quad (10.2.14)$$

10.2.3 The properties of the Legendre polynomial and recurrence formula

We have known that the first kind Legendre polynomial of order n is

$$P_n(x) = \sum_{k=0}^{N} \frac{(-1)^k (2n-2k)!}{2^n k!(n-k)!(n-2k)!} x^{n-2k},$$

where $N = \begin{cases} \dfrac{n}{2}, & n=2k, \ k=0, 1, \cdots, \\ \dfrac{n-1}{2}, & n=2k+1, \ k=0, 1, \cdots. \end{cases}$

Now we will prove that the **differential representation** of the Legendre polynomial is

$$P_n(x) = \frac{1}{2^n n!} \frac{d^n}{dx^n} (x^2 - 1)^n \quad (10.2.15)$$

which is also called **Rodrigues formula**.

Proof If we expand $(x^2-1)^n$ by the binomial formula, we can have

$$\frac{1}{2^n n!}\frac{d^n}{dx^n}(x^2-1)^n = \frac{1}{2^n n!}\frac{d^n}{dx^n}\sum_{k=0}^{n}\frac{n!}{k!(n-k)!}(x^2)^{n-k}(-1)^k.$$

Then

$$\frac{1}{2^n n!}\frac{d^n}{dx^n}(x^2-1)^n = \sum_{k=0}^{\frac{n}{2}}\frac{(-1)^k n!(2n-2k)\cdots(n-2k+1)}{2^n k!(n-k)!}x^{n-2k}$$

$$= \sum_{k=0}^{\frac{n}{2}}\frac{(-1)^k n!(2n-2k)!}{2^n k!(n-k)!(n-2k)!}x^{n-2k}.$$

By the differential representation, the analytical expression of the Legendre polynomials with lower orders are given as follows:

$$P_0(x)=1;$$
$$P_1(x)=\frac{1}{2}\frac{d}{dx}(x^2-1)=x;$$
$$P_2(x)=\frac{1}{2^2 \cdot 2!}\frac{d^2}{dx^2}(x^2-1)^2=\frac{1}{2}(3x^2-1);$$
$$P_3(x)=\frac{1}{2}(5x^3-3x);$$
$$P_4(x)=\frac{1}{8}(35x^4-30x^2+3);$$
$$P_5(x)=\frac{1}{8}(63x^5-70x^3+15x).$$

(10.2.16)

The integral representation of Legendre polynomial is

$$P_n(x) = \frac{1}{\pi}\int_0^{\pi}(x+i(1-x^2)^{\frac{1}{2}}\cos\varphi)^n d\varphi \qquad (10.2.17)$$

and it is concluded from this formula that $P_n(1)=1$, $P_n(-1)=(-1)^n$.

Similar to the Bessel function, Legendre polynomials satisfy some important recurrence formulas as follows:

(1) $(n+1)P_{n+1}(x)-(2n+1)xP_n(x)+nP_{n-1}(x)=0$, $(n\geqslant 1)$, (10.2.18)
(2) $(x^2-1)P'_n(x)=nxP_n(x)-nP_{n-1}(x)$, $(n\geqslant 1)$, (10.2.19)
(3) $nP_n(x)+P'_{n-1}(x)-xP'_n(x)=0$, $(n\geqslant 1)$, (10.2.20)
(4) $P'_{n+1}(x)=xP'_n(x)+(n+1)P_n(x)$, $(n\geqslant 0)$. (10.2.21)

It is clear that we can give the expression of Legendre polynomial of any order with two lower Legendre polynomials by (10.2.18).

Example 10.2.1 Write down the explicit expression of $P_2(x)$, $P_3(x)$ with $P_0(x)=1$, $P_1(x)=x$.

Solution According to the recurrence formula

$$(n+1)P_{n+1}(x)-(2n+1)xP_n(x)+nP_{n-1}(x)=0,$$

we have

$$n=1, \quad 2P_2(x)-3xP_1(x)+P_0(x)=0 \Rightarrow P_2(x)=\frac{3}{2}x^2-\frac{1}{2},$$

$$n=2, \quad 3P_3(x)-5xP_2(x)+2P_1(x)=0 \Rightarrow P_3(x)=\frac{5}{2}x^3-\frac{3}{2}x,$$

$$n=3, \quad 4P_4(x)-7xP_3(x)+3P_2(x)=0 \Rightarrow P_4(x)=\frac{35}{8}x^4-\frac{30}{8}x^2+\frac{3}{8}.$$

When we solve some defining problem applying the Legendre polynomials, we need to expand some function defined on $[-1,1]$ with respect to the Legendre polynomials. So it is necessary to prove that Legendre polynomials construct an orthogonal function system on $[-1,1]$. Now we will prove that Legendre polynomials satisfy the following relation

$$\int_{-1}^{1} P_n(x)P_m(x)\,dx = \begin{cases} 0, & n \ne m, \\ \dfrac{2}{2n+1}, & n = m. \end{cases} \quad (10.2.22)$$

Proof Since $P_n(x)$ satisfies the Legendre equation of order n, then

$$\frac{d}{dx}\left[(1-x^2)\frac{dP_n(x)}{dx}\right]+n(n+1)P_n(x)=0.$$

Multiplying the left side of this equation by $P_m(x)$ and integrating it on $[-1,1]$, we have

$$\int_{-1}^{1} P_m(x)\frac{d}{dx}\left[(1-x^2)\frac{dP_n(x)}{dx}\right]dx$$

$$= P_m(x)\left[(1-x^2)\frac{dP_n(x)}{dx}\right]\bigg|_{-1}^{1} - \int_{-1}^{1}(1-x^2)\frac{dP_n(x)}{dx}dP_m(x)$$

$$= -\int_{-1}^{1}(1-x^2)\frac{dP_n(x)}{dx}\frac{dP_m(x)}{dx}dx,$$

and then

$$-\int_{-1}^{1}(1-x^2)\frac{dP_n(x)}{dx}\frac{dP_m(x)}{dx}dx + \int_{-1}^{1} n(n+1)P_n(x)P_m(x)dx = 0. \quad (10.2.23)$$

On the other hand, $P_m(x)$ satisfies Legendre equation of order m, then

$$\frac{d}{dx}\left[(1-x^2)\frac{dP_m(x)}{dx}\right]+m(m+1)P_m(x)=0.$$

Similar to the above process, we have

$$-\int_{-1}^{1}(1-x^2)\frac{dP_n(x)}{dx}\frac{dP_m(x)}{dx}dx + \int_{-1}^{1} m(m+1)P_n(x)P_m(x)dx = 0. \quad (10.2.24)$$

Subtract (10.2.23) by (10.2.24), we can obtain

$$[n(n+1)-m(m+1)]\int_{-1}^{1} P_n(x)P_m(x)dx = 0.$$

Then it is clear that when $n \ne m$,

$$\int_{-1}^{1} P_n(x)P_m(x)dx = 0.$$

When $m = n$, we can prove that $N_n^2 = \int_{-1}^{1} P_n^2(x)dx = \dfrac{2}{2n+1}$.

With formula (10.2.22), we can expand some function with respect to Legendre polynomial. That is, if $f(x)$ is defined on $[-1,1]$, and it has continuous second order

derivative, then
$$f(x) = \sum_{n=0}^{\infty} C_n P_n(x),$$
where $C_n = \dfrac{2n+1}{2}\int_{-1}^{1} f(x) P_n(x) dx$.

In the following, we will give some examples.

Example 10.2.2 Expand $f(x)=x^3$ as a series of $P_n(x)$.

Solution Let $f(x)=x^3=C_0 P_0(x)+C_1 P_1(x)+C_2 P_2(x)+C_3 P_3(x)$, then $x^3=C_0+C_1 x+\dfrac{1}{2}C_2(3x^2-1)+\dfrac{C_3}{2}(5x^3-3x)$, that is $x^3=C_0+(C_1-\dfrac{3}{2}C_3)x+\dfrac{3}{2}C_2 x^2+\dfrac{5C_3}{2}x^3-\dfrac{1}{2}C_2$.

Thus we have the following equations
$$\begin{cases} C_0 - \dfrac{1}{2}C_2 = 0, \\ C_1 - \dfrac{3}{2}C_3 = 0, \\ \dfrac{3}{2}C_2 = 0, \\ \dfrac{5}{2}C_3 = 1. \end{cases}$$

Solving this equation, we obtain
$$C_0 = C_2 = 0, \quad C_3 = \dfrac{2}{5}, \quad C_1 = \dfrac{3}{5}.$$

Then $f(x)=x^3=\dfrac{3}{5}P_1(x)+\dfrac{2}{5}P_3(x)$.

Example 10.2.3 Evaluate the integral $\int_{-1}^{1} x P_n(x) dx$.

Solution Since $P_1(x)=x$, then
$$\int_{-1}^{1} x P_n(x) dx = \int_{-1}^{1} P_1(x) P_n(x) dx = \begin{cases} 0, & n \neq 1, \\ \dfrac{2}{3}, & n = 1. \end{cases}$$

10.2.4 Application of Legendre polynomial

In this subsection, we consider some defining problem which is related with the Legendre polynomial.

Example 10.2.4 Potential distribution in a sphere.

Find potential distribution $u(r,\theta)$ in the sphere with the center 0 and the radius a, which is determined by the following problem
$$\begin{cases} \dfrac{1}{r^2}\dfrac{\partial}{\partial r}\left(r^2 \dfrac{\partial u}{\partial r}\right) + \dfrac{1}{r^2 \sin\theta}\dfrac{\partial}{\partial \theta}\left(\sin\theta \dfrac{\partial u}{\partial \theta}\right) = 0, & 0<r<a,\ 0<\theta<\pi, \\ u(r,\theta)|_{r=a} = c, & 0 \leqslant \theta \leqslant \pi, \\ u(r,\theta)|_{r\to\infty} = -E_0 r \cos\theta, & 0 \leqslant \theta \leqslant \pi. \end{cases}$$

Solution Let $u(r,\theta)=R(r)\Theta(\theta)$.

Substituting it into the above equation, we have
$$\frac{r^2 \frac{d^2 R}{dr^2}+2r \frac{dR}{dr}}{R(r)} = -\frac{\frac{d^2 \Theta}{d\theta^2}+\frac{\cos\theta}{\sin\theta}\frac{d\Theta}{d\theta}}{\Theta(\theta)} = \lambda.$$

Setting $\lambda = v(v+1)$, we obtain two ordinary differential equation as follows

$$r^2 \frac{d^2 R}{dr^2}+2r \frac{dR}{dr}-v(v+1)R(r)=0, \tag{10.2.25}$$

$$\frac{d^2 \Theta}{d\theta^2}+\frac{\cos\theta}{\sin\theta}\frac{d\Theta}{d\theta}+v(v+1)\Theta(\theta)=0. \tag{10.2.26}$$

In (10.2.26), let $\Theta(\theta) = \Theta(\arccos x) \triangleq P(x)$, ($x = \cos\theta$), then (10.2.26) becomes

$$(1-x^2)\frac{d^2 P}{dx^2} - 2x \frac{dP}{dx} + v(v+1)P(x)=0. \tag{10.2.27}$$

This is a Legendre equation. According to the physical meaning of this problem, $u(r,\theta)$ is bounded and then $\Theta(\theta)$ is bounded. According to the results in subsection 10.2.2, only when $v = n$, the solution of (10.2.27) is bounded and the solution is $\Theta_n(\theta) = P_n(\cos\theta)$, $n = 0, 1, \cdots$. In fact, $\Theta_n(\theta) = P_n(\cos\theta)$ is the eigenfunction of (10.2.27) with the natural boundary condition $|P(\pm 1)| < +\infty$.

Then equation (10.2.25) becomes
$$r^2 R''(r) + 2r R'(r) - n(n+1)R(r) = 0.$$

This is an Euler equation, and its general solution is
$$R_n(r) = A_n r^n + B_n r^{-(n+1)}.$$

Hence the solution of this problem is
$$u(r,\theta) = \sum_{n=0}^{\infty}(A_n r^n + B_n r^{-(n+1)})P_n(\cos\theta).$$

From $u(r,\theta)|_{r \to +\infty} = -E_0 r \cos\theta$, we have
$$\sum_{n=0}^{\infty} A_n r^n P_n(\cos\theta) = -E_0 r \cos\theta = -E_0 r P_1(\cos\theta).$$

Then
$$A_0 = 0, \quad A_1 = -E_0, \quad A_2 = A_3 = \cdots = A_n = \cdots = 0.$$

Then the solution becomes
$$u(r,\theta) = -E_0 r P_1(\cos\theta) + \sum_{n=0}^{\infty} B_n r^{-(n+1)} P_n(\cos\theta).$$

From $u(a,\theta) = c$, we have
$$-E_0 a P_1(\cos\theta) + \sum_{n=0}^{\infty} B_n a^{-(n+1)} P_n(\cos\theta) = c.$$

That is
$$\sum_{n=0}^{\infty} B_n a^{-(n+1)} P_n(\cos\theta) = c + E_0 a P_1(\cos\theta) = c P_0(\cos\theta) + E_0 a P_1(\cos\theta),$$

which means
$$B_0 a^{-1} = c, \quad B_1 a^{-2} = E_0 a, \quad B_2 = B_3 = \cdots = 0.$$

Finally, the solution of this defining problem is

$$u(r,\theta) = -E_0 r P_1(\cos\theta) + acr^{-1} + E_0 a^3 r^{-2} \cos\theta$$
$$= acr^{-1} + E_0(a^3 r^{-2} - r)\cos\theta.$$

Exercises 10.2

1. Expand the following functions according to the Legendre polynomial:
   (1) $f(x) = 2x^3 + 3x + 4$;
   (2) $f(x) = x^4$.

2. Evaluate $\int_{-1}^{1} x^5 P_5(x) \, dx$.

3. Evaluate $\int_{-1}^{1} [P_4(x)]^2 \, dx$.

4. Solve the following defining problem:
$$\begin{cases} \dfrac{1}{r^2}\dfrac{\partial}{\partial r}\left(r^2 \dfrac{\partial u}{\partial r}\right) + \dfrac{1}{r^2 \sin\theta}\dfrac{\partial}{\partial \theta}\left(\sin\theta \dfrac{\partial u}{\partial \theta}\right) = 0, & 0 < r < 1, \ 0 < \theta < \pi, \\ u(r,\theta)|_{r=1} = \cos^2\theta, & 0 \leqslant \theta \leqslant \pi. \end{cases}$$

Chapter 11 Integral Transformations

The method for integral transformation can be used to solve the problems in the unbounded region.

11.1 Fourier integral transformation

The Fourier Transform provides a way of expanding functions on the whole line $R = (-\infty, +\infty)$ as (continuous) superposition of the basic oscillatory functions $e^{i\xi x}$ ($\xi \in R$) in much the same way that Fourier series are used to expand functions on a finite interval.

Consider the Fourier series of the function f on $[-l, l]$. Suppose that f is a function on R. For any $l > 0$, we can expand f on the interval $[-l, l]$ in Fourier series, and we wish to see what happens to this expansion as we let $l \to \infty$.

To this end, we write the Fourier expansion as follows: for $x \in [-l, l]$,

$$f(x) = \frac{1}{\sqrt{2l}} \sum_{n=-\infty}^{\infty} \tilde{C}_{n,l} e^{i\frac{n\pi x}{l}},$$

where

$$\tilde{C}_{n,l} = \frac{1}{\sqrt{2l}} \int_{-l}^{l} f(y) e^{-i\frac{n\pi y}{l}} dy.$$

Let $\Delta\xi = \frac{\pi}{l}$ and $\xi_n = n\Delta\xi = \frac{n\pi}{l}$. Then these formulas become

$$f(x) = \frac{1}{\sqrt{2\pi}} \sum_{n=-\infty}^{\infty} C_{n,l} e^{i\xi_n x} \Delta\xi, \quad C_{n,l} = \frac{1}{\sqrt{2\pi}} \int_{-l}^{l} f(y) e^{-i\xi_n y} dy.$$

Let us suppose that $f(x)$ vanishes rapidly as $x \to \pm\infty$. Then $C_{n,l}$ will not change much if we extend the region of integration from $[-l, l]$ to $[-\infty, +\infty]$.

$$C_{n,l} \approx \frac{1}{\sqrt{2\pi}} \int_{-\infty}^{\infty} f(y) e^{-i\xi_n y} dy.$$

This last integral is a function of ξ_n, which we call $\hat{f}(\xi_n)$, and we now have

$$f(x) \approx \frac{1}{\sqrt{2\pi}} \sum_{n=-\infty}^{\infty} \hat{f}(\xi_n) e^{i\xi_n x} \Delta\xi, \quad (|x| < l).$$

This looks very much like a Riemann sum. If we now let $l \to \infty$, so that $\Delta\xi \to 0$, the "$\approx$"

should become "=" and the sum should turn into an integral, thus

$$f(x) = \frac{1}{\sqrt{2\pi}} \int_{-\infty}^{\infty} \hat{f}(\xi) e^{i\xi x} \, d\xi,$$

where

$$\hat{f}(\xi) = \frac{1}{\sqrt{2\pi}} \int_{-\infty}^{\infty} f(y) e^{-i\xi y} \, dy.$$

These limiting calculation are utterly non-rigorous as they stand; nonetheless, the final result is correct under suitable condition on f, as we shall prove in due course.

11.1.1 Definition of Fourier integral transformation

Definition 11.1.1 Assume that the function $f(x)$ is piecewise smooth (piecewise continuously derivable) and absolutely integrable in $(-\infty, +\infty)$, then the integral

$$F(\lambda) = \frac{1}{\sqrt{2\pi}} \int_{-\infty}^{+\infty} f(x) e^{i\lambda x} \, dx \equiv \mathscr{F}[f(x)] \qquad (11.1.1)$$

is called the **Fourier integral transformation** of $f(x)$, and $f(x)$ is called the **Fourier inverse transformation** of $F(\lambda)$, that is

$$f(x) = \frac{1}{\sqrt{2\pi}} \int_{-\infty}^{+\infty} F(\lambda) e^{-i\lambda x} \, d\lambda. \qquad (11.1.2)$$

Remark As a supplement, there are some results concerned with the Fourier transformation as follows.

Theorem 11.1.1 (Dirichlet conditions) Assume that $f(x)$ satisfies the Dirichlet conditions:

(1) $f(x)$ is bounded and absolutely integrable for all $x \in (-\infty, +\infty)$;

(2) $f(x)$ has at most finite number of extreme points and discontinuities of the first kind.

Then, for any $x \in (-\infty, +\infty)$,

$$\frac{f(x+0) + f(x-0)}{2} = \frac{1}{2\pi} \int_{-\infty}^{+\infty} \left[\int_{-\infty}^{+\infty} f(y) e^{i y \lambda} \, dy \right] e^{-i x \lambda} \, d\lambda. \qquad (11.1.3)$$

Let $f(x)$ be an odd function satisfying the Dirichlet condition, then the Fourier transformation becomes the sine Fourier transformation, such that

$$F_s(\lambda) = \sqrt{\frac{2}{\pi}} \int_0^{+\infty} f(y) \sin(\lambda y) \, dy,$$

and

$$f(x) = \sqrt{\frac{2}{\pi}} \int_0^{+\infty} F_s(\lambda) \sin(x\lambda) \, d\lambda.$$

Together with these, we have

$$\frac{f(x+0) + f(x-0)}{2} = \frac{1}{2\pi} \int_{-\infty}^{+\infty} \left[\int_{-\infty}^{+\infty} f(y) \sin \lambda (x-y) \, dy \right] d\lambda.$$

If $f(x)$ is an even function satisfying the Dirichlet condition, then the Fourier transformation becomes the cosine Fourier transformation, such that

$$F_c(\lambda) = \sqrt{\frac{2}{\pi}} \int_0^{+\infty} f(y)\cos(\lambda y)\,dy,$$

$$f(x) = \sqrt{\frac{2}{\pi}} \int_0^{+\infty} F_c(\lambda)\cos(x\lambda)\,d\lambda,$$

and together with these, we have

$$\frac{f(x+0)+f(x-0)}{2} = \frac{1}{2\pi} \int_{-\infty}^{+\infty} \left[\int_{-\infty}^{+\infty} f(y)\cos\lambda(x-y)\,dy\right]d\lambda.$$

Example 11.1.1 Find the Fourier integral transformation of $f(x) = \begin{cases} 1, & |x| \leqslant 1, \\ 0, & |x| > 1. \end{cases}$

Solution $F(\lambda) = \mathscr{F}[f(x)] = \frac{1}{\sqrt{2\pi}} \int_{-\infty}^{+\infty} f(x)e^{i\lambda x}\,dx = \frac{1}{\sqrt{2\pi}} \int_{-1}^{1} e^{i\lambda x}\,dx$

$$= \frac{1}{\sqrt{2\pi}} \frac{1}{i\lambda}(e^{i\lambda} - e^{-i\lambda}) = \sqrt{\frac{2}{\pi}} \frac{\sin\lambda}{\lambda}.$$

Example 11.1.2

(1) Find the Fourier transformation of $f(x) = e^{-|x|}$.

(2) And then prove that

$$\int_0^{+\infty} \frac{\cos\lambda x}{1+\lambda^2}\,d\lambda = \frac{\pi}{2}e^{-|x|}.$$

Solution (1) $F(\lambda) = \mathscr{F}[f(x)] = \frac{1}{\sqrt{2\pi}} \int_{-\infty}^{+\infty} e^{-|x|} e^{i\lambda x}\,dx$

$$= \frac{1}{\sqrt{2\pi}}\left(\int_{-\infty}^{0} e^{x} e^{i\lambda x}\,dx + \int_0^{+\infty} e^{-x} e^{i\lambda x}\,dx\right)$$

$$= \frac{1}{\sqrt{2\pi}}\left[\frac{1}{1+i\lambda} + \frac{1}{1-i\lambda}\right] = \sqrt{\frac{2}{\pi}} \frac{1}{1+\lambda^2}.$$

(2) By the Fourier inverse transformation, we then have

$$f(x) = \frac{1}{\sqrt{2\pi}} \int_{-\infty}^{+\infty} F(\lambda) e^{-i\lambda x}\,d\lambda$$

$$= \frac{1}{\sqrt{2\pi}}\left(\int_{-\infty}^{+\infty} \sqrt{\frac{2}{\pi}} \frac{1}{1+\lambda^2}(\cos\lambda x - i\sin\lambda x)\,d\lambda\right)$$

$$= \frac{1}{\pi} \int_{-\infty}^{+\infty} \frac{1}{1+\lambda^2}\cos\lambda x\,d\lambda$$

$$= \frac{2}{\pi} \int_0^{+\infty} \frac{\cos\lambda x}{1+\lambda^2}\,d\lambda.$$

Thus, $\int_0^{+\infty} \frac{\cos\lambda x}{1+\lambda^2}\,dx = \frac{\pi}{2}f(x) = \frac{\pi}{2}e^{-|x|}.$

Example 11.1.3 Compute Fourier transformations of step function $u_a(x) = \begin{cases} 0, & x < a, \\ 1, & x \geqslant a, \end{cases}$

for $a > 0$.

Solution It is obvious that $u_a(x)$ is not absolutely integrable. Take $0 < \beta < 1$, then we consider the function

$$u_a(x,\beta) = \begin{cases} 0, & x<a, \\ e^{-\beta x}, & x>a. \end{cases}$$

Note that $\lim\limits_{\beta \to 0^+} u_a(x,\beta) = u_a(x)$.

In the following, let us compute the Fourier transformations of $u_a(x,\beta)$.

$$\mathscr{F}[u_a(x,\beta)] = \frac{1}{\sqrt{2\pi}} \int_{-\infty}^{+\infty} u_a(x,\beta) e^{i\lambda x} dx = \frac{1}{\sqrt{2\pi}} \int_a^{+\infty} e^{-\beta x} e^{i\lambda x} dx$$

$$= \frac{1}{\sqrt{2\pi}} \int_a^{+\infty} e^{-(\beta-i\lambda)x} dx = \frac{1}{\sqrt{2\pi}} \left\{ \frac{-e^{-(\beta-i\lambda)x}}{\beta-i\lambda} \Big|_a^{+\infty} \right\}$$

$$= \frac{1}{\sqrt{2\pi}} \frac{e^{-(\beta-i\lambda)a}}{(\beta-i\lambda)}.$$

Now, we define

$$\mathscr{F}[u_a(x)] = \lim_{\beta \to 0^+} \mathscr{F}[u_a(x,\beta)] = \frac{1}{\sqrt{2\pi}} \frac{e^{i\lambda a}}{-i\lambda} = \frac{i e^{i\lambda a}}{\sqrt{2\pi}\lambda}.$$

And, if $a=0$, then $\mathscr{F}[u_0(x)] = \dfrac{i}{\sqrt{2\pi}\lambda}$.

Example 11.1.4 Compute the Fourier transformation of the impulse function $\delta(x)$, where

$$\delta(x) = \begin{cases} 0, & x \neq 0, \\ \int_{-\infty}^{+\infty} \delta(x)dx = 1. \end{cases}$$

Solution As we know that the impulse function (Dirac function) $\delta(x)$ is not a normal function. And $\delta(x)$ is called a generalized function or function of distribution, where $\int_{-\infty}^{+\infty} \delta(x)dx = 1$.

Firstly, let us consider the following function:

$$P_\varepsilon(x,a) = \begin{cases} h, & |x-a|<\varepsilon, \\ 0, & |x-a|\geqslant\varepsilon, \end{cases} \quad \text{for } \varepsilon>0.$$

In the following, let us compute the Fourier transformations of $P_\varepsilon(x,a)$.

$$\mathscr{F}[P_\varepsilon(x,a)] = \frac{1}{\sqrt{2\pi}} \int_{a-\varepsilon}^{a+\varepsilon} h e^{i\lambda x} dx = \frac{h}{\sqrt{2\pi}} \frac{e^{i\lambda x}}{i\lambda} \Big|_{a-\varepsilon}^{a+\varepsilon}$$

$$= \frac{h}{\sqrt{2\pi}} \frac{e^{i\lambda a}}{i\lambda} (e^{i\lambda\varepsilon} - e^{-i\lambda\varepsilon})$$

$$= \frac{2h\varepsilon}{\sqrt{2\pi}} e^{i\lambda a} \left(\frac{\sin \lambda\varepsilon}{\lambda\varepsilon} \right).$$

Setting $h = \dfrac{1}{2\varepsilon}$, and defining $\mathscr{F}[\delta(x-a)] = \lim\limits_{\varepsilon \to 0} \mathscr{F}[P_\varepsilon(x,a)] = \dfrac{1}{\sqrt{2\pi}} e^{i\lambda a}$, where

$$\delta(x-a) = \begin{cases} 0, & x \neq a, \\ \int_{-\infty}^{+\infty} \delta(x-a) = 1. \end{cases}$$

As $a=0$, we have $\mathscr{F}[\delta(x)] = \dfrac{1}{\sqrt{2\pi}}$.

11.1.2 The properties of Fourier integral transformation

Assume $\mathscr{F}[f] = F(\lambda) = \dfrac{1}{\sqrt{2\pi}} \displaystyle\int_{-\infty}^{+\infty} f(\xi) e^{i\lambda\xi} d\xi$.

In the following, we give some properties of Fourier integral transformation and then we can calculate Fourier integral transformation of some functions by these properties easily.

(1) Linear property

Assume that f, g are absolutely integrable in $(-\infty, +\infty)$. Let a, b be two constants, then
$$\mathscr{F}[af(x)+bg(x)] = a\mathscr{F}[f(x)] + b\mathscr{F}[g(x)].$$

Proof
$$\mathscr{F}[af(x)+bg(x)] = \frac{1}{\sqrt{2\pi}} \int_{-\infty}^{+\infty} [af(x)+bg(x)] e^{i\lambda x} dx$$
$$= a \frac{1}{\sqrt{2\pi}} \int_{-\infty}^{+\infty} f(x) e^{i\lambda x} dx + b \frac{1}{\sqrt{2\pi}} \int_{-\infty}^{+\infty} g(x) e^{i\lambda x} dx$$
$$= a\mathscr{F}[f(x)] + b\mathscr{F}[g(x)].$$

(2) Displacement theorem

Assume that f is absolutely integral in $(-\infty, +\infty)$, then for any constant number $c \in R$,
$$\mathscr{F}[f(x-c)] = e^{i\lambda c} F(\lambda).$$

Proof
$$\mathscr{F}[f(x-c)] = \frac{1}{\sqrt{2\pi}} \int_{-\infty}^{+\infty} f(x-c) e^{i\lambda x} dx$$
$$\xrightarrow{y=x-c} \frac{1}{\sqrt{2\pi}} \int_{-\infty}^{+\infty} f(y) e^{i\lambda(y+c)} dy$$
$$= e^{i\lambda c} F(\lambda).$$

(3) Similarity theorem

Assume that f is absolutely integral in $(-\infty, +\infty)$, and let $c \neq 0$ be a constant, then
$$\mathscr{F}[f(cx)] = \frac{1}{|c|} F\left(\frac{\lambda}{c}\right).$$

Proof
$$\mathscr{F}[f(cx)] = \frac{1}{\sqrt{2\pi}} \int_{-\infty}^{+\infty} f(cx) e^{i\lambda x} dx$$
$$\xrightarrow{y=cx} \begin{cases} \dfrac{1}{\sqrt{2\pi}} \displaystyle\int_{-\infty}^{+\infty} f(y) e^{i\frac{\lambda}{c}y} \dfrac{1}{c} dy, & c > 0, \\ \dfrac{1}{\sqrt{2\pi}} \displaystyle\int_{+\infty}^{-\infty} f(y) e^{i\frac{\lambda}{c}y} \dfrac{1}{c} dy, & c < 0, \end{cases}$$
$$= \frac{1}{|c|} \frac{1}{\sqrt{2\pi}} \int_{-\infty}^{+\infty} f(y) e^{i\frac{\lambda}{c}y} dy = \frac{1}{|c|} F\left(\frac{\lambda}{c}\right).$$

(4) Delay property

Assume that f is absolutely integrable in $(-\infty, +\infty)$, then
$$\mathscr{F}[e^{-i\lambda_0 x}f(x)] = F[\lambda - \lambda_0].$$

Proof
$$\mathscr{F}[e^{-i\lambda_0 x}f(x)] = \frac{1}{\sqrt{2\pi}}\int_{-\infty}^{+\infty} e^{-i\lambda_0 x}f(x)e^{i\lambda x}\,\mathrm{d}x$$
$$= \frac{1}{\sqrt{2\pi}}\int_{-\infty}^{+\infty} f(x)e^{i(\lambda-\lambda_0)x}\,\mathrm{d}x = F[\lambda - \lambda_0].$$

(5) Differential theorem

Assume $f(x)$ and $f'(x)$ are piecewise smooth and absolutely integrable in $(-\infty, +\infty)$, and $\lim\limits_{|x| \to +\infty} f(x) = 0$, then
$$\mathscr{F}[f'(x)] = (-i\lambda)\mathscr{F}[f(x)].$$

Proof
$$\mathscr{F}[f'(x)] = \frac{1}{\sqrt{2\pi}}\int_{-\infty}^{+\infty} f'(x)e^{i\lambda x}\,\mathrm{d}x$$
$$= \frac{1}{\sqrt{2\pi}}\left[f(x)e^{i\lambda x}\Big|_{-\infty}^{+\infty} - i\lambda\int_{-\infty}^{+\infty} f(x)e^{i\lambda x}\,\mathrm{d}x\right]$$
$$= (-i\lambda)\mathscr{F}[f(x)].$$

Corollary Assume $f(x)$ and $f^{(k)}(x)$ $(k=1, 2, \cdots, n)$ can be operated by the Fourier transformation, and $f^{(k)}(\pm\infty) = 0$, $k = 0, 1, \cdots, n-1$, where $f^{(0)}(x) = f(x)$, then
$$\mathscr{F}[f^{(n)}(x)] = (-i\lambda)^n \mathscr{F}[f(x)].$$

(6) Integral property

Assume that f is absolutely integrable in $(-\infty, +\infty)$, then
$$\mathscr{F}\left[\int_{x_0}^{x} f(t)\,\mathrm{d}t\right] = \frac{1}{(-i\lambda)}\mathscr{F}[f(x)].$$

Proof Firstly,
$$\mathscr{F}\left[\frac{\mathrm{d}}{\mathrm{d}x}\int_{x_0}^{x} f(t)\,\mathrm{d}t\right] = \mathscr{F}[f(x)].$$

Secondly,
$$\mathscr{F}\left[\frac{\mathrm{d}}{\mathrm{d}x}\int_{x_0}^{x} f(t)\,\mathrm{d}t\right] = (-i\lambda)\mathscr{F}\left[\int_{x_0}^{x} f(t)\,\mathrm{d}t\right].$$

Thus, we have $\mathscr{F}\left[\int_{x_0}^{x} f(t)\,\mathrm{d}t\right] = \dfrac{1}{(-i\lambda)}\mathscr{F}[f(x)].$

Example 11.1.5 Please compute $\mathscr{F}[e^{-ax^2}]$ by the formula $\int_0^{+\infty} e^{-x^2}\,\mathrm{d}x = \dfrac{\sqrt{\pi}}{2}$.

Solution Noting that $\int_0^{+\infty} e^{-x^2}\,\mathrm{d}x = \dfrac{\sqrt{\pi}}{2}$, we have

$$\mathscr{F}[e^{-ax^2}] = \frac{1}{\sqrt{2\pi}} \int_{-\infty}^{+\infty} e^{-ax^2} e^{i\lambda x} dx$$

$$= \frac{1}{\sqrt{2\pi}} \int_{-\infty}^{+\infty} e^{-a(x-\frac{i\lambda}{2a})^2 - \frac{\lambda^2}{4a}} dx$$

$$= \frac{1}{\sqrt{2\pi a}} e^{-\frac{\lambda^2}{4a}} \int_{-\infty}^{+\infty} e^{-[\sqrt{a}(x-\frac{i\lambda}{2a})]^2} d\left(\sqrt{a}\left(x - \frac{i\lambda}{2a}\right)\right)$$

$$= \frac{1}{\sqrt{2a}} e^{-\frac{\lambda^2}{4a}}.$$

Example 11.1.6 Use the result of Example 11.1.5 to compute $\mathscr{F}[xe^{-ax^2}]$.

Solution From Example 11.1.5, we have

$$\mathscr{F}[e^{-ax^2}] = \frac{1}{\sqrt{2a}} e^{-\frac{\lambda^2}{4a}}.$$

Moreover, $\dfrac{d}{dx}[e^{-ax^2}] = -2ax e^{-ax^2} \Rightarrow xe^{-ax^2} = \dfrac{1}{-2a}\dfrac{d}{dx}[e^{-ax^2}].$

So,

$$\mathscr{F}[xe^{-ax^2}] = \frac{1}{-2a} \mathscr{F}\left[\frac{d}{dx}[e^{-ax^2}]\right]$$

$$= \frac{1}{-2a}(-i\lambda) \mathscr{F}[e^{-ax^2}] = \frac{i\lambda}{(\sqrt{2a})^3} e^{-\frac{\lambda^2}{4a}}.$$

11.1.3 Convolution and its Fourier transformation

Let f and g be functions on R. Assume that there exist $\mathscr{F}[f]$ and $\mathscr{F}[g]$, then the integral

$$\frac{1}{\sqrt{2\pi}} \int_{-\infty}^{+\infty} f(x-\xi) g(\xi) d\xi = \frac{1}{\sqrt{2\pi}} \int_{-\infty}^{+\infty} g(x-\xi) f(\xi) d\xi$$

is called the convolution of f and g, and it can be denoted by $f * g(x)$ or $g * f(x)$.

Similarly, let $F(\lambda) = \mathscr{F}[f(x)], G(\lambda) = \mathscr{F}[g(x)]$, then the integral

$$\frac{1}{\sqrt{2\pi}} \int_{-\infty}^{+\infty} F(\lambda - s) G(s) ds = \frac{1}{\sqrt{2\pi}} \int_{-\infty}^{+\infty} G(\lambda - s) F(s) ds$$

is called the convolution of $F(\lambda)$ and $G(\lambda)$, and is denoted by $F * G(\lambda)$ or $G * F(\lambda)$.

Theorem 11.1.2 Convolution obeys the same algebraic laws as ordinary multiplication.

(1) $f * (ag + bh) = a(f * g) + b(f * h)$ for any constants a and b;

(2) $f * g = g * f$;

(3) $f * (g * h) = (f * g) * h$.

Proof

(1) It is easy to see that the equation is true since integration is a linear operation.

(2) Make the change of variable $z = x - y$,

$$f * g(x) = \frac{1}{\sqrt{2\pi}} \int_{-\infty}^{+\infty} f(x-y) g(y) dy = \frac{1}{\sqrt{2\pi}} \int_{-\infty}^{+\infty} f(z) g(x-z) dz = g * f(x).$$

(3) Using (2) and interchanging the order of integration, we have

$$(f * g) * h = \frac{1}{\sqrt{2\pi}} \int_{-\infty}^{+\infty} f * g(x-y)h(y) \mathrm{d}y$$

$$= \frac{1}{2\pi} \int_{-\infty}^{+\infty} \int_{-\infty}^{+\infty} f(z) g(x-y-z) h(y) \mathrm{d}z \mathrm{d}y$$

$$= \frac{1}{\sqrt{2\pi}} \int_{-\infty}^{+\infty} f(z) \left[\frac{1}{\sqrt{2\pi}} \int_{-\infty}^{+\infty} g(x-z-y) h(y) \mathrm{d}y \right] \mathrm{d}z$$

$$= \frac{1}{\sqrt{2\pi}} \int_{-\infty}^{+\infty} f(z) g * h(x-z) \mathrm{d}z$$

$$= f * (g * h)(x).$$

Theorem 11.1.3 Assume that there exit Fourier transformations of $f(x)$ and $g(x)$, and $\mathscr{F}[f]=F(\lambda)$, $\mathscr{F}[g]=G(\lambda)$, then
(1) $\mathscr{F}[f * g] = F(\lambda) \cdot G(\lambda)$;
(2) $\mathscr{F}[f \cdot g] = F * G(\lambda)$.

Proof

(1) $\mathscr{F}[f * g] = \frac{1}{\sqrt{2\pi}} \int_{-\infty}^{+\infty} \left[\frac{1}{\sqrt{2\pi}} \int_{-\infty}^{+\infty} f(x-\xi) g(\xi) \mathrm{d}\xi \right] \mathrm{e}^{\mathrm{i}\lambda x} \mathrm{d}x$

$$= \frac{1}{\sqrt{2\pi}} \int_{-\infty}^{+\infty} \left[\frac{1}{\sqrt{2\pi}} \int_{-\infty}^{+\infty} f(x-\xi) \mathrm{e}^{\mathrm{i}\lambda x} \mathrm{d}x \right] g(\xi) \mathrm{d}\xi$$

$$\underline{\eta = x - \xi} \quad \frac{1}{\sqrt{2\pi}} \int_{-\infty}^{+\infty} \left[\frac{1}{\sqrt{2\pi}} \int_{-\infty}^{+\infty} f(\eta) \mathrm{e}^{\mathrm{i}\lambda(\eta+\xi)} \mathrm{d}\eta \right] g(\xi) \mathrm{d}\xi$$

$$= \left[\frac{1}{\sqrt{2\pi}} \int_{-\infty}^{+\infty} f(\eta) \mathrm{e}^{\mathrm{i}\lambda \eta} \mathrm{d}\eta \right] \cdot \left[\frac{1}{\sqrt{2\pi}} \int_{-\infty}^{+\infty} g(\xi) \mathrm{e}^{\mathrm{i}\lambda \xi} \mathrm{d}\xi \right]$$

$$= F(\lambda) \cdot G(\lambda).$$

(2) $F * G(\lambda) = \frac{1}{\sqrt{2\pi}} \int_{-\infty}^{+\infty} F(\lambda - s) G(s) \mathrm{d}s$

$$= \frac{1}{\sqrt{2\pi}} \int_{-\infty}^{+\infty} \left[\frac{1}{\sqrt{2\pi}} \int_{-\infty}^{+\infty} f(x) \mathrm{e}^{\mathrm{i}(\lambda-s)x} \mathrm{d}x \right] G(s) \mathrm{d}s$$

$$= \frac{1}{\sqrt{2\pi}} \int_{-\infty}^{+\infty} f(x) \left[\frac{1}{\sqrt{2\pi}} \int_{-\infty}^{+\infty} G(s) \mathrm{e}^{-\mathrm{i}sx} \mathrm{d}s \right] \mathrm{e}^{\mathrm{i}\lambda x} \mathrm{d}x$$

$$= \frac{1}{\sqrt{2\pi}} \int_{-\infty}^{+\infty} f(x) g(x) \mathrm{e}^{\mathrm{i}\lambda x} \mathrm{d}x = \mathscr{F}[f(x) \cdot g(x)].$$

11.1.4 Application of Fourier integral transformation

Example 11.1.7 Solve the Cauchy problem of string oscillation equation:

$$\begin{cases} u_{tt} - a^2 u_{xx} = 0, & -\infty < x < \infty, \ t > 0, \\ u(x,0) = f(x), \ u_t(x,0) = 0, & -\infty < x < \infty. \end{cases} \quad (11.1.4) \\ (11.1.5)$$

Solution Denote $V(\lambda,t) = \mathscr{F}[u(x,t)]$, $F(\lambda) = F[f(x)]$, then

$$\mathscr{F}[u_{tt} - a^2 u_{xx}] = \mathscr{F}[u_{tt}] - a^2 \mathscr{F}[u_{xx}]$$

$$= \frac{d^2}{dt^2}\mathscr{F}[u] - a^2 (i\lambda)^2 \mathscr{F}[u]$$

$$= \frac{d^2 V}{dt^2} + a^2 \lambda^2 V = 0.$$

The equations of (11.1.5) give that
$$\mathscr{F}[u(x,0)] = \mathscr{F}[f(x)] \Rightarrow V(\lambda,0) = F[\lambda].$$
$$\mathscr{F}[u_t(x,0)] = 0 \Rightarrow V_t(\lambda, 0) = 0.$$

Thus the problem becomes
$$\begin{cases} \dfrac{d^2 V}{dt^2} + a^2 \lambda^2 V = 0, & (11.1.6) \\ V(\lambda,0) = F[\lambda], & (11.1.7) \\ V_t(\lambda,0) = 0. & (11.1.8) \end{cases}$$

The general solution of the ordinary differential equation (ODE) (11.1.6) is
$$V(\lambda,t) = C_1(\lambda)\cos a\lambda t + C_2(\lambda)\sin a\lambda t.$$
From the initial conditions (11.1.7) and (11.1.8), we have
$$V(\lambda,0) = F[\lambda] \Rightarrow C_1(\lambda) = F[\lambda],$$
and
$$V_t(\lambda,0) = 0, \Rightarrow a\lambda C_2(\lambda) = 0, \Rightarrow C_2(\lambda) = 0.$$
Thus the solution of (11.1.6)~(11.1.8) is
$$V(\lambda,t) = F(\lambda)\cos a\lambda t.$$
Finally, we take the Fourier inverse transformation to get the solution of this problem.
$$u(x,t) = \mathscr{F}^{-1}[V(\lambda,t)] = \mathscr{F}^{-1}[F(\lambda)\cos a\lambda t]$$
$$= \frac{1}{\sqrt{2\pi}} \int_{-\infty}^{+\infty} F(\lambda)\cos a\lambda t\, e^{-i\lambda x}\, d\lambda$$
$$= \frac{1}{\sqrt{2\pi}} \int_{-\infty}^{+\infty} F(\lambda) \frac{e^{ia\lambda t} + e^{-ia\lambda t}}{2} e^{-i\lambda x}\, d\lambda$$
$$= \frac{1}{2\sqrt{2\pi}} \left[\int_{-\infty}^{+\infty} F(\lambda)e^{i\lambda(at-x)}\, d\lambda + \int_{-\infty}^{+\infty} F(\lambda)e^{-i\lambda(at+x)}\, d\lambda \right]$$
$$= \frac{1}{2}\left[\frac{1}{\sqrt{2\pi}} \int_{-\infty}^{+\infty} F(\lambda)e^{-i\lambda(x-at)}\, d\lambda + \frac{1}{\sqrt{2\pi}} \int_{-\infty}^{+\infty} F(\lambda)e^{-i\lambda(x+at)}\, d\lambda \right]$$
$$= \frac{1}{2}[f(x-at) + f(x+at)].$$

This concise with D'Alembert's formula.

Example 11.1.8 Solve the heat conduction problem for an infinite beam:
$$\begin{cases} u_t - a^2 u_{xx} = 0, & -\infty < x < +\infty, \quad (11.1.9) \\ u(x,0) = f(x), & -\infty < x < +\infty. \quad (11.1.10) \end{cases}$$

Solution Let $V(\lambda,t) = \mathscr{F}[u(x,t)]$, $F(\lambda) = \mathscr{F}[f(x)]$, then by (11.1.9) and (11.1.10), we have
$$\begin{cases} \dfrac{dV}{dt} + a^2 \lambda^2 V = 0, & (11.1.11) \\ V(\lambda,0) = F(\lambda). & (11.1.12) \end{cases}$$

$\Rightarrow V(\lambda,t) = F(\lambda) \cdot e^{-a^2 \lambda^2 t}$.

Then

$$u(x,t) = \mathscr{F}^{-1}[V(\lambda,t)]$$
$$= \mathscr{F}^{-1}[F(\lambda)e^{-a^2\lambda^2 t}]$$
$$= \mathscr{F}^{-1}\{\mathscr{F}(f(x))\mathscr{F}[\mathscr{F}^{-1}(e^{-a^2\lambda^2 t})]\}.$$

Let $\mathscr{F}^{-1}(e^{-a^2\lambda^2 t}) = g(x,t)$, thus we get
$$u(x,t) = f * g(x,t).$$

In the following, let us compute $g(x,t)$. We find,
$$g(x,t) = \frac{1}{\sqrt{2\pi}}\int_{-\infty}^{+\infty} e^{-a^2\lambda^2 t} \cdot e^{-i\lambda x}\,\mathrm{d}\lambda = \frac{1}{\sqrt{2\pi}}\int_{-\infty}^{+\infty} e^{-a^2\lambda^2 t}\cos\lambda x\,\mathrm{d}\lambda.$$

And,
$$\frac{\mathrm{d}g}{\mathrm{d}x} = \frac{1}{\sqrt{2\pi}}\int_{-\infty}^{+\infty} e^{-a^2\lambda^2 t}(-\lambda\sin\lambda x)\,\mathrm{d}\lambda$$
$$= \frac{1}{\sqrt{2\pi}}\left\{\int_{-\infty}^{+\infty}\frac{1}{2a^2 t}(e^{-a^2\lambda^2 t})'_\lambda \sin\lambda x\,\mathrm{d}\lambda\right\}$$
$$= \frac{1}{2a^2 t}\frac{1}{\sqrt{2\pi}}\left\{e^{-a^2\lambda^2 t}\sin\lambda x\,\Big|_{-\infty}^{+\infty} - \int_{-\infty}^{+\infty} e^{-a^2\lambda^2 t}(x\cos\lambda x)\,\mathrm{d}\lambda\right\}$$
$$= -\frac{x}{2a^2 t}\left[\frac{1}{\sqrt{2\pi}}\int_{-\infty}^{+\infty} e^{-a^2\lambda^2 t}\cos\lambda x\,\mathrm{d}\lambda\right] = -\frac{x}{2a^2 t}g.$$

thus $\frac{\mathrm{d}g}{\mathrm{d}x} + \frac{x}{2a^2 t}g = 0$. Solving this equation, we have
$$g(x,t) = c(t)e^{-\frac{x^2}{4a^2 t}}.$$

Let $\eta = a\lambda\sqrt{t}$, $\mathrm{d}\lambda = \frac{\mathrm{d}\eta}{a\sqrt{t}}$. Noting that
$$g(0,t) = \frac{1}{\sqrt{2\pi}}\int_{-\infty}^{+\infty} e^{-a^2\lambda^2 t}\,\mathrm{d}\lambda \xlongequal{\eta = a\lambda\sqrt{t}} \frac{1}{a\sqrt{t}}\cdot\frac{1}{\sqrt{2\pi}}\int_{-\infty}^{+\infty} e^{-\eta^2}\,\mathrm{d}\eta = \frac{1}{a\sqrt{2t}},$$

then we have
$$c(t) = \frac{1}{a\sqrt{2t}} \quad \text{and} \quad g(x,t) = \frac{1}{a\sqrt{2t}}e^{-\frac{x^2}{4a^2 t}}.$$

Hence, $u(x,t) = f * g(x,t)$
$$= \frac{1}{\sqrt{2\pi}}\int_{-\infty}^{+\infty} f(\xi)g(x-\xi,t)\,\mathrm{d}\xi$$
$$= \frac{1}{2a\sqrt{\pi t}}\int_{-\infty}^{+\infty} f(\xi)e^{-\frac{(x-\xi)^2}{4a^2 t}}\,\mathrm{d}\xi.$$

Specially, if $f(x) = \begin{cases} a, & x > 0, \\ 0, & x < 0, \end{cases}$ then
$$u(x,t) = \frac{a}{2a\sqrt{\pi t}}\int_0^{+\infty} e^{-\frac{(\xi-x)^2}{4a^2 t}}\,\mathrm{d}\xi \quad (\text{let } \eta = \frac{\xi - x}{2a\sqrt{t}})$$
$$= \frac{a}{\sqrt{\pi}}\int_{\frac{-x}{2a\sqrt{t}}}^{+\infty} e^{-\eta^2}\,\mathrm{d}\eta$$
$$= \frac{a}{\sqrt{\pi}}\left\{\int_0^{+\infty} e^{-\eta^2}\,\mathrm{d}\eta + \int_{\frac{-x}{2a\sqrt{t}}}^0 e^{-\eta^2}\,\mathrm{d}\eta\right\}$$
$$= \frac{a}{2}\left[1 + \frac{2}{\sqrt{\pi}}\int_0^{\frac{x}{2a\sqrt{t}}} e^{-\eta^2}\,\mathrm{d}\eta\right] = \frac{a}{2}\left[1 + \mathrm{erf}\left(\frac{x}{2a\sqrt{t}}\right)\right],$$

where
$$\mathrm{erf}(x) = \frac{2}{\sqrt{\pi}}\int_0^x e^{-\eta^2}\,\mathrm{d}\eta = \frac{2}{\sqrt{\pi}}\left(x - \frac{x^3}{1!\cdot 3} + \frac{x^5}{2!\cdot 5} - \frac{x^7}{3!\cdot 7} + \cdots\right) \text{—Error function,}$$

$$\mathrm{erf}(\infty) = \mathrm{erf}(z) + \mathrm{erfc}(z)$$
$$= \frac{2}{\sqrt{\pi}}\int_0^z e^{-\eta^2}\,\mathrm{d}\eta + \frac{2}{\sqrt{\pi}}\int_z^{+\infty} e^{-\eta^2}\,\mathrm{d}\eta = \frac{2}{\sqrt{\pi}}\int_0^{+\infty} e^{-\eta^2}\,\mathrm{d}\eta \equiv 1.$$

Example 11.1.9 Solve the problems

$$\begin{cases} u_{xx} + u_{yy} = 0, & -\infty < x < +\infty,\ y > 0, \quad (11.1.13)\\ u(x,0) = f(x), & -\infty < x < +\infty, \quad (11.1.14)\\ \lim_{|x|\to +\infty} u(x,y) = 0,\ \lim_{|x|\to +\infty} u_x(x,y) = 0, & \quad (11.1.15)\\ \lim_{y\to +\infty} |u(x,y)| < +\infty. & \quad (11.1.16) \end{cases}$$

Solution Let $V(\lambda,y) = \mathscr{F}[u(x,y)]$, $F(\lambda) = \mathscr{F}[f(x)]$. Then, using (11.1.15) and taking the Fourier transformation for (11.1.13), we have

$$\mathscr{F}[u_{xx} + u_{yy}] = \mathscr{F}[u_{xx}] + \mathscr{F}[u_{yy}] = -\lambda^2 F[u] + \frac{\mathrm{d}^2}{\mathrm{d}y^2}F[u] = -\lambda^2 V + \frac{\mathrm{d}^2 V}{\mathrm{d}y^2} = 0.$$

By (11.1.14) and (11.1.16), we get

$$\mathscr{F}[u(x,0)] = V(\lambda,0) = \mathscr{F}[f(x)] = F(\lambda),$$
$$\lim_{y\to +\infty}|\mathscr{F}[u(x,y)]| = \lim_{y\to +\infty}|V(\lambda,y)| < +\infty.$$

Thus, we have

$$\begin{cases} \dfrac{\mathrm{d}^2 V}{\mathrm{d}y^2} - \lambda^2 V = 0, & (11.1.17)\\ V(\lambda,0) = F(\lambda), & (11.1.18)\\ \lim_{y\to +\infty}|V(\lambda,y)| < +\infty. & (11.1.19) \end{cases}$$

Solving (11.1.17), we have
$$V(\lambda,y) = C_1(\lambda)e^{\lambda y} + C_2(\lambda)e^{-\lambda y}.$$

By (11.1.19), $\begin{cases} \text{if } \lambda > 0, & \text{then } C_1(\lambda) = 0,\\ \text{if } \lambda < 0, & \text{then } C_2(\lambda) = 0, \end{cases}$ or

$$V(\lambda,y) = \begin{cases} C_2(\lambda)e^{-\lambda y}, & \text{if } \lambda > 0\\ C_1(\lambda)e^{-\lambda y}, & \text{if } \lambda < 0 \end{cases} = C(\lambda)e^{-|\lambda|y}.$$

By (11.1.18), we get $C(\lambda) = F(\lambda)$, then
$$V(\lambda,y) = F(\lambda)e^{-|\lambda|y}.$$

Using the inverse transformation, we have

$$u(x,y) = \mathscr{F}^{-1}[F(\lambda)e^{-|\lambda|y}] = \frac{1}{\sqrt{2\pi}}\int_{-\infty}^{+\infty} F(\lambda)e^{-|\lambda|y}e^{-i\lambda x}d\lambda$$

$$= \frac{1}{\sqrt{2\pi}}\int_{-\infty}^{+\infty}\left[\frac{1}{\sqrt{2\pi}}\int_{-\infty}^{+\infty} f(\xi)e^{i\lambda\xi}d\xi\right]e^{-|\lambda|y-i\lambda x}d\lambda$$

$$= \frac{1}{2\pi}\int_{-\infty}^{+\infty} f(\xi)\left[\int_{-\infty}^{+\infty} e^{-|\lambda|y+i(\xi-x)\lambda}d\lambda\right]d\xi$$

$$= \frac{1}{2\pi}\int_{-\infty}^{+\infty} f(\xi)\left[\int_{0}^{+\infty} e^{-[y-i(\xi-x)]\lambda}d\lambda + \int_{-\infty}^{0} e^{[y+i(\xi-x)]\lambda}d\lambda\right]d\xi$$

$$= \frac{1}{2\pi}\int_{-\infty}^{+\infty} f(\xi)\left[\left.\frac{-e^{-[y-i(\xi-x)]\lambda}}{y-i(\xi-x)}\right|_{0}^{+\infty} + \left.\frac{e^{[y+i(\xi-x)]\lambda}}{y+i(\xi-x)}\right|_{-\infty}^{0}\right]d\xi$$

$$= \frac{1}{2\pi}\int_{-\infty}^{+\infty} f(\xi)\left[\frac{1}{y-i(\xi-x)} + \frac{1}{y+i(\xi-x)}\right]d\xi$$

$$= \frac{y}{\pi}\int_{-\infty}^{+\infty} \frac{f(\xi)}{y^2+(\xi-x)^2}d\xi.$$

Example 11.1.10 Solve the problem on temperature distribution for a half infinite rod:

$$\begin{cases} u_t = u_{xx}, & 0 < x < +\infty, \ t > 0, & (11.1.20) \\ u(x,0) = 0, & 0 \leqslant x < +\infty, & (11.1.21) \\ u_x(0,t) = g(t), & t > 0, & (11.1.22) \\ \lim_{x \to +\infty} u(x,t) = \lim_{x \to +\infty} u_x(x,t) = 0. & (11.1.23) \end{cases}$$

Solution Let $U(\lambda,t) = F_c[u(x,t)] = \sqrt{\frac{2}{\pi}}\int_{0}^{+\infty} u(x,t)\cos\lambda x\, dx$,

then, by (11.1.20)~(11.1.23), we have

$$\begin{cases} \dfrac{dU}{dt} + \lambda^2 U = -\sqrt{\dfrac{2}{\pi}}g(t), & (11.1.24) \\ U(\lambda,0) = 0. & (11.1.25) \end{cases}$$

Thus, by (11.1.24) and (11.1.25), we get

$$U(\lambda,t) = -\sqrt{\frac{2}{\pi}}\int_{0}^{t} g(\tau)e^{-\lambda^2(t-\tau)}d\tau.$$

And then

$$u(x,t) = -\frac{2}{\pi}\int_{0}^{+\infty}\left[\int_{0}^{t} g(\tau)e^{-\lambda^2(t-\tau)}d\tau\right]\cos\lambda x\, d\lambda$$

$$= -\frac{2}{\pi}\int_{0}^{t} g(\tau)\left[\int_{0}^{+\infty} e^{-\lambda^2(t-\tau)}\cos\lambda x\, d\lambda\right]d\tau$$

$$= -\frac{1}{\sqrt{\pi}}\int_{0}^{t} \frac{g(\tau)}{\sqrt{t-\tau}}e^{-\frac{x^2}{4(t-\tau)}}d\tau.$$

Exercises 11.1

1. Find the Fourier integral transformation of the following functions:

(a) $f(x) = \begin{cases} 0, & x > 0 \\ e^{2x}, & x \leqslant 0 \end{cases}$;

(b) $f(x) = \begin{cases} 1-x, & |x| \leqslant 1 \\ 0, & |x| > 1 \end{cases}$.

2. Let $F(\lambda)=\mathscr{F}[f(x)]$. Please show that

$$\mathscr{F}[f(x)\cos\lambda_0 x]=\frac{1}{2}[F(\lambda+\lambda_0)+F(\lambda-\lambda_0)],$$

$$\mathscr{F}[f(x)\sin\lambda_0 x]=\frac{1}{2i}[F(\lambda+\lambda_0)-F(\lambda-\lambda_0)].$$

3. Prove that $\mathscr{F}[f(at-b)]=\dfrac{1}{|a|}e^{i\frac{b}{a}\lambda}F\left(\dfrac{\lambda}{a}\right)$.

4. Prove the convolution formula $\mathscr{F}^{-1}[F(\lambda)G(\lambda)]=f*g(t)$, where $f(t)=\mathscr{F}^{-1}[F(\lambda)]$ and $g(t)=\mathscr{F}^{-1}[F(\lambda)]$.

5. Assume $f(x)$ and $f^{(k)}(x)$ ($k=1, 2, \cdots, n$) can be operated by the Fourier transformation, and $f^{(k)}(\pm\infty)=0$, $k=0, 1, \cdots, n-1$, where $f^{(0)}(x)=f(x)$, then show that

$$\mathscr{F}[f^{(n)}(x)]=(-i\lambda)^n\mathscr{F}[f(x)].$$

6. Solve the Neumann problem

$$\begin{cases} u_{xx}+u_{yy}=0, & -\infty<x<+\infty,\ y>0, \\ u_y(x,0)=g(x), & -\infty<x<+\infty, \\ \lim_{|x|\to+\infty} u_y(x,y)=0,\ \lim_{|x|\to+\infty} u_{yx}(x,y)=0, \\ \lim_{y\to+\infty}|u_y(x,y)|<+\infty. \end{cases}$$

7. Solve the problem

$$\begin{cases} u_{xx}+u_{yy}=0, & 0<x<+\infty,\ 0<y<2, \\ u(0,y)=0, & 0<y<2, \\ u(x,0)=f(x),\ u(x,2)=0. \end{cases}$$

11.2 Laplace integral transformation

11.2.1 Definition of Laplace transformation

Suppose that $\begin{cases} f(t)\neq 0, & t>0, \\ f(t)=0, & t\leq 0. \end{cases}$ Then, consider the Fourier transform of f,

$$\mathscr{F}[f]=F(\lambda)=\frac{1}{\sqrt{2\pi}}\int_0^{+\infty} f(t)e^{-i\lambda t}\,dt.$$

But, f could not be absolutely integrable. In this case, if $|f(t)|\leq e^{s_0 t}$ ($s_0>0$), we choose $f_1(t)$ as follows:

$$f_1(t)=\begin{cases} e^{-st}f(t), & t\geq 0,\ \text{for } s>s_0>0, \\ 0, & t<0. \end{cases}$$

Then

$$f_1(t) = \frac{1}{\sqrt{2\pi}} \int_{-\infty}^{+\infty} \left[\frac{1}{\sqrt{2\pi}} \int_{-\infty}^{+\infty} f_1(\tau) e^{i\lambda\tau} d\tau \right] e^{-i\lambda t} d\lambda$$

$$= \frac{1}{\sqrt{2\pi}} \int_{-\infty}^{+\infty} \left[\frac{1}{\sqrt{2\pi}} \int_{0}^{+\infty} f(\tau) e^{-s\tau} e^{i\lambda\tau} d\tau \right] e^{-i\lambda t} d\lambda$$

$$= \frac{1}{2\pi} \int_{-\infty}^{+\infty} \left[\int_{0}^{+\infty} f(\tau) e^{-(s-i\lambda)\tau} d\tau \right] e^{-i\lambda t} d\lambda.$$

From the above formula, it is easy to see that

$$f(t) = f_1(t) e^{st} = \frac{1}{2\pi} \int_{-\infty}^{+\infty} \left[\int_{0}^{+\infty} f(\tau) e^{-(s-i\lambda)\tau} d\tau \right] e^{(s-i\lambda)t} d\lambda.$$

Let $p = s - i\lambda$, $dp = -id\lambda$, then

$$f(t) = \frac{1}{2\pi i} \int_{s-i\infty}^{s+i\infty} \left[\int_{0}^{+\infty} f(\tau) e^{-p\tau} d\tau \right] e^{pt} dp.$$

Definition 11.2.1 Assume that $f(t)$ satisfies the following conditions (denoted by L-(A)):

(1) $f(t)$ is piecewise smooth on $[0, +\infty)$;

(2) There exist constants $M > 0$ and $s_0 \geq 0$ such that
$$|f(t)| \leq M e^{s_0 t} \quad \text{for} \quad 0 \leq s_0 < s.$$

Then the integral

$$L(p) = \int_{0}^{+\infty} f(t) e^{-pt} dt$$

is called **Laplace integral transformation** of $f(t)$ in $(0, +\infty)$, and denote $\mathscr{L}[f(t)] = L(p)$, for all p with $\operatorname{Re} p = s > s_0$.

The integral

$$f(t) = \frac{1}{2\pi i} \int_{s-i\infty}^{s+i\infty} L(p) e^{pt} dp$$

is called the **inverse Laplace transformation** of $L(p)$, and it is denoted as $\mathscr{L}^{-1}[L(p)] = f(t)$.

Example 11.2.1 Find the Laplace transformation of the following functions.

(1) $f(t) = 1$;
(2) $f(t) = t^n$, $(n \geq 0, n \in Z)$;
(3) $f(t) = e^{kt}$, $(k \in R)$;
(4) $f(t) = \sin kt$, $(k \in R)$.

Solution

(1) $\mathscr{L}[f(t)] = \int_{0}^{+\infty} f(t) e^{-pt} dt = \int_{0}^{+\infty} e^{-pt} dt = \left. \frac{e^{-pt}}{-p} \right|_{0}^{+\infty} = \frac{1}{p}$, $(\operatorname{Re} p > 0)$.

(2) $\mathscr{L}[f(t)] = \int_{0}^{+\infty} f(t) e^{-pt} dt = \int_{0}^{+\infty} t^n e^{-pt} dt$

$$= -\frac{1}{p} \int_{0}^{+\infty} t^n d e^{-pt}$$

$$= -\frac{1}{p} t^n e^{-pt} \Big|_{0}^{+\infty} + \frac{n}{p} \int_{0}^{+\infty} t^{n-1} e^{-pt} dt$$

$$= \frac{n}{p} \mathscr{L}[t^{n-1}], \quad (\operatorname{Re} p > 0).$$

When $n=1$, $\mathscr{L}[t^{n-1}]=\mathscr{L}[1]=\dfrac{1}{p}$, thus by recursion formula we have

$$\mathscr{L}[t^n]=\frac{n!}{p^{n+1}}, \quad (\mathrm{Re}\,p>0).$$

(3) $\mathscr{L}[f(t)] = \displaystyle\int_0^{+\infty} f(t)e^{-pt}\,dt = \int_0^{+\infty} e^{kt}e^{-pt}\,dt = \dfrac{1}{k-p}e^{(k-p)t}\Big|_0^{+\infty} = \dfrac{1}{p-k}, \quad (\mathrm{Re}\,p>k).$

(4) $\mathscr{L}[f(t)] = \displaystyle\int_0^{+\infty} f(t)e^{-pt}\,dt = \int_0^{+\infty} \sin kt\, e^{-pt}\,dt$

$$= \int_0^{+\infty} \frac{e^{ikt}-e^{-ikt}}{2i} e^{-pt}\,dt$$

$$= \frac{1}{2i}\left(\frac{1}{p-ik} - \frac{1}{p+ik}\right)$$

$$= \frac{k}{p^2+k^2}, \quad (\mathrm{Re}\,p>0).$$

Similarly, $\mathscr{L}[\cos kt]=\dfrac{p}{p^2+k^2}$, $(\mathrm{Re}\,p>0)$.

Theorem 11.2.1 (Inversion theorem) Assume that $\mathscr{L}[f(t)]=L(p)$, then

$$\frac{1}{2\pi i}\int_{s-i\infty}^{s+i\infty} L(p)e^{pt}\,dp = \begin{cases} f(t), & t\geq 0,\ t \text{ is a point of continuity,} \\ 0, & t<0. \end{cases}$$

11.2.2 Properties of Laplace transformation

1. Linear property

Let $f(t)$ and $g(t)$ satisfy L-(A). Then

$$\mathscr{L}[c_1 f(t)+c_2 g(t)] = c_1\mathscr{L}[f(t)] + c_2\mathscr{L}[g(t)],$$

for any two constants c_1 and c_2.

Example 11.2.2 Find the Laplace transformation of function $f(t)=\sin 2t\cos 3t$.

Solution

$$\mathscr{L}[\sin 2t\cos 3t] = \mathscr{L}\left[\frac{1}{2}(\sin 5t - \sin t)\right]$$

$$= \frac{1}{2}[\mathscr{L}[\sin 5t] - \mathscr{L}[\sin t]]$$

$$= \frac{1}{2}\left(\frac{5}{p^2+5^2} - \frac{1}{p^2+1^2}\right), \quad (\mathrm{Re}\,p>0).$$

2. Delay theorem

Assume $f(t)$ satisfies L-(A), $L(p)=\mathscr{L}[f(t)]$, then for $c>0$,

$$\mathscr{L}[f(t-c)] = e^{-pc}\mathscr{L}[f(t)] = e^{-pc}L(p).$$

Proof

$$\mathscr{L}[f(t-c)] = \int_0^{+\infty} f(t-c)e^{-pt}\,dt$$

$$= \int_0^{+\infty} f(\eta)e^{-p(\eta+c)}\,d\eta = e^{-pc}\mathscr{L}[f(t)].$$

Example 11.2.3 (1) Find the Laplace transformation of $f(t)=\sin(t-t_0)$, $(t_0>0)$.

(2) And, please point out that the following formula is true or not, why?
$$\mathscr{L}[\sin(t-t_0)] = e^{-p t_0} \mathscr{L}[\sin t] = e^{-p t_0} L(p) = e^{-p t_0} \frac{1}{1+p^2}.$$

Solution (1) By trigonometric formula,
$$\begin{aligned}\mathscr{L}[\sin(t-t_0)] &= \mathscr{L}[\sin t \cos t_0 - \cos t \sin t_0] \\ &= \cos t_0 \mathscr{L}[\sin t] - \sin t_0 \mathscr{L}[\cos t] \\ &= \frac{\cos t_0 - p \sin t_0}{p^2 + 1}, \quad (\text{Re} p > 0).\end{aligned}$$

(2) The formula is not true. Since when $t \leqslant t_0$, $\sin(t-t_0) \neq 0$.

3. Displacement theorem

Assume $f(t)$ satisfies L-(A) and $L(p) = \mathscr{L}[f(t)]$, then
$$L(p - p_0) = \mathscr{L}[e^{p_0 t} f(t)], \quad (\text{Re}(p - p_0) > 0).$$

Proof
$$L(p - p_0) = \int_0^{+\infty} f(t) e^{-(p - p_0)t} dt = \int_0^{+\infty} e^{p_0 t} f(t) e^{-pt} dt = \mathscr{L}[e^{p_0 t} f(t)].$$

Example 11.2.4 Find the Laplace transformation of the function $f(t) = e^{-at} t^m$, ($m \geqslant 1$, $m \in Z$).

Solution $\mathscr{L}[t^m] = \frac{m!}{p^{n+1}}$, (Re $p > 0$), thus by the displacement theorem
$$\mathscr{L}[e^{-at} t^m] = L(p + a) = \frac{m!}{(p+a)^{n+1}}, \quad (\text{Re}(p+a) > 0).$$

4. Similar theorem

Assume $f(t)$ satisfies L-(A) and $L(p) = \mathscr{L}[f(t)]$, then for $a > 0$, we have
$$\mathscr{L}[f(at)] = \frac{1}{a} L\left(\frac{p}{a}\right).$$

Proof
$$\begin{aligned}\mathscr{L}[f(at)] &= \int_0^{+\infty} f(at) e^{-pt} dt \\ &= \int_0^{+\infty} f(\eta) e^{-\frac{p}{a} \eta} \frac{1}{a} d\eta = \frac{1}{a} L\left(\frac{p}{a}\right).\end{aligned}$$

5. Differential theorem

(1) Assume $f(t)$ and $f'(t)$ satisfy L-(A), then
$$\mathscr{L}[f'(t)] = p\mathscr{L}[f(t)] - f(0^+).$$

Proof
$$\begin{aligned}\mathscr{L}[f'(t)] &= \int_0^{+\infty} f'(\tau) e^{-p\tau} d\tau \\ &= f(\tau) e^{-p\tau} \Big|_0^{+\infty} + \int_0^{+\infty} f(\tau) p e^{-p\tau} d\tau \\ &= p\mathscr{L}[f(t)] - f(0^+).\end{aligned}$$

Furthermore, we have the following corollary.

Corollary Assume $f(t)$ and $f^{(k)}(t)$, ($k = 1, 2, \cdots, n$) satisfy L-(A), then
$$\mathscr{L}[f^{(n)}(t)] = p^n \left(\mathscr{L}[f(t)] - \frac{f(0)}{p} - \frac{f'(0)}{p^2} - \cdots - \frac{f^{(n-1)}(0)}{p^n}\right),$$

where $f(0)=f(0^+)$, $f^{(k)}(0)=f^{(k)}(0^+)$, $k=1, 2, \cdots, n-1$.

(2) Assume $f(t)$ satisfies L-(A), then
$$\frac{d}{dp}\mathscr{L}[f(t)]=\mathscr{L}[-tf(t)],$$
and then
$$\frac{d^n \mathscr{L}[f(t)]}{dp^n}=\mathscr{L}[(-t)^n f(t)].$$

Proof
$$\frac{d}{dp}\mathscr{L}[f(t)] = \frac{d}{dp}\int_0^{+\infty} f(\tau)e^{-p\tau}d\tau = \int_0^{+\infty} f(\tau)(-\tau)e^{-p\tau}d\tau = \mathscr{L}[-tf(t)].$$

Example 11.2.5 Please use the differential theorem to find the Laplace transformation of the following functions.

(1) $f(t)=t\sin kt$; (2) $f(t)=t\cos kt$.

Solution

(1) $\mathscr{L}[t\sin kt]=-\dfrac{d}{dp}\mathscr{L}[\sin kt]=-\dfrac{d}{dp}\left(\dfrac{k}{p^2+k^2}\right)=\dfrac{2kp}{(p^2+k^2)^2}$, (Re$p>0$).

(2) $\mathscr{L}[t\cos kt]=-\dfrac{d}{dp}\mathscr{L}[\cos kt]=-\dfrac{d}{dp}\left(\dfrac{p}{p^2+k^2}\right)=\dfrac{p^2-k^2}{(p^2+k^2)^2}$, (Re$p>0$).

Example 11.2.6 Please use the differential theorem to compute the inverse Laplace transformation
$$\mathscr{L}^{-1}\left[\ln\frac{p+1}{p-1}\right].$$

Solution Let $\mathscr{L}[f(t)]=L(p)=\ln\dfrac{p+1}{p-1}$, $L'(p)=\dfrac{1}{p+1}-\dfrac{1}{p-1}$, then by the differential theorem (2),
$$-tf(t)=\mathscr{L}^{-1}[\mathscr{L}[-tf(t)]]=\mathscr{L}^{-1}\left[\frac{d}{dp}\mathscr{L}[f(t)]\right].$$

Thus,
$$f(t)=-\frac{1}{t}\mathscr{L}^{-1}[L'(p)]=-\frac{1}{t}\mathscr{L}^{-1}\left[\frac{1}{p+1}-\frac{1}{p-1}\right]$$
$$=-\frac{1}{t}\left\{\mathscr{L}^{-1}\left[\frac{1}{p+1}\right]-\mathscr{L}^{-1}\left[\frac{1}{p-1}\right]\right\}$$
$$=-\frac{1}{t}(e^{-t}-e^{t}).$$

6. Integral theorem

(1) Assume $f(t)$ satisfies L-(A) and $\varphi(t)=\displaystyle\int_0^t f(\tau)d\tau$, then
$$\mathscr{L}[\varphi(t)]=\frac{1}{p}\mathscr{L}[f(t)]=\frac{1}{p}L(p).$$

Proof Since $\varphi'(t)=f(t)$, $\varphi(0)=0$, then
$$\mathscr{L}[\varphi'(t)]=\mathscr{L}[f(t)]=p\mathscr{L}[\varphi(t)]-\varphi(0)$$
$$\Rightarrow \mathscr{L}[\varphi(t)]=\frac{1}{p}\mathscr{L}[f(t)]=\frac{1}{p}L(p).$$

(2) Assume $f(t)$ satisfies L-(A), and $L(p)=\mathscr{L}[f(t)]$, $\int_{p}^{+\infty}|L(s)|ds<+\infty$, then
$$\int_{p}^{+\infty} L(s)ds = \mathscr{L}\left[\frac{f(t)}{t}\right].$$

Proof
$$\int_{p}^{+\infty} L(s)ds = \int_{p}^{+\infty}\left(\int_{0}^{+\infty} f(\tau)e^{-s\tau}d\tau\right)ds$$
$$= \int_{0}^{+\infty} f(\tau)\left(\int_{p}^{+\infty} e^{-s\tau}ds\right)d\tau = \int_{0}^{+\infty} f(\tau)\left(\frac{-e^{-s\tau}}{\tau}\right)\bigg|_{p}^{+\infty}d\tau$$
$$= \int_{0}^{+\infty} \frac{f(\tau)}{\tau}e^{-p\tau}d\tau = \mathscr{L}\left[\frac{f(t)}{t}\right].$$

Example 11.2.7 Please use the integral theorem to compute
$$\mathscr{L}\left[\int_{0}^{t}\frac{\sin t}{t}dt\right] \quad \text{and} \quad \int_{0}^{+\infty}\frac{\sin t}{t}dt.$$

Solution (1) By $\mathscr{L}[\sin t]=\dfrac{1}{p^{2}+1}$ and the integral theorem (2), we have
$$\mathscr{L}\left[\frac{\sin t}{t}\right]=\int_{p}^{+\infty}\frac{1}{p^{2}+1}dp = \arctan p \bigg|_{p}^{+\infty} = \frac{\pi}{2}-\arctan p.$$

Thus using the integral theorem (1), we have
$$\mathscr{L}\left[\int_{0}^{t}\frac{\sin t}{t}dt\right]=\frac{1}{p}\left(\frac{\pi}{2}-\arctan p\right).$$

(2) Let $L(p)=\mathscr{L}[\sin t]$, then by $\int_{0}^{+\infty}\frac{f(t)}{t}dt = \mathscr{L}\left[\frac{f(t)}{t}\right]\bigg|_{p=0}=\int_{0}^{+\infty}L(p)dp$, we get
$$\int_{0}^{+\infty}\frac{\sin t}{t}dt=\int_{0}^{+\infty}\frac{1}{p^{2}+1}dp=\arctan p\bigg|_{0}^{+\infty}=\frac{\pi}{2}.$$

11.2.3 Convolution and its Laplace transformation

Definition 11.2.2 Assume $f(t)$ and $g(t)$ satisfy L-(A), then the integral
$$\int_{0}^{t} f(t-\tau)g(\tau)d\tau \quad \text{or} \quad \int_{0}^{t} g(t-\tau)f(\tau)d\tau$$
is called the convolution of $f(t)$ and $g(t)$, denoted by $f*g(t)$ or $g*f(t)$.

Let $F(p)=\mathscr{L}[f(t)]$ and $G(p)=\mathscr{L}[g(t)]$, then the integral
$$\frac{1}{2\pi i}\int_{s-i\infty}^{s+i\infty} F(p-q)G(q)dq \quad \text{or} \quad \frac{1}{2\pi i}\int_{s-i\infty}^{s+i\infty} G(p-q)F(q)dq$$
is called the convolution of $F(p)$ and $G(p)$, denoted by $F*G(p)$ or $G*F(p)$.

Properties Assume that $f(t)$ and $g(t)$ satisfy L-(A), $F(p)=\mathscr{L}[f(t)]$, $G(p)=\mathscr{L}[g(t)]$, then

(1) $\mathscr{L}[f*g(t)]=F(p)\cdot G(p)$;

(2) $\mathscr{L}[f(t)\cdot g(t)]=F*G(p)$.

Proof (1) $\mathscr{L}[f*g(t)] = \int_{0}^{+\infty}\left[\int_{0}^{t} f(t-\tau)g(\tau)d\tau\right]e^{-pt}dt$
$$= \int_{0}^{+\infty}\left[\int_{\tau}^{+\infty} f(t-\tau)e^{-pt}dt\right]g(\tau)d\tau$$

$$= \int_0^{+\infty} \left[\int_0^{+\infty} f(\eta) e^{-p(\eta+\tau)} d\eta\right] g(\tau) d\tau$$

$$= \left[\int_0^{+\infty} f(\eta) e^{-p\eta} d\eta\right]\left[\int_0^{+\infty} g(\tau) e^{-p\tau} d\tau\right]$$

$$= F(p) \cdot G(p).$$

(2) $F * G(p) = \dfrac{1}{2\pi i} \int_{s-i\infty}^{s+i\infty} F(p-q) G(q) dq$

$$= \dfrac{1}{2\pi i} \int_{s-i\infty}^{s+i\infty} \left[\int_0^{+\infty} f(t) e^{-(p-q)t} dt\right] G(q) dq$$

$$= \int_0^{+\infty} f(t) \left[\dfrac{1}{2\pi i} \int_{s-i\infty}^{s+i\infty} G(q) e^{qt} dq\right] e^{-pt} dt$$

$$= \int_0^{+\infty} f(t) g(t) e^{-pt} dt = \mathscr{L}[f(t) \cdot g(t)].$$

Note that $\mathscr{L}[1] = \int_0^{+\infty} e^{-pt} dt = \dfrac{1}{p}$ and $\mathscr{L}[1] = \mathscr{L}\left[\left(\dfrac{t^n}{n!}\right)^{(n)}\right]$.

Since $f(t) = \dfrac{t^n}{n!}$, $f^{(n)}(t) = 1$, $f^{(k)}(0) = 0$, $k = 0, 1, \cdots, n-1$, by the corollary of 5, we have

$$\mathscr{L}[1] = \dfrac{1}{p} = \mathscr{L}[f^{(n)}(t)] = p^n \mathscr{L}[f(t)],$$

$$\Rightarrow \mathscr{L}[f(t)] = \mathscr{L}\left[\dfrac{t^n}{n!}\right] = \dfrac{1}{p^{n+1}} \quad \text{or} \quad \mathscr{L}[t^n] = \dfrac{n!}{p^{n+1}}.$$

11.2.4 Application of Laplace integral transformation

Example 11.2.8 Find the Laplace transformation of step function

$$u_a(t) = \begin{cases} 1, & t > a \geq 0, \\ 0, & t < a. \end{cases}$$

Solution $\mathscr{L}[u_a(t)] = \int_0^{+\infty} u_a(t) e^{-pt} dt = \int_a^{+\infty} e^{-pt} dt = \left.\dfrac{-e^{-pt}}{p}\right|_0^{+\infty} = \dfrac{e^{-pa}}{p}.$

As $a = 0$, then $\mathscr{L}[u_0(t)] = \mathscr{L}[1] = \dfrac{1}{p}$.

Example 11.2.9 Find the Laplace transformation of impulse function

$$P_\varepsilon(t,a) = \begin{cases} h, & |t-a| < \varepsilon, \\ 0, & |t-a| > \varepsilon, \end{cases} \quad (a > 0).$$

Solution $\mathscr{L}[P_\varepsilon(t,a)] = \int_0^\infty P_\varepsilon(t,a) e^{-pt} dt = \int_{a-\varepsilon}^{a+\varepsilon} h e^{-pt} dt$

$$= h \left.\dfrac{-e^{-pt}}{p}\right|_{a-\varepsilon}^{a+\varepsilon} = \dfrac{h}{p} e^{-pa} (e^{p\varepsilon} - e^{-p\varepsilon})$$

$$= 2h\varepsilon e^{-pa} \dfrac{e^{p\varepsilon} - e^{-p\varepsilon}}{2p\varepsilon} = 2h\varepsilon e^{-pa} \left(\dfrac{\sinh(p\varepsilon)}{p\varepsilon}\right).$$

Let $h = \dfrac{1}{2\varepsilon}$, then

$$P_\varepsilon(t, a) = \begin{cases} \dfrac{1}{2\varepsilon}, & |t-a| < \varepsilon, \\ 0, & |t-a| \geq \varepsilon. \end{cases}$$

And,
$$\lim_{\varepsilon \to 0^+} P_\varepsilon(t,a) = \delta(t-a) = \begin{cases} 0, & t \neq a, \\ \int_{-\infty}^{+\infty} \delta(t-a)\,dt = a, & t = a \end{cases}$$

is the δ-function.

Since $\int_{-\infty}^{+\infty} P_\varepsilon(t,a)\,dt = \int_{a-\varepsilon}^{a+\varepsilon} \dfrac{1}{2\varepsilon}\,dt \equiv 1$, we have $\int_{-\infty}^{+\infty} \delta(t-a)\,dt \equiv 1$, and

$$\mathscr{L}[p_\varepsilon(t-a)] = e^{-pa}\left(\frac{\sinh p\varepsilon}{p\varepsilon}\right) \xrightarrow{\varepsilon \to 0^+} e^{-pa}.$$

Thus,
$$\mathscr{L}[\delta(t-a)] = \lim_{\varepsilon \to 0^+} \mathscr{L}[P_\varepsilon(t, a)] = e^{-pa}.$$

Specially, if $a=0$, then $\mathscr{L}[\delta(t)]=1$.

Example 11.2.10 Find the Laplace inverse transformation of $\dfrac{1}{p(p+1)^2}$.

Solution Since $\dfrac{1}{p(p+1)^2} = \dfrac{1}{p} - \dfrac{1}{p+1} - \dfrac{1}{(p+1)^2}$, and

$$\mathscr{L}^{-1}\left[\frac{1}{p}\right]=1, \quad \mathscr{L}^{-1}\left[\frac{1}{p+1}\right]=e^{-t},$$

$$\mathscr{L}^{-1}\left[\frac{1}{(p+1)^2}\right] = \mathscr{L}^{-1}\left[\frac{1}{p+1} \cdot \frac{1}{p+1}\right] = \mathscr{L}^{-1}[\mathscr{L}(e^{-t} * e^{-t})] = e^{-t} * e^{-t} = \int_0^t e^{-\tau}e^{-(t-\tau)}\,d\tau = te^{-t}.$$

Thus, $\mathscr{L}^{-1}\left[\dfrac{1}{p(p+1)^2}\right] = 1 - e^{-t} - te^{-t}$.

Example 11.2.11 Solve the ordinary differential equation (ODE)
$$\begin{cases} x'''(t) + 3x''(t) + 3x'(t) + x(t) = 1, \\ x(0) = x'(0) = x''(0) = 0. \end{cases}$$

Solution Let $L(p) = \mathscr{L}[x(t)]$. Taking Laplace transformation on the ODE, we have
$$p^3 L(p) + 3p^2 L(p) + 3p L(p) + L(p) = \frac{1}{p},$$

So, $L(p) = \mathscr{L}[x(t)] = \dfrac{1}{p(p+1)^3} = \dfrac{1}{p} - \dfrac{1}{p+1} - \dfrac{1}{(p+1)^2} - \dfrac{1}{(p+1)^3}$.

Thus, $x(t) = \mathscr{L}^{-1}[L(p)] = 1 - e^{-t} - te^{-t} - \dfrac{1}{2}t^2 e^{-t}$.

Example 11.2.12 Consider the vibration problem of a half infinite string
$$\begin{cases} u_{tt} = c^2 u_{xx}, & 0 < x < +\infty,\ t > 0, \quad (11.2.1) \\ u(x,0) = 0,\ u_t(x,0) = 0, & (11.2.2) \\ u(0,t) = f(t),\ \lim_{x \to +\infty} u_x(x,t) = 0. & (11.2.3) \end{cases}$$

Solution Let $V(x,p) = \mathscr{L}[u(x,t)]$, $L(p) = \mathscr{L}[f(t)]$.
Then by (11.2.1), we have
$$\mathscr{L}[u_{tt}] - c^2 \mathscr{L}[u_{xx}] = 0. \quad (11.2.4)$$

Moreover, by (11.2.2)
$$\mathscr{L}[u_{tt}] = p^2 L[u] - p u(x,0) - u_t(x,0) = p^2 V. \quad (11.2.5)$$
Thus, combining (11.2.4) and (11.2.5), we obtain
$$\frac{d^2 V}{dx^2} - \frac{p^2}{c^2} V = 0.$$

Solving this equation, then
$$V(x,p) = C_1(p) e^{-\frac{p}{c}x} + C_2(p) e^{\frac{p}{c}x}. \quad (11.2.6)$$

From (11.2.3), it is obvious that
$$V(0,p) = L(p), \quad \lim_{x \to +\infty} \frac{dV(x,p)}{dx} = 0. \quad (11.2.7)$$

Thus, substituting (11.2.7) into (11.2.6), we have
$$C_1(p) = L(p), \quad C_2(p) = 0.$$

Then
$$V(x,p) = L(p) e^{-\frac{p}{c}x}.$$

Finally, by the Laplace inverse transformation we have
$$u(x,t) = \mathscr{L}^{-1}[V(x,p)] = \mathscr{L}^{-1}[L(p) e^{-\frac{p}{c}x}]$$
$$= \mathscr{L}^{-1}\left[\int_0^{+\infty} f(t) e^{-pt} e^{-\frac{p}{c}x} dt\right]$$
$$= \mathscr{L}^{-1}\left[\mathscr{L}\left[f\left(t - \frac{x}{c}\right)\right]\right] = f\left(t - \frac{x}{c}\right).$$

Exercises 11.2

1. Find the Laplace transformation of the following functions:
   (1) $f(t) = t^2 + 3t + 2$;
   (2) $f(t) = 1 - t e^t$;
   (3) $f(t) = (t-1)^2 e^t$;
   (4) $f(t) = e^{-2t} \sin 6t$.

2. Find the Laplace transformation of the following functions:
   (1) $f(t) = t e^{-3t} \sin 2t$;
   (2) $f(t) = \dfrac{\sin kt}{t}$.

3. Find the Laplace inverse transformation of the following functions:
   (1) $L(p) = \dfrac{1}{p^2 + 4^2}$;
   (2) $L(p) = \dfrac{1}{p^4}$;
   (3) $L(p) = \dfrac{1}{p+3}$;
   (4) $L(p) = \ln \dfrac{p-2}{p+1}$.

4. Solve the problem:
$$\begin{cases} u_t = k u_{xx} - hu, & 0 < x < +\infty, \ t > 0, \\ u(x,0) = T_0, & 0 \leqslant x < +\infty, \\ u(0,t) = 0, & \\ \lim\limits_{x \to +\infty} u_x(x,t) = 0. & \end{cases}$$

References

[1] Brown J W, Churchill R V. Complex Variables and Application [M]. Beijing: China Machine Press, 2008.
[2] 谢鸿政. Equations of Classical Mathematical Physics [M]. 北京:科学出版社,2006.
[3] 郭玉翠. 工程数学——复变函数与数学物理方法[M]. 北京:清华大学出版社,2014.
[4] 盖云英,邢宇明. Functions of a Complex Variable and Integral Transforms [M]. 北京:科学出版社,2007.
[5] Evans G, Blackledge J, Yardley P. Analytic Methods for Partial Differential Equations [M]. Beijing: Beijing World Publishing Corporation, 2004.
[6] 王元明. 数学物理方程与特殊函数 [M]. 北京:高等教育出版社,2004.